Association for Women in Mathematics Series

Volume 37

Focusing on the groundbreaking work of women in mathematics past, present, and future, Springer's Association for Women in Mathematics Series presents the latest research and proceedings of conferences worldwide organized by the Association for Women in Mathematics (AWM). All works are peer-reviewed to meet the highest standards of scientific literature, while presenting topics at the cutting edge of pure and applied mathematics, as well as in the areas of mathematical education and history. Since its inception in 1971, The Association for Women in Mathematics has been a non-profit organization designed to help encourage women and girls to study and pursue active careers in mathematics and the mathematical sciences and to promote equal opportunity and equal treatment of women and girls in the mathematical sciences. Currently, the organization represents more than 3000 members and 200 institutions constituting a broad spectrum of the mathematical community in the United States and around the world.

Titles from this series are indexed by Scopus.

Cristina Garcia-Cardona • Harlin Lee
Editors

Advances in Data Science

Women in Data Science and Mathematics (WiSDM) 2023

Springer

ASSOCIATION FOR WOMEN IN MATHEMATICS

Editors
Cristina Garcia-Cardona
Los Alamos National Laboratory
Los Alamos, NM, USA

Harlin Lee
School of Data Science and Society
University of North Carolina
Chapel Hill, NC, USA

ISSN 2364-5733 ISSN 2364-5741 (electronic)
Association for Women in Mathematics Series
ISBN 978-3-031-87803-9 ISBN 978-3-031-87804-6 (eBook)
https://doi.org/10.1007/978-3-031-87804-6

Mathematics Subject Classification: 68P01, 68R12, 62-08, 15A69, 68R10, 68T07, 68T09, 68T10, 68T45, 68T50, 05C62

This Springer imprint is published by the registered company Springer Nature Switzerland AG
The registered company address is: Gewerbestrasse 11, 6330 Cham, Switzerland

Preface

This special issue of *Advances in Data Science*, part of the Association for Women in Mathematics (AWM) Springer Series, offers a diverse collection of papers in data science. Data science is a cross-disciplinary field relying on statistics, computer science, and mathematics that is driven by problems from a wide range of disciplines. This volume aims to make more visible the role of theoretical and applied mathematics in data science.

Some contributions are products of the collaborations initiated during the third "Women in Data Science and Mathematics" (WiSDM) Research Workshop that took place between August 7 to August 11, 2023, at the Institute for Pure & Applied Mathematics, a National Science Foundation math institute at the University of California, Los Angeles. The goal of WiSDM is to bring small interdisciplinary teams to work on real-world problems and focused open research questions for a one-week workshop. Participants typically range from senior researchers to early graduate students, collaborating as equals and building relationships centered on shared research interests and complementary technical skills. This third edition congregated six different groups and generated six contributions to the proceedings.[1]

However, works included in the collection go beyond the WiSDM workshop. Calls for contributions were extended to a network of WiSDM affiliates, a vibrant group of previous workshop participants that is growing with each WiSDM edition, and to a wider mathematical community composed by past participants' institutions. Ten additional submissions were collected, representing a diversity of research advancements in applied mathematics.

Note that all the works were subjected to a single-blind peer-review process where two to three experts reviewed the submission and provided recommendations for the authors to improve their manuscript. We take this opportunity to thank the anonymous reviewers for their constructive feedback and their valuable suggestions.

[1] For further details about the projects and participants in WiSDM 2023 please refer to Appendix "Appendix A WiSDM 2023: Projects and Participants".

Overall, we hope that this volume will constitute a useful resource for researchers working in data science and applied mathematics. The in-depth discussions included about complex data problems and cutting-edge methodologies from mathematics, statistics, and computer science will help researchers to keep up with state-of-the-art tools in data analysis. Furthermore, the interdisciplinary nature of the contributions, some showcasing applications to real-world problems, will promote collaborations among different fields and inspire novel practical demonstrations of the presented research. Additionally, since the final papers are the result of collaborations between researchers with different level of expertise, such as graduate students, post-doctoral fellows, and junior and senior faculty members, we expect that this volume will be accessible and particularly beneficial to junior researchers, providing insights, guidance, and motivation to tackle new data science projects.

Chapter Summary

These proceedings cover a broad array of theory and applications illustrating the wide use of tensor decompositions, graph-based algorithms, dimension reduction techniques, and mathematically constrained machine learning. Works range from early research results, to new algorithms inspired by related fields, to objective evaluation of published methods. It also includes promising theoretical developments as well as novel application of techniques to practical problems. A special section highlighting how data science tools can be used to examine aspects of higher education concludes the volume.

Part I: Matrix and Tensor Methods

Part I includes novel contributions in the areas of tensor and matrix methods. Essential to both chapters is the exploitation of the problem structure to produce more efficient algorithms.

Chapter “Randomized Iterative Methods for Tensor Regression Under the t-product” develops new algorithms that generalize iterative methods for matrix-vector regression problems to the tensor regime. Tensor representations naturally arise when dealing with complex multi-modal data (e.g., multidimensional arrays such as video with spatial and time dimensions). This chapter presents a comprehensive survey of iterative methods for tensor regression (under the t-product) and arrives at novel insights by extending variants of the Kaczmarz and Gauss-Seidel methods to tensor regression settings. Results demonstrate the empirical efficacy and accuracy of tensor methods while interesting follow-up theoretical directions are delineated.

Chapter “Matrix exponentials: Lie-Trotter-Suzuki fractal decomposition, Gauss Runge-Kutta polynomial formulation, and compressible features” compares two existing numerical approaches for efficient computation of the matrix exponential. The matrix exponential is an important component when performing numerical simulation of science or engineering time-dependent systems. The methods assume that the system matrix can be expressed as the sum of two simple structured matrices, specifically a diagonal matrix and a low-rank matrix, but use different solution decompositions, yielding distinct numerical formulations and achieving different levels of performance.

Part II: Graph Algorithms

Part II contains contributions related to graph methods. These range from new theoretical definitions of graph properties, to new problem formulations based on graph regularizations or graph representations, to algorithms for making graph-based computations more efficient.

Chapter "An exploration of graph distances, graph curvature, and applications to network analysis" evaluates how notions of graph curvature induced by different definitions of distance on a graph correlate with graph centrality measures. The motivation behind the study is to investigate if some of these graph curvature definitions are able to capture analogous intuitions from the continuous space, where geometrical curvature characterizes how much the space differs from a flat Euclidean space. The correlations computed in synthetic and real-world graphs constitute a first step toward a better understanding of curvature on graphs, which may prove fruitful for advancing new analysis tools.

Chapter "Time-Varying Graph Signal Recovery Using High-Order Smoothness and Adaptive Low-rankness" proposes two new algorithms for recovering signals in a graph, specifically time-varying signals. These are useful descriptions for problems such as predicting sea surface temperatures, i.e., problems involving time-series with given additional relationships among them, e.g., geographical locations, which often can be represented via graphs. The proposed methods combine high-order temporal smoothness and graph structures, and include a novel low-rank regularization. A generalized recovery framework that encompasses the new methods, as well as methods previously published, is presented too.

Chapter "Graph-Directed Topic Models of Text Documents" develops a new topic modeling formulation that incorporates a graph-based regularization term. A text document in a corpus can be represented via a bag-of-words model that captures the distribution of corpus words in the document. However, extracting knowledge requires more than representing individual documents. Topic modeling is a methodology used to summarize the corpus in terms of topics, i.e., documents are linear combinations of topics, where each topic corresponds to a specific word distribution. Typically, a sparsity constraint is used. This work demonstrates how a similarity graph between documents can be used as an alternative regularization.

Chapter "Linear independent component analysis in Wasserstein space" introduces a framework for performing independent component analysis, i.e., identifying the set of independent random variables from observations of the (linearly) mixed components, when the observations consist of probability measures or point-clouds, not vectors in an Euclidean space. The framework proposed uses a Wasserstein-based graph Laplacian. The work studies under which theoretical conditions the eigenvectors of this Laplacian approximate the independent components and shows preliminary results for data that is isometric and almost isometric to Euclidean data.

Chapter "Faster HodgeRank Approximation Algorithm for Statistical Ranking and User Recommendation Problems" proposes a new algorithm to accelerate the solution of the problem of statistical ranking on graphs. A ranking algorithm tries to order entities by a given measure (e.g., preference, votes, etc.). The HodgeRank algorithm is able to compute a global ranking from datasets with incomplete or inconsistent scores, and is based on pairwise differences represented as edge flows on a graph. This work develops a new method to run the HodgeRank algorithm on smaller subgroups, to improve the scalability when the number of entities to be ranked increases.

Chapter "A Comparison Study of Graph Laplacian Computation" compares existing approaches for accelerating the numerical computation of the eigendecomposition of a graph Laplacian matrix. In data analysis, the eigendecomposition of the graph Laplacian can be used to solve clustering or classification problems. However, the methods become computationally demanding when the size of the data, and consequently the size of the graph, is large. The methods evaluated are based on different approximated eigendecompositions that only use a subsample of the dataset, reducing their computational burden. This work performs extensive numerical comparisons and reports trade-offs incurred by the different methods.

Part III: Dimensionality Reduction

Part III focuses on novel developments on dimensionality reduction for improving the efficiency on different data analysis tasks or for decreasing the complexity of neural network model representations so that they can be deployed in resource-limited settings.

Chapter "Supervised Dimension Reduction via Local Gradient Elongation" proposes a novel geometric approach to perform nonlinear supervised dimensionality reduction, i.e., obtain a low-dimensional representation of the input data features via embeddings guided by the response variable (label). The method developed uses a new metric that elongates the standard Euclidean distance in the direction of the (univariate) label gradient. Also, it is able to consider different supervision levels in the proposed local metric, i.e., different weights between feature distance and label gradient. Demonstrations focus on visualization (of embedded features) and prediction (of output variables to new input data) tasks, for synthetic datasets and for real-world data from biology.

Chapter "Reducing NLP Model Embeddings for Deployment in Embedded Systems" aims to reduce the number of parameters required to represent natural language processing (NLP) models while maintaining a tolerable level of performance. NLP models often involve a large number of parameters (hundreds of millions to billions) which makes for a problematic deployment in resource-limited environments (e.g., embedded systems such as field-programmable gate arrays FPGAs). This work applies dimensionality reduction methods to the token embedding layer of the BERT model, a state-of-the-art large language model, to produce new embedding vectors that maximize the variance of its components in a smaller vector space.

Part IV: Data Analysis and Machine Learning

Part IV demonstrates how tailored data analysis and machine learning algorithms can improve applications in road safety, pharmacokinetics, inverse problems in imaging, and speech recognition.

Chapter "Automated extraction of roadside slope from aerial LiDAR data in rural North Carolina" devises a new Python-based, open-source, computational pipeline for processing aerial LiDAR data with the goal of calculating slope grades adjacent to roads on rural North Carolina. The slope of the roadside terrain is a significant component in crash prediction but this data is scarce, particularly in rural regions where physical surveys are impractical. LiDAR is a compelling alternative since it has widespread availability, allowing for cost-effective and large-scale hazard assessments. The processing pipeline includes roadway segmentation, road segment identification, and linear regression fitting.

Chapter "A non-parametric optimal design algorithm for population pharmacokinetics" describes how the non-parametric estimation of the joint distribution of the model parameters, from model observations, can be accelerated by using directional derivatives of the log-likelihood function. This is a principled approach that replaces the ad-hoc exploration of previous methods, allowing for less time spent evaluating nonrelevant points and yielding a significant boost in computation speed. Pharmacokinetics modeling aims to describe the evolution of the amount of drug on a subject, given different conditions of elimination rate, input dose, apparent distribution volume, etc. By reducing computation time, this approach may enable faster therapeutic decisions.

Chapter "Unrolling Deep Learning End-to-End Method for Phase Retrieval" proposes a deep learning framework for the phase retrieval problem. The approach unfolds an iterative algorithm used for regularized optimization, specifically an alternating direction method of multipliers (ADMM) formulation, into a feed-forward network structure, yielding a framework that is interpretable and more amenable to theoretical analysis. A convolutional neural network and a graph convolutional network are learned as part of the unfolded structure, in order to incorporate local and non-local smoothing regularizations. Recovering phase information from intensity data is crucial in fields like coherent diffraction imaging and crystallography.

Chapter "Performance Analysis of MFCC and wav2vec on Stuttering Data" develops machine learning models based on a Siamese neural network to accurately identify types of dysfluency in speech. Scarcity of labeled data and inconsistency of labels make it difficult to train accurate models and improve assistive technologies for individuals with speech disorders. The approach focuses on evaluating different feature representations for audio clips and on adding auxiliary classification tasks to increase the generalization power of the model. Different settings are compared: a single task to identify if the audio pairs provided belong to the same class, and multitask configurations including classification of the six stuttering types and/or binary classification of normal vs. stuttering.

Part V: Data Science and Higher Education

Part V illustrates how data analysis tools can be used to assess gender disparities in STEM, and in some cases, serve as guidance for adjustments that may be beneficial on different stages of the higher education pipeline.

Chapter "Active Learning for Reducing Gender Gaps in Undergraduate Computing and Data Science" reports two instructors' attempt to increase female students' confidence in computing via active learning and collaborative project-based learning. The efforts focused on adapting the course design from a purely introductory programming material to a more diverse offering including data science and machine learning related topics. Surveys were administered during different terms to assess change in student perceptions. The findings from the survey analysis partially align with research suggesting that early educational interventions can enhance comfort and interest in STEM for women.

Chapter "Quantifying and Documenting Gender-Based Inequalities in the Mathematical Sciences in the United States" examines gender inequality at the institutional level within US PhD-granting math departments, where structural barriers may exist in funding and hiring practices. The analysis is based on public data, coming from a census of tenured or tenure-track faculty employed at PhD-granting institutions in the United States and from the National Science Foundation, and focuses on estimating quantitative relations between department's gender composition, the funding received, and the department's perceived prestige. This work illustrates the usefulness of data science tools for diagnosing issues in the mathematical sciences itself.

Together, these studies highlight that while classroom-level interventions can boost early interest and confidence, broader institutional changes are needed to support long-term gender equity in STEM.

Los Alamos, NM, USA Cristina Garcia-Cardona
Chapel Hill, NC, USA Harlin Lee

Acknowledgments

Part of this research was performed while the authors were visiting the Institute for Pure and Applied Mathematics (IPAM), which is supported by the National Science Foundation (Grant No. DMS-1925919). We would also like to thank the staff at IPAM for hosting WiSDM from August 7 to August 11, 2023.

We thank the anonymous reviewers for their valuable feedback to individual chapters. Finally, we acknowledge Linda Ness and Kathryn Leonard for their inspiring example, for their commitment to the community and for their guidance.

Contents

Part I
Matrix and Tensor Methods

Randomized Iterative Methods for Tensor Regression Under the t-Product

Alejandra Castillo, Jamie Haddock, Iryna Hartsock, Paulina Hoyos, Lara Kassab, Alona Kryshchenko, Kamila Larripa, Deanna Needell, Shambhavi Suryanarayanan, and Karamatou Yacoubou Djima

1 Introduction

The extreme challenges of modern data analysis can be caused not just by the size of data sets but also by the inherent complexity of this data. Often, this data is *multimodal*, with modes representing measurements along different dimensions, e.g., spatial and temporal dimensions of video data or word and document dimensions of text corpora data. Such data is naturally formatted as a *tensor*, a multidimensional array, or, in other words, a higher-order generalization of a matrix. In a tensor, the number of dimensions (or *modes*) is called the *order* of the tensor; the higher the tensor order, the higher the complexity. Because the development of data analytic methods for tensor data is far behind that for matrices, practitioners must frequently

A. Castillo
Pomona College, Claremont, CA, USA
e-mail: alejandra.castillo@pomona.edu

J. Haddock (✉)
Harvey Mudd College, Claremont, CA, USA
e-mail: jhaddock@g.hmc.edu

I. Hartsock
H. Lee Moffitt Cancer Center & Research Institute, Tampa, FL, USA
e-mail: iryna.hartsock@moffitt.org

P. Hoyos
University of Texas at Austin, Austin, TX, USA
e-mail: paulinah@utexas.edu

L. Kassab · D. Needell
University of California, Los Angeles, CA, USA
e-mail: lkassab@math.ucla.edu; deanna@math.ucla.edu

C. Garcia-Cardona, H. Lee (eds.), *Advances in Data Science*, Association for Women in Mathematics Series 37, https://doi.org/10.1007/978-3-031-87804-6_1

first transform their tensor data and apply inadequate matrix-based methods that ignore the natural, unified structure of the data as tensor.

Many data analytic problems in the realm of tensors come with a unique set of challenges not encountered for analogous tasks for matrix (lower-order tensor) data. For example, the notion of tensor rank is not uniquely defined [33], unlike the rank of matrices. Further, the various definitions of tensor rank are not computable in polynomial time [27], as opposed to the case of matrices. The landscape is far more complex for tensor data; computations with tensor data often remain challenging even when their matrix counterparts can be taught in introductory linear algebra courses.

Solving large-scale systems of linear equations or linear regressions is one of the most commonly encountered problems across the data-rich sciences. This problem arises in machine learning, as subroutines of several optimization methods [7], in medical imaging [18, 26], in sensor networks [61], and in statistical analysis, to name only a few. In the matrix-vector and matrix-matrix regime, this problem is very well-understood with many highly efficient methods with provable guarantees in the literature. For example, the Kaczmarz method and the Gauss-Seidel method are popular and memory-efficient iterative methods that have been well-studied. Iterative methods are algorithms that begin with an approximation to the solution $\boldsymbol{x}^{(\mathbf{0})}$ and then provide a series of improved approximations that converge to the solution set. If a system of equations is large, iterative methods are advantageous because they allow control of round-off error in contrast to elimination methods, such as Gaussian elimination. Additionally, if one can make an accurate initial guess (e.g., based on the physical context of the problem), this can lead to faster convergence than seen in elimination methods.

Recently, iterative methods have been proposed for a variety of tensor linear systems and regression problems [10, 39, 67]. Tensor regression problems arise organically in settings in which the model inputs or outputs are naturally formulated as a multidimensional tensor array, and the tensor product governs the dependence structure between input and output tensors. Examples include weather and climate forecasting, age estimation from medical imaging data or other biomarker information, and many others; see [36] for more details and an excellent survey of tensor regression models. In this paper, we will be concerned with tensor regression under the *tensor t-product*. Iterative methods for tensor regression can be applied for a variety of deblurring [31], denoising [43], and dictionary representation-learning

A. Kryshchenko
California State University Channel Islands, Camarillo, CA, USA

K. Larripa
Cal Poly Humboldt, Arcata, CA, USA

S. Suryanarayanan
Princeton University, Princeton, NJ, USA

K. Y. Djima
Wellesley College, Wellesley, MA, USA

imaging application [50]; each of these can be formulated as a possibly regularized tensor regression problem:

$$\min_{\mathcal{X}\in\mathfrak{X}} \|\mathcal{B} - \mathcal{A}\mathcal{X}\|_F^2 + \Phi(\mathcal{X}) \tag{1}$$

where $\mathcal{A} \in \mathbb{R}^{m\times n\times p}$ is the measurement operator or dictionary, $\mathcal{B} \in \mathbb{R}^{m\times l\times p}$ represents the measurements or data, $\mathcal{X} \in \mathfrak{X} \subset \mathbb{R}^{n\times l\times p}$ is the signal of interest, $\mathcal{A}\mathcal{X}$ is the t-product between $\mathcal{A}$, and $\mathcal{X}$ [32] defined in Sect. 2, and Φ is a choice of regularizer.

1.1 Contributions

Our main contributions in this paper are twofold. The first main contribution is to provide a survey of the growing area of literature dedicated to iterative methods for tensor regression and related problems. This survey is presented in Sect. 3. Our second main contribution is to provide new generalizations of iterative methods for matrix-vector regression problems to the tensor regime. We first generalize the randomized Gauss-Seidel method [34] to tensor linear systems and explore its application to consistent systems; see Sect. 4.1. We additionally consider the regime in which the linear system is defined by an operator with a given factorization and generalize the randomized Kaczmarz variant of [41] to the tensor linear system regime; see Sect. 4.2. Finally, we consider tensor linear systems that may be corrupted by adversarial perturbations and generalize the *quantile randomized Kaczmarz* method of [22] to the tensor regime; see Sect. 4.3. These three new directions are important to explore as the complexity of tensor linear systems demands methods tailored to new settings, such as systems defined by "wide" operators, systems defined by operators with known factorization, or systems with measurements perturbed by corruption. Existing methods in the literature are not yet amenable to such domains; our work provides initial methods that can tackle these challenging tensor systems.

We note that our work, and much of the work dealing with iterative methods for tensor linear regression, parallels and generalizes the literature dealing with iterative methods for matrix-vector linear regression. To help illustrate the parallels in the literature, we provide Table 1 and include citations to some relevant literature that has motivated our work.

Table 1 Related literature summary

	Factorized system	Column-action	Corruption-robust
Matrix-vector	[41]	[34]	[22]
Tensor-tensor (t-product)	This paper		

1.2 Organization

We begin with the necessary background, definitions, and notation in Sect. 2. In Sect. 3, we present our first main contribution, a survey of iterative methods for tensor regression and related problems. In Sect. 4, we present our second main contribution, our proposed new iterative methods for tensor regression problems in a variety of challenging settings. In Sect. 4.1, we generalize the randomized Gauss-Seidel method to the tensor setting and present initial numerical experiments illustrating the behavior of the method on systems of a variety of sizes. In Sect. 4.2, we generalize the RK-RK method of [41] to tensor linear systems defined with factorized operator and present initial numerical experiments illustrating the method on a variety of factorized operators. In Sect. 4.3, we present the generalization of the quantile randomized Kaczmarz method of [22] to the regime of tensor linear systems possibly corrupted adversarially and experiment with this method on synthetic corrupted systems. We end with some illustrative numerical experiments applying these methods to a simple tensor linear system formulation of the video deconvolution problem in Sect. 5. Finally, we present some conclusions and discussion of future directions in Sect. 6.

2 Background and Notation

2.1 Notation

We use boldfaced lowercase Latin letters (e.g., $\boldsymbol{x}$) to denote vectors, boldfaced uppercase Latin letters (e.g., $\boldsymbol{A}$) to denote matrices, and boldfaced uppercase calligraphic Latin letters (e.g., $\mathcal{A}$) to denote higher-order tensors. We use lightfaced lowercase Latin and Roman letters (e.g., q and β) to denote scalars. We let $[m]$ denote the set $\{1, 2, \cdots, m\}$. We utilize "MATLAB" notation; e.g., $\boldsymbol{A}_{i:}$ is the ith row of matrix $\boldsymbol{A}$ and $\mathcal{A}_{:j:}$ is the jth column-slice of tensor $\mathcal{A}$. We use $\mathcal{A}^*$ to denote the conjugate transpose of the tensor $\mathcal{A} \in \mathbb{C}^{m\times n\times p}$, which is obtained by taking the conjugate transpose of each of the frontal slices and then reversing the order of transposed frontal slices 2 through p.

The notation $\|\boldsymbol{v}\|$ denotes the Euclidean norm of a vector $\boldsymbol{v}$ and $\|\cdot\|_F$ the Frobenius norm of any tensor input. Throughout, we denote by $\sigma_{\min}(\boldsymbol{A})$ the smallest singular values of the matrix $\boldsymbol{A}$ (i.e., the smallest eigenvalue of the matrix $\sqrt{\boldsymbol{A}^\top \boldsymbol{A}}$). We use $\boldsymbol{A} \otimes \boldsymbol{B}$ to denote the Kronecker product of matrices $\boldsymbol{A}$ and $\boldsymbol{B}$.

2.2 Background on Kaczmarz Methods

Before we begin our survey of iterative methods for tensor regression, we remind the reader of one of the most popular and well-studied iterative methods for matrix-vector and matrix-matrix regression: the Kaczmarz method. The Kaczmarz algorithm aims to find a solution $\boldsymbol{x}$ of a system of equations $\boldsymbol{A}\boldsymbol{x} = \boldsymbol{b}$, where $\boldsymbol{A}$ is an $m \times n$ matrix, $\boldsymbol{x}$ is a vector of unknowns, and $\boldsymbol{b}$ is a vector of constants. The algorithm iteratively updates the approximation by selecting one equation at a time and projecting the current approximation onto the hyperplane defined by that equation. This process continues until convergence is reached or a maximum number of iterations is reached. The update is defined by

$$\boldsymbol{x}^{(k)} = \boldsymbol{x}^{(k-1)} + \frac{b_{i_k} - \boldsymbol{A}_{i_k:}\boldsymbol{x}^{(k-1)}}{\|\boldsymbol{A}_{i_k:}\|^2}\boldsymbol{A}_{i_k:}^\top, \tag{2}$$

where $\boldsymbol{A}_{i_k:}$ is the selected i_kth row of $\boldsymbol{A}$ and b_{i_k} is the selected i_kth entry of $\boldsymbol{b}$ at iteration k. The Kaczmarz method is particularly useful when the system of equations is large and sparse, meaning that most entries in the matrix $\boldsymbol{A}$ are zero. It allows for efficient computation by updating the solution one equation at a time, making it suitable for problems with a large number of equations.

In [66], the authors introduced a randomized variant of the Kaczmarz method where the probability that the ith row of $\boldsymbol{A}$ is sampled in the kth iteration is $\|\boldsymbol{A}_{i:}\|^2/\|\boldsymbol{A}\|_F^2$. The authors showed that for a consistent system with unique solution $\boldsymbol{x}^*$, the randomized Kaczmarz (RK) method converges at least linearly in expectation with the guarantee:

$$\mathbb{E}\|\boldsymbol{x}^{(k)} - \boldsymbol{x}^*\|^2 \le \left(1 - \frac{\sigma_{\min}^2(\boldsymbol{A})}{\|\boldsymbol{A}\|_F^2}\right)^k \|\boldsymbol{x}^{(0)} - \boldsymbol{x}^*\|^2. \tag{3}$$

Many variants and extensions followed, including convergence analyses for inconsistent and random linear systems [9, 46], connections to other popular iterative algorithms [13, 40, 47, 53, 54], block approaches [48, 55], acceleration and parallelization strategies [14, 35, 44, 45], and techniques for reducing noise and corruption [21, 81]. Lastly, it is worth noting that there is a clear relationship between row-action methods like randomized Kaczmarz and column-action methods like Gauss-Siedel. It is often the case, however, that when row-action methods are more efficient (i.e., the system is over-determined), the methods do not guarantee convergence to the least-squares solution without modification, and vice versa for column-action methods, which do not converge to the least-norm solution in the under-determined case. While not the focus of this paper, the interested reader can refer to [25] for a thorough discussion of this relationship and trade-off.

2.3 Background on Tensor t-Product Algebra

Before we may discuss the generalization of the Kaczmarz method, and other iterative methods, to the tensor regression regime, we require some definitions and will discuss the tensor t-product. The first definition is the *block-circulant* matrix of a tensor $\mathcal{T}$, bcirc$(\mathcal{T})$. For $\mathcal{T} \in \mathbb{R}^{m\times n\times p}$, we denote by bcirc$(\mathcal{T}) \in \mathbb{R}^{mp\times np}$ the matrix

$$\operatorname{bcirc}(\mathcal{T}) = \begin{bmatrix} \mathcal{T}_0 & \mathcal{T}_{p-1} & \mathcal{T}_{p-2} & \cdots & \mathcal{T}_1 \\ \mathcal{T}_1 & \mathcal{T}_0 & \mathcal{T}_{p-1} & \cdots & \mathcal{T}_2 \\ \vdots & & & \cdots & \vdots \\ \mathcal{T}_{p-1} & \mathcal{T}_{p-2} & \mathcal{T}_{p-3} & \cdots & \mathcal{T}_0 \end{bmatrix}$$

We use $\mathcal{T}_k$ to denote the kth frontal slice of $\mathcal{T}$ (i.e., $\mathcal{T}_k := \mathcal{T}_{::k}$ is a $m \times n$ matrix).

Additionally, the t-product is defined in terms of tensor fiber transformation by the discrete Fourier transform (DFT). The discrete Fourier transform (DFT) matrix of size $p \times p$ is defined as

$$\mathbf{F}_p = \frac{1}{\sqrt{p}} \begin{bmatrix} 1 & 1 & 1 & \cdots & 1 \\ 1 & \omega & \omega^2 & \cdots & \omega^{p-1} \\ 1 & \omega^2 & \omega^4 & \cdots & \omega^{2(p-1)} \\ \vdots & \vdots & \vdots & \ddots & \vdots \\ 1 & \omega^{p-1} & \omega^{2(p-1)} & \cdots & \omega^{(p-1)(p-1)} \end{bmatrix}$$

where $\omega = e^{-\frac{2\pi i}{p}}$ is the principal p-th root of unity.

The tensor *t-product* [32] of tensors $\mathcal{T} \in \mathbb{R}^{m\times n\times p}$ and $\mathcal{S} \in \mathbb{R}^{n\times l\times p}$ is the tensor $\mathcal{T}\mathcal{S}$ of dimension $m \times l \times p$ given by

$$(\mathcal{T}\mathcal{S})_{[2]} := (\boldsymbol{F}_p^* \otimes \boldsymbol{I}_{n\times n})\widehat{\mathcal{T}}(\boldsymbol{F}_p \otimes \boldsymbol{I}_{n\times n})\mathcal{S}_{[2]}, \tag{4}$$

where

$$\mathcal{S}_{[2]} := \begin{bmatrix} \mathcal{S}_1 \\ \vdots \\ \mathcal{S}_p \end{bmatrix} \in \mathbb{R}^{np\times l}$$

is the unfolding of $\mathcal{S}$ along its second mode, and

$$\widehat{\mathcal{T}} := (\boldsymbol{F}_p \otimes \boldsymbol{I}_{m\times m})\operatorname{bcirc}(\mathcal{T})(\boldsymbol{F}_p^* \otimes \boldsymbol{I}_{n\times n}) = \operatorname{diag}\left(\widehat{\mathcal{T}}_1, \ldots, \widehat{\mathcal{T}}_p\right) \tag{5}$$

is the circular discrete Fourier transform (DFT) of $\mathcal{T}$ with the $p \times p$ DFT matrix $\boldsymbol{F}_p$.

As with matrix-vector linear system iterative methods, the row space and range of a given tensor are important concepts for defining and analyzing tensor linear system iterative methods. The *row space* of $\mathcal{A}$ is defined as

$$R(A) = \{\mathcal{A}^T \mathcal{Y} \colon \mathcal{Y} \in \mathbb{R}^{m \times k \times p}\}$$

If $p, k = 1$, then $R(\mathcal{A})$ coincides with the row space of the $m \times n$ matrix $\mathcal{A}$. A related space is the *k-range space* of tensor $\mathcal{A}$, which is defined as

$$range_k(\mathcal{A}) = \{\mathcal{A}\mathcal{Y} \colon \mathcal{Y} \in \mathbb{R}^{m \times k \times p}\}.$$

Finally, we note that the definition of the t-product implies that a tensor linear system may be reformulated as an equivalent linear system.

Fact 1 *The tensor linear system*

$$\mathcal{A}\mathcal{X} = \mathcal{B}$$

is equivalent to the matrix-matrix linear system

$$\text{bcirc}(\mathcal{A})\mathcal{X}_{[2]} = \mathcal{B}_{[2]};$$

that is, solutions to the tensor linear system, $\mathcal{X}$, after unfolding, $\mathcal{X}_{[2]}$, are solutions to the matrix linear system and vice versa.

We will exploit this fact to compare iterative methods for tensor linear systems to their counterpart iterative methods for matrix linear systems. We note that the data in a single row slice of the tensor system $\mathcal{A}\mathcal{X} = \mathcal{B}$ is distributed among a block of rows of the equivalent matricized system $\text{bcirc}(\mathcal{A})\mathcal{X}_{[2]} = \mathcal{B}_{[2]}$. Row-slice-action methods on the tensor system $\mathcal{A}\mathcal{X} = \mathcal{B}$ are equivalent to carefully constructed row-block-action methods on the matricized system $\text{bcirc}(\mathcal{A})\mathcal{X}_{[2]} = \mathcal{B}_{[2]}$. In the matrix setting, there has been interest in understanding and solving structured linear regression problems defined with block-circulant matrices [4]. It is also pertinent to note that block circulant matrices naturally pop up in many applications including computer vision, vibration analysis [51], and time series analysis [52].

3 Survey of Related Work for Tensor Regression

In this section, we survey existing literature on methods and models for tensor regression. This survey is certainly not exhaustive and is focused on iterative methods for tensor regression and related problems under the t-product. The t-product was defined in the foundational work [32] as the product between two three-order tensors. The authors subsequently derived formulations of the associated tensor identity, inverse, pseudoinverse, and transpose and extended orthogonal matrix factorizations such as the SVD and QR factorizations to tensors. The t-product has gained significant traction and is now being applied in dictionary learning [50, 64, 76], low-rank tensor completion [62, 77, 78, 80], low-rank tensor recovery [10], facial recognition [23, 74], and neural networks [49, 71]. The t-product-based decompositions have been shown to be more efficient than their equivalent matrix formulations in many multimodal settings [38, 78].

3.1 Consistent Tensor Linear Systems

In [39], A. Ma and D. Molitor proposed the generalization of the randomized Kaczmarz method for tensor linear systems defined under the t-product, called tensor randomized Kaczmarz (TRK). This method begins with an initial approximation $\mathcal{X}^{(0)}$ to the solution $\mathcal{X}^*$ to the tensor linear system $\mathcal{A}\mathcal{X} = \mathcal{B}$ and iteratively samples a row slice of the system defined by $\mathcal{A}_{i_k::}$ and $\mathcal{B}_{i_k::}$ and projects the previous iterate onto the space of solutions to this sampled subsystem. The pseudocode is provided in Algorithm 1. The authors prove that this method converges at least linearly in expectation to the unique solution $\mathcal{X}^*$ of the system.

Algorithm 1 Tensor Randomized Kaczmarz (TRK) [39]

1: **procedure** TRK($\mathcal{A}$, $\mathcal{B}$, K)
2: $\quad\mathcal{X}^{(0)} = 0$
3: $\quad$**for** $k = 1, \dots, K$ **do**
4: $\qquad$Sample $i_k \in [m]$.
5: $\qquad\mathcal{X}^{(k)} = \mathcal{X}^{(k-1)} - \mathcal{A}^*_{i_k::}(\mathcal{A}_{i_k::}\mathcal{A}^*_{i_k::})^{-1}(\mathcal{A}_{i_k::}\mathcal{X}^{(k-1)} - \mathcal{B}_{i_k::})$
6: $\quad$**end for**
7: **return** $\mathcal{X}^{(K)}$
8: **end procedure**

Further, the authors observed that, while the TRK method can equivalently be viewed as a block Kaczmarz method applied to a matrix-matrix system for a specific choice of blocks, a naive application of the block Kaczmarz method would give us weaker theoretical guarantees than the one that TRK provides. This highlights the advantage of exploiting the intrinsic structure of the data by using row-slice or column-slice-action-based updates over their matrix counterparts.

In [3], W. Bao, F. Zhang, W. Li, Q. Wang, and Y. Gao compute the least Frobenius-norm solution for consistent tensor linear systems of the form

$$\mathcal{A}\mathcal{X} = \mathcal{B}, \tag{6}$$

where $\mathcal{A} \in \mathbb{R}^{m\times n\times p}$, $\mathcal{X} \in \mathbb{R}^{n\times l\times p}$, and $\mathcal{B} \in \mathbb{R}^{m\times l\times p}$. The authors propose the tensor randomized average Kaczmarz (TRAK) method, which is pseudoinverse- and inverse-free and offers a speed-up over the TRK method for solving tensor linear systems (6); see Algorithm 2 for the pseudocode of the TRAK method. The authors prove that the iterates $\mathcal{X}^{(k)}$ converge at least linearly in expectation to the unique least Frobenius-norm solution $\mathcal{X}^* = \mathcal{A}^\dagger \mathcal{B}$.

Algorithm 2 Tensor Randomized Average Kaczmarz (TRAK) [3]

---

procedure TRAK($\mathcal{A}$, $\mathcal{B}$, K, stepsize $\alpha > 0$, partition $\{J_i\}_{i=1}^s$ of $[m]$)
  $\mathcal{X}^{(0)} \in range_l(\mathcal{A}^T)$
  for $k = 1, \ldots, K$ **do**
    Sample $i_k \in [s]$ with probability $\|\mathcal{A}_{J_{i_k}::}\|_F^2 / \|\mathcal{A}\|_F^2$.
    $\mathcal{X}^{(k)} = \mathcal{X}^{(k-1)} - \frac{\alpha}{\|\mathcal{A}_{J_{i_k}::}\|_F^2} (\mathcal{A}_{J_{i_k}::})^T (\mathcal{A}_{i_k::} \mathcal{X}^{(k-1)} - \mathcal{B}_{i_k::})$
  end for return $\mathcal{X}^{(K)}$
end procedure

In [68], L. Tang, Y. Yu, Y. Zhang, and H. Li. consider the consistent tensor linear equation:

$$\mathcal{A}\mathcal{X}\mathcal{B} = \mathcal{C} \tag{7}$$

for given third-order tensors $\mathcal{A} \in \mathbb{R}^{m \times r \times p}$, $\mathcal{B} \in \mathbb{R}^{s \times n \times p}$, and $\mathcal{C} \in \mathbb{R}^{m \times n \times p}$. They propose the tensor regular sketch-and-project method, TESP for short. In this approach, one takes the point that is closest to the current iteration $\mathcal{X}^{(k)}$ and solves a sketched version of the original tensor equation (7) as the next iteration $\mathcal{X}^{(k+1)}$ as described in Algorithm 3. It is shown that the iterates $\mathcal{X}^{(k)}$ converge linearly in expectation to a solution $\mathcal{X}^\star$ of the system $\mathcal{A}\mathcal{X}^\star\mathcal{B} = \mathcal{C}$.

Algorithm 3 Tensor Regular Sketch-and-Project Algorithm (TESP) [68]

---

procedure TESP($\mathcal{A}$, $\mathcal{B}$, $\mathcal{C}$, K)
  $\mathcal{X}^{(0)} \in \mathbb{R}^{r \times s \times p}$
  for $k = 1, \ldots, K$ **do**
    Sample independent copies $\mathcal{S} \sim \mathcal{D}_\mathcal{S}$ and $\mathcal{V} \sim \mathcal{D}_\mathcal{V}$
    Compute $\mathcal{E} = \mathcal{S}(\mathcal{S}^T \mathcal{A} \mathcal{M}^{-1} \mathcal{A}^T \mathcal{S})^\dagger \mathcal{S}^T$ and $\mathcal{G} = \mathcal{V}(\mathcal{V}^T \mathcal{B}^T \mathcal{N}^{-1} \mathcal{B} \mathcal{V})^\dagger \mathcal{V}^T$
    $\mathcal{X}^{(k)} = \mathcal{X}^{(k-1)} - \mathcal{M}^{-1} \mathcal{A}^T \mathcal{E} (\mathcal{A}\mathcal{X}^{(k-1)}\mathcal{B} - \mathcal{C}) \mathcal{G} \mathcal{B}^T \mathcal{N}^{-1}$
  end for return $\mathcal{X}^{(K)}$
end procedure

3.2 *Inconsistent Tensor Linear Systems*

In [31], M. Kilmer, K. Braman, N. Hao, and R. Hoover generalize a number of linear algebraic algorithms to the t-product tensor algebra. They define the conjugate gradient method for this regime and illustrate a number of applications for this method, including image deconvolution. The pseudocode for this method is provided in Algorithm 4. This method can, of course, be applied to consistent linear systems defined by positive definite operators and is additionally appropriate for arbitrary (inconsistent) tensor systems by passing to the normal equations, $\mathcal{A}^\top \mathcal{A}\mathcal{X} = \mathcal{A}^\top \mathcal{B}$.

Algorithm 4 Tensor Conjugate Gradient (TCG) [31]

procedure TCG($\mathcal{A}$, $\mathcal{B}$, K)
    $\mathcal{X}^{(0)} = 0$ ▷ $\mathcal{A}$ assumed positive definite
    $\mathcal{R} = \mathcal{B}$
    $\mathcal{P} = \mathcal{R}$
    for $k = 1, \ldots, K$ **do**
        $\mathcal{R}' = \mathcal{R}$
        $\mathcal{C} = (\mathcal{P}^\top \mathcal{A}\mathcal{P})^{-1}\mathcal{R}^\top \mathcal{R}$
        $\mathcal{X}^{(k)} = \mathcal{X}^{(k-1)} + \mathcal{P}\mathcal{C}$
        $\mathcal{R} = \mathcal{R}' - \mathcal{A}\mathcal{P}\mathcal{C}$
        $\mathcal{D} = ((\mathcal{R}')^\top \mathcal{R}')^{-1}\mathcal{R}^\top \mathcal{R}$
        $\mathcal{P} = \mathcal{R} + \mathcal{P}\mathcal{D}$
    end for
return $\mathcal{X}^{(K)}$
end procedure

In [28], G.-X. Huang and S.-Y. Zhong propose an extended variant of TRK, which they refer to as the tensor randomized extended Kaczmarz (TREK) method; see Algorithm 5 for the method pseudocode. The authors prove that the iterates converge at least linearly in expectation to the solution of the unperturbed tensor system.

Algorithm 5 Tensor Randomized Extended Kaczmarz (TREK) [28]

procedure TREK($\mathcal{A}$, $\mathcal{B}$, K)
    $\mathcal{X}^{(0)} = 0$
    $\mathcal{Z}^{(0)} = \mathcal{B}$
    for $k = 1, \ldots, K$ **do**
        Sample $j_k \in [n]$
        $\mathcal{Z}^{(k)} = \mathcal{Z}^{(k-1)} - \mathcal{A}_{:j_k:}(\mathcal{A}^*_{:j_k:}\mathcal{A}_{:j_k:})^\dagger \mathcal{A}^*_{:j_k:}\mathcal{Z}^{(k-1)}$
        Sample $i_k \in [m]$.
        $\mathcal{X}^{(k)} = \mathcal{X}^{(k-1)} - \mathcal{A}^*_{i_k::}(\mathcal{A}_{i_k::}\mathcal{A}^*_{i_k::})^\dagger(\mathcal{A}_{i_k::}\mathcal{X}^{(k-1)} - \mathcal{B}_{i_k::} + \mathcal{Z}^{(k)}_{i_k::})$
    end for
return $\mathcal{X}^{(K)}$
end procedure

3.3 Convex Optimization over Linear System Constraints

In [10], X. Chen and J. Qin study how to recover $\mathcal{X}$ in the consistent and underdetermined system $\mathcal{A}\mathcal{X} = \mathcal{B}$, where $\mathcal{A} \in \mathbb{R}^{m\times n\times p}$, $\mathcal{X} \in \mathbb{R}^{n\times l\times p}$, and $\mathcal{B} \in \mathbb{R}^{m\times l\times p}$, by solving the minimization:

$$\underset{\mathcal{X}\in\mathbb{R}^{n\times l\times p}}{\operatorname{argmin}} f(\mathcal{X}), \qquad \text{s.t.} \quad \mathcal{A}\mathcal{X} = \mathcal{B}, \tag{8}$$

where f is an α_f-strongly convex function. To solve this linear constrained minimization problem, they propose a general regularized Kaczmarz tensor algorithm, which involves random projections onto the solution space of individual equations, and a gradient calculation on the convex conjugate function of f, f^*. The pseudocode for their method is given in Algorithm 6. The authors prove Algorithm 6 enjoys a linear convergence rate in expectation if the objective function f is α_f strongly convex and admissible. We note that the authors additionally provide a deterministic algorithm, but we do not present this algorithm in detail. They additionally consider a variety of problems and applications, including sparse and low-rank recovery and image deconvolution or deblurring.

Algorithm 6 Tensor Randomized Regularized Kaczmarz (TRRK) [10]

procedure TRRK($\mathcal{A}$, $\mathcal{B}$, K, tolerance tol, stepsize t)
  $\mathcal{Z}^{(0)} \in R(\mathcal{A})$
  $\mathcal{X}^{(0)} = \nabla f^*(\mathcal{Z}^{(0)})$
  for $k = 1, \ldots, K$ **do**
    Sample $i(k)$ with probability $\|\mathcal{A}(i(k))\|_F^2 / \|\mathcal{A}\|_F^2$.
    $\mathcal{Z}^{(k)} = \mathcal{Z}^{(k-1)} + t\mathcal{A}(i(k))^T \frac{\mathcal{B}(i(k)) - \mathcal{A}(i(k))\mathcal{X}^{(k-1)}}{\|\mathcal{A}(i(k))\|_F^2}$
    $\mathcal{X}^{(k)} = \nabla f^*(\mathcal{Z}^{(k-1)})$
    If $\|\mathcal{X}^{(k)} - \mathcal{X}^{(k-1)}\|_F / \|\mathcal{X}^{(k-1)}\|_F < tol$, return $\mathcal{X}^{(k-1)}$.
  end for return $\mathcal{X}^{(K)}$
end procedure

In [12], Kui Du and Xiao-Hui Sun assume that the linear system $\mathcal{A}\mathcal{X} = \mathcal{B}$ is inconsistent and therefore considers the constrained minimization problem:

$$\hat{\mathcal{X}} = \underset{\mathcal{X}\in\mathbb{R}^{n\times l\times p}}{\arg\min} f(\mathcal{X}) \quad \text{s.t.} \quad \mathcal{A}^T\mathcal{A}\mathcal{X} = \mathcal{A}^T\mathcal{B}. \tag{9}$$

The authors propose the tensor randomized regularized extended Kaczmarz (TRREK) method; see Algorithm 7 for the pseudocode. If f is a strongly admissible and γ-strongly convex function, then the TRREK algorithm converges linearly in expectation to the solution $\hat{\mathcal{X}}$ of the least-squares problem (9). Their method is an extended generalization of that in [10], which allows it to pass below the usual convergence horizon of a Kaczmarz-type method.

Algorithm 7 Tensor Randomized Regularized Extended Kaczmarz (TRREK) [12]

procedure TRREK($\mathcal{A}$, $\mathcal{B}$, K, stepsizes $\alpha_\Gamma, \alpha_c > 0$)
  $\mathcal{Z}^{(0)} = \mathcal{B}, \mathcal{Y}^{(0)} \in range_l(\mathcal{A}^T), \mathcal{X}^{(0)} = \nabla f^*(\mathcal{Y}^{(0)})$.
  for $k = 1, \ldots, K$ **do**
    Sample $j_k \in [n]$ with probability $\|\mathcal{A}_{:j_k:}\|_F^2 / \|\mathcal{A}\|_F^2$.
    $\mathcal{Z}^{(k)} = \mathcal{Z}^{(k-1)} - \frac{\alpha_c}{\|\mathcal{A}_{:j_k:}\|_F^2}\mathcal{A}_{:j_k:}(\mathcal{A}_{:i_k:})^T \mathcal{Z}^{(k-1)}$
    Sample $i_k \in [m]$ with probability $\|\mathcal{A}_{:i_k:}\|_F^2 / \|\mathcal{A}\|_F^2$.
    $\mathcal{Y}^{(k)} = \mathcal{Y}^{(k-1)} - \frac{\alpha_\Gamma}{\|\mathcal{A}_{i_k::}\|_F^2}(\mathcal{A}_{:j_k:})^T(\mathcal{A}_{i_k::}\mathcal{X}^{(k-1)} - \mathcal{B}_{i_k::} + \mathcal{Z}^{(k)}_{i_k::})$
    $\mathcal{X}^{(k)} = \nabla f^*(\mathcal{Y}^{(k)})$
  end for return $\mathcal{X}^{(K)}$
end procedure

3.4 Regression Under Other Tensor Products

While the focus of this paper is linear tensor regression under the t-product, there has been work dealing with tensor regression under other tensor products. Broadly, this work tends to be focused either on frameworks or models of tensor regression, on algorithmic approaches for training or fitting a given tensor regression model, on applications of these models, or combinations of these. In [37], the authors propose a framework for the linear tensor regression problem under the contracted tensor product. In [73], the authors introduce and analyze a subsampled tensor projected gradient method for tensor regression. In [79], the authors propose a family of rank-R generalized linear tensor regression models and suggest training this model with an alternating block relaxation method. The survey [36] contains a good overview of the broad class of tensor regression models, training methods, and application of these techniques. More generally, many tensor decomposition training methods utilize an alternating optimization scheme in which they solve subproblems holding all but one factor fixed. These subproblems are special instances of tensor regression problems that can occasionally be rewritten to ordinary least-squares problems; see, e.g., [8, 24, 33].

4 Proposed Iterative Methods for Tensor Linear Systems

In this section, we provide new approaches for tensor linear systems in a variety of scenarios. First, in Sect. 4.1, we propose randomized column-slice-action iterative methods for tensor linear systems, which are especially important for systems where the measurement tensor is smaller than the signal tensor. In Sect. 4.2, we provide randomized iterative methods for tensor linear systems with a factorized measurement operator. Finally, in Sect. 4.3, we propose randomized iterative methods for tensor linear systems that are robust to sparse adversarial corruptions in the measurement tensor.

4.1 Column-Slice-Action Methods

While the row-oriented Kaczmarz-type methods are being actively explored in the tensor regression setting, coordinate-wise or column-action methods have been considered far less in the literature. The benefit of developing a column-action method is clear in situations where row slices of the measurement tensor are extremely large and cannot be stored in active memory, or the tensor data is naturally stored in column-slice components (e.g., distributed across computational servers or priority indexed by column). In this setting, accessing column-slices of the tensor may be the only reliable form of data access available. We will generalize classical column-action methods for matrix-vector systems, the Jacobi and Gauss-Seidel methods. First, we will describe how the Jacobi and Gauss-Seidel methods solve a linear system by operating on columns and show that randomized Gauss-Seidel (RGS) can be viewed as a variant of these methods applied to the normal equations. In [34], the authors prove that the residuals of randomized RGS converge linearly in expectation, yielding an improvement over the convergence rate in the classical settings.

Classical Gauss-Seidel Method The classical Jacobi and Gauss-Seidel methods are iterative methods used to solve a system of linear equations $\boldsymbol{A}\boldsymbol{x} = \boldsymbol{b}$, where the square matrix $\boldsymbol{A}$ and the vector $\boldsymbol{b}$ are known and the goal is to approximate $\boldsymbol{x}$. The Gauss-Seidel method was developed in the 1800s and is considered one of the first iterative methods developed [60]. It is taught in undergraduate numerical method courses and is similar to the Jacobi method, with the main difference being when updates are applied.

When considering $\boldsymbol{A}\boldsymbol{x} = \boldsymbol{b}$, the matrix $\boldsymbol{A}$ is decomposed into the sum of a strictly lower triangular matrix $\boldsymbol{L}$, a diagonal matrix $\boldsymbol{D}$, and a strictly upper triangular matrix $\boldsymbol{U}$, $\boldsymbol{A} = \boldsymbol{D} + \boldsymbol{L} + \boldsymbol{U}$. This allows the system of linear equations to be rewritten as $\boldsymbol{D}\boldsymbol{x} + \boldsymbol{L}\boldsymbol{x} + \boldsymbol{U}\boldsymbol{x} = \boldsymbol{b}$. The Jacobi method exploits this rewritten system and produces a fixed-point iterative method on the fixed-point equation $\boldsymbol{D}\boldsymbol{x} = -(\boldsymbol{L} + \boldsymbol{U})\boldsymbol{x} + \boldsymbol{b}$ of the form

$$\boldsymbol{x}^{(k)} = -\boldsymbol{D}^{-1}(\boldsymbol{L} + \boldsymbol{U})\boldsymbol{x}^{(k-1)} + \boldsymbol{D}^{-1}\boldsymbol{b}.$$

Entry-wise, this takes the form

$$\boldsymbol{x}_i^{(k)} = -\frac{1}{A_{ii}} \sum_{j \neq i} A_{ij} \boldsymbol{x}_j^{(k-1)} + \frac{1}{A_{ii}} \boldsymbol{b}_i.$$

The Gauss-Seidel method, meanwhile, uses the fixed-point equation: $(\boldsymbol{D} + \boldsymbol{L})\boldsymbol{x} = -\boldsymbol{U}\boldsymbol{x} + \boldsymbol{b}$ to construct the fixed-point iterative method

$$\boldsymbol{x}^{(k)} = -(\boldsymbol{D} + \boldsymbol{L})^{-1}\boldsymbol{U}\boldsymbol{x}^{(k-1)} + (\boldsymbol{D} + \boldsymbol{L})^{-1}\boldsymbol{b}.$$

This is equivalent to $\boldsymbol{x}^{(k)} = -\boldsymbol{D}^{-1}\boldsymbol{L}\boldsymbol{x}^{(k)} - \boldsymbol{D}^{-1}\boldsymbol{U}\boldsymbol{x}^{(k-1)} + \boldsymbol{D}^{-1}\boldsymbol{b}$, which takes the following form entry-wise:

$$\boldsymbol{x}_i^{(k)} = -\frac{1}{A_{ii}}\sum_{j=1}^{i-1} A_{ij}\boldsymbol{x}_j^{(k)} - \frac{1}{A_{ii}}\sum_{j=i+1}^{n} A_{ij}\boldsymbol{x}_j^{(k-1)} + \frac{1}{A_{ii}}\boldsymbol{b}_i.$$

The convergence properties of the Jacobi and Gauss-Seidel method are dependent on the properties of the matrix $\boldsymbol{A}$, specifically upon the spectral radius of the matrices involved in the Jacobi and Gauss-Seidel updates, $-\boldsymbol{D}^{-1}(\boldsymbol{L}+\boldsymbol{U})$ and $-(\boldsymbol{D}+\boldsymbol{L})^{-1}\boldsymbol{U}$, respectively.

Randomized Gauss-Seidel (RGS) Method In the recent literature, the method referred to as *randomized Gauss-Seidel* is, in fact, a variant of randomized coordinate descent applied to the least-squares objective. We note that this can be viewed as a variant of either Jacobi's method or Gauss-Seidel applied to the normal equations $\boldsymbol{A}^\top\boldsymbol{A}\boldsymbol{x} = \boldsymbol{A}^\top\boldsymbol{b}$ in which only a single coordinate is updated. This method iterates by minimizing a subset of the residual error with respect to a single coordinate; the kth iterate is

$$\boldsymbol{x}^{(k)} = \boldsymbol{x}^{(k-1)} - \frac{\boldsymbol{A}_{:i_k}^T(\boldsymbol{A}\boldsymbol{x}^{(k-1)} - \boldsymbol{b})}{\|\boldsymbol{A}_{:i_k}\|^2}\boldsymbol{e}_{i_k}, , \tag{10}$$

where $\boldsymbol{A}_{:i_k}$ is the i_kth column of $\boldsymbol{A}$. These methods have found success in subroutines for multigrid methods [59, 69], high-performance computing [72], and PDEs [16, 42]. Here, the probability that the jth column of $\boldsymbol{A}$ is sampled in the kth iteration is $\|\boldsymbol{A}_{:j}\|^2/\|\boldsymbol{A}\|_F^2$; that is, the probability is proportional to the square of the Euclidean norm of the column [34]. The algorithm has an expected linear convergence rate [34] given by

$$\mathbb{E}\|\boldsymbol{x}^{(k)} - \boldsymbol{x}^*\|^2 \le \left(1 - \frac{\sigma_{\min}^2(\boldsymbol{A})}{\|\boldsymbol{A}\|_F^2}\right)^k \|\boldsymbol{x}^{(0)} - \boldsymbol{x}^*\|^2. \tag{11}$$

This method and variants have additionally been analyzed in [15, 40, 57].

Extension to Tensor Settings We now consider the consistent tensor linear system $\mathcal{A}\mathcal{X} = \mathcal{B}$, where $\mathcal{B} \in \mathbb{C}^{m\times l\times p}$, $\mathcal{A} \in \mathbb{C}^{m\times n\times p}$, and $\mathcal{X} \in \mathbb{C}^{n\times l\times p}$. We have formulated the tensor version of the randomized Gauss-Seidel for this setting:

$$\mathcal{X}^{(k)} = \mathcal{X}^{(k-1)} - \mathcal{E}_j(\mathcal{A}_{:j:}^*\mathcal{A}_{:j:})^{-1}\mathcal{A}_{:j:}^*(\mathcal{A}\mathcal{X}^{(k-1)} - \mathcal{B}) \tag{12}$$

with $\mathcal{X}^{(k)}$ and $\mathcal{X}^{(k-1)}$ in $\mathbb{C}^{n\times l\times p}$, $\mathcal{E}_j$ in $\mathbb{C}^{n\times 1\times p}$ a vertical slice tensor with the first frontal slice being a standard basis vector e_j with a 1 in the jth coordinate and 0's elsewhere and the rest of the frontal slices are all vectors of zeros, $\mathcal{A}_{:j:}^*$ in $\mathbb{C}^{1\times m\times p}$,

$\mathcal{A}_{:j:}$ in $\mathbb{C}^{m\times 1\times p}$, and j is a uniformly sampled column index from the set of indices $[n]$. We give the pseudocode for this method in Algorithm 8.

Algorithm 8 Tensor Randomized Gauss-Seidel (TRGS)

procedure TRGS($\mathcal{A}$, $\mathcal{B}$, K)
 $\mathcal{X}^{(0)} = 0, \mathcal{R}^{(0)} = \mathcal{A}\mathcal{X}^{(0)} - \mathcal{B}$
 for $k = 1, 2, ..., K$ **do**
  Sample $j_k \in [n]$
  $\mathcal{X}^{(k)} = \mathcal{X}^{(k-1)} - \mathcal{E}_{j_k}(\mathcal{A}^*_{:j_k:}\mathcal{A}_{:j_k:})^{-1}\mathcal{A}^*_{:j_k:}\mathcal{R}^{(k-1)}$
  $\mathcal{R}^{(k)} = \mathcal{A}\mathcal{X}^{(k)} - \mathcal{B}$
 end for return $\mathcal{X}^{(K)}$
end procedure

Comparing TRGS to Matrix Method and TRK The performance of our proposed TRGS algorithm was studied empirically on synthetic data. In the first experiment, we compared the performance of the TRGS to TRK method. In Fig. 1, we consider tensor $\mathcal{A} \in \mathbb{R}^{50\times 20\times 30}$ generated with i.i.d. random Gaussian entries and tensor $\mathcal{X}^*$ the same as described above. It looks like TRGS and TRK perform in a very similar manner.

In the second suite of experiments, tensors $\mathcal{A} \in \mathbb{R}^{m\times 20\times 30}$ and $\mathcal{X}^* \in \mathbb{R}^{20\times 10\times 30}$ were generated with i.i.d. random Gaussian entries. Different cases spanning the under-determined and over-determined setting of the problem were considered by varying m, the dimension of the first mode of tensor $\mathcal{A}$ as tabulated in Table 2. In Figs. 2, 3, 4, 5, 6, we compare the performance of our TRGS algorithm with that of matrix RGS on the equivalent system, $\text{bcirc}(\mathcal{A})\mathcal{X}_{[2]} = \mathcal{B}_{[2]}$, defined in Fact 1. Throughout these experiments, the matricized error is $\|(\mathcal{X}^{(k)} - \mathcal{X}^*)_{[2]}\|/\|\mathcal{X}^*_{[2]}\|$; the matricized least-norm error is $\|(\mathcal{X}^{(k)} - \mathcal{X}_{LN})_{[2]}\|/\|(\mathcal{X}_{LN})_{[2]}\|$, where $\mathcal{X}_{LN}$ is the least-norm solution; and the matricized residual error is $\|\text{bcirc}(\mathcal{A})\mathcal{X}^{(k)}_{[2]} - \mathcal{B}_{[2]}\|$. In the under-determined setting (Cases 1–2), we see that TRGS performs better, as evidenced by the rapid convergence of the residual error. This suggests that the TRGS converges to a solution of the system, but the nature of this solution is unclear and a potential direction of our future work. In the over-determined setting (Cases

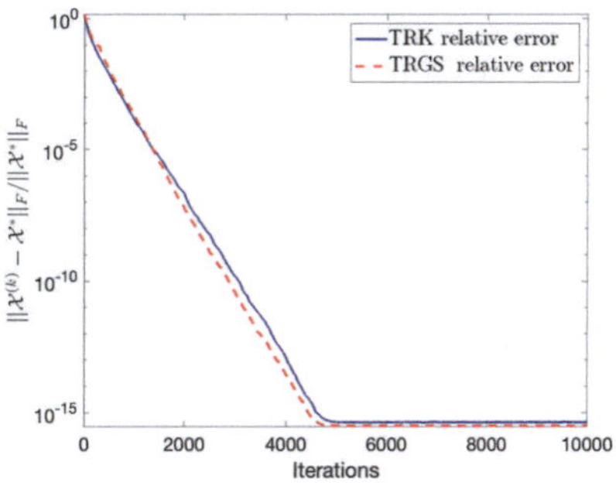

Fig. 1 Relative error of solution, $\|\mathcal{X}^{(k)} - \mathcal{X}^*\|_F/\|\mathcal{X}^*\|_F$, over all iterations of TRK and TRGS for $\mathcal{A} \in \mathbb{R}^{50\times 20\times 30}$ where $\mathcal{X}^{(k)}$ are iterates and $\mathcal{X}^*$ is the exact solution

Table 2 Cases of the numerical experiments of RGS algorithm on various sizes of tensor $\mathcal{A}$. The dimensions of the $\mathcal{X}$ were $20 \times 10 \times 30$, the same for all the cases

	Under-determined		Over-determined
Case 1: Fig. 2	$\mathcal{A} \in \mathbb{R}^{15\times 20\times 30}$	Case 3: Fig. 4	$\mathcal{A} \in \mathbb{R}^{40\times 20\times 30}$
Case 2: Fig. 3	$\mathcal{A} \in \mathbb{R}^{10\times 20\times 30}$	Case 4: Fig. 5	$\mathcal{A} \in \mathbb{R}^{30\times 20\times 30}$
		Case 5: Fig. 6	$\mathcal{A} \in \mathbb{R}^{20\times 20\times 30}$

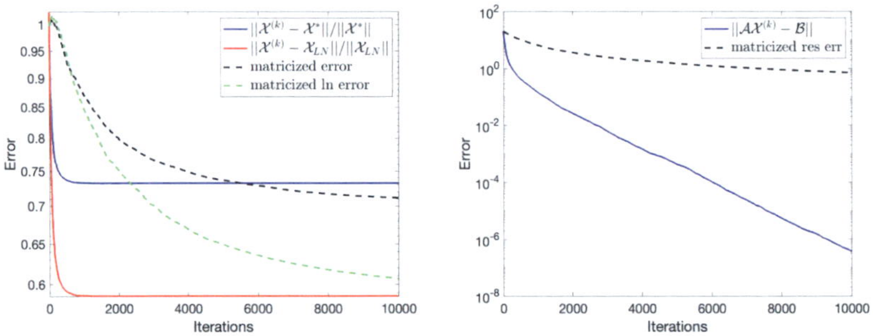

Fig. 2 Relative (left) and residual (right) solution error with iterations for TRGS and matrix RGS (on bcirc($\mathcal{A}$)$\mathcal{X}_{[2]} = \mathcal{B}_{[2]}$) for tensor system from Case 1 from Table 2 ($\mathcal{A} \in \mathbb{R}^{15\times 20\times 30}$ under-determined), where $\mathcal{X}^{(k)}$ are iterates, $\mathcal{X}^*$ is the exact solution, and $\mathcal{X}_{LN}$ is the least-norm solution

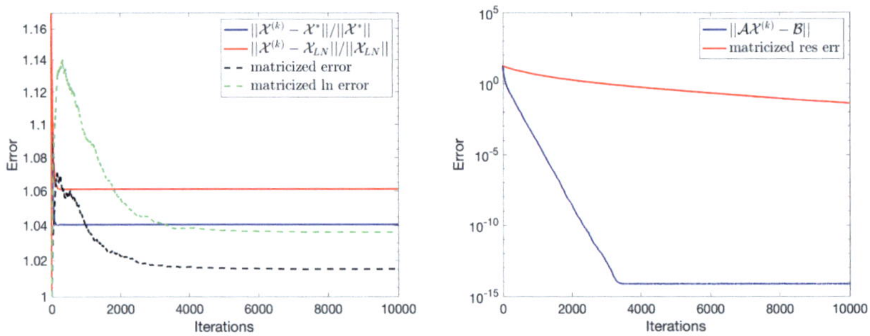

Fig. 3 Relative (left) and residual (right) solution error with iterations for TRGS and matrix RGS (on bcirc($\mathcal{A}$)$\mathcal{X}_{[2]} = \mathcal{B}_{[2]}$) for tensor system from Case 2 from Table 2 ($\mathcal{A} \in \mathbb{R}^{10\times 20\times 30}$ under-determined), where $\mathcal{X}^{(k)}$ are iterates, $\mathcal{X}^*$ is the exact solution, and $\mathcal{X}_{LN}$ is the least-norm solution

3–5), TRGS exhibits faster convergence to the true solution over matrix RGS. In fact, it was able to recover the exact solution (up to machine precision) in Case 3, which corresponds to a highly over-determined system.

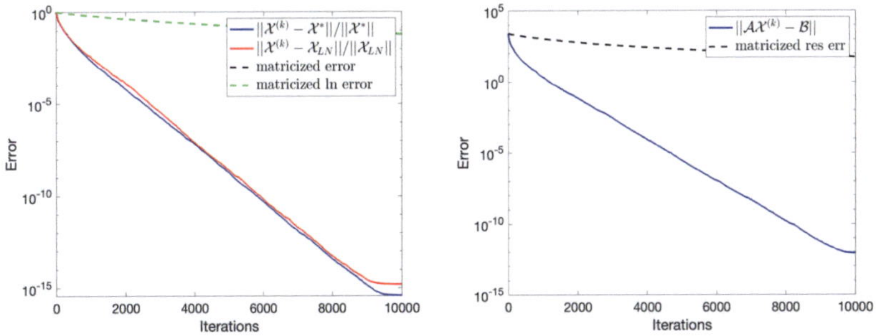

Fig. 4 Relative (left) and residual (right) solution error with iterations for TRGS and matrix RGS (on bcirc($\mathcal{A}$)$X_{[2]}$ = $\mathcal{B}_{[2]}$) for tensor system from Case 3 from Table 2 ($\mathcal{A} \in \mathbb{R}^{40\times20\times30}$ over-determined), where $\mathcal{X}^{(k)}$ are iterates, $\mathcal{X}^*$ is the exact solution, and $\mathcal{X}_{LN}$ is the least-norm solution

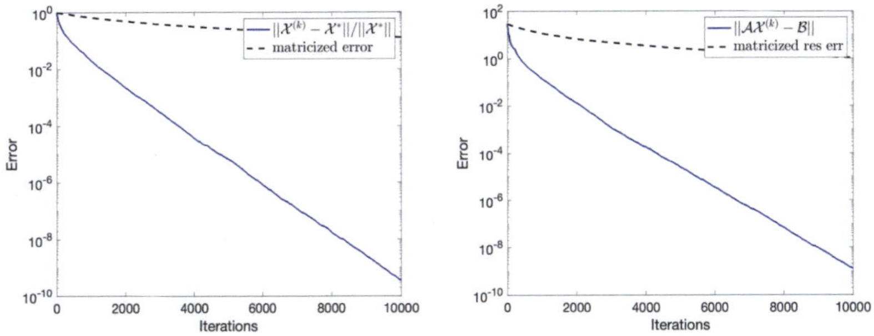

Fig. 5 Relative (left) and residual (right) solution error with iterations for TRGS and matrix RGS (on bcirc($\mathcal{A}$)$X_{[2]}$ = $\mathcal{B}_{[2]}$) for tensor system from Case 4 from Table 2 ($\mathcal{A} \in \mathbb{R}^{30\times20\times30}$ over-determined), where $\mathcal{X}^{(k)}$ are iterates and $\mathcal{X}^*$ is the exact unique solution

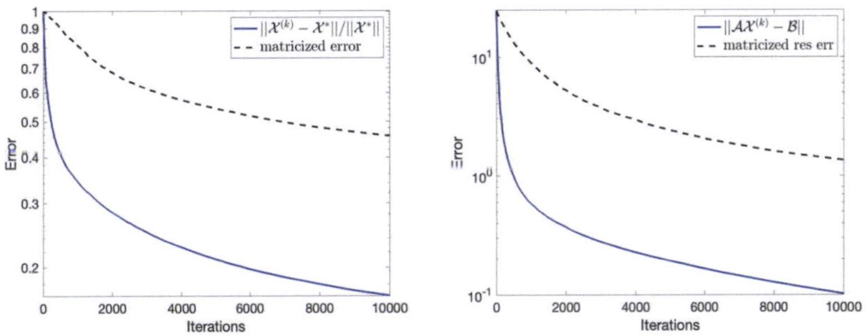

Fig. 6 Relative (left) and residual (right) solution error with iterations for TRGS and matrix RGS (on bcirc($\mathcal{A}$)$X_{[2]}$ = $\mathcal{B}_{[2]}$) for tensor system from Case 5 from Table 2 ($\mathcal{A} \in \mathbb{R}^{20\times20\times30}$ over-determined), where $\mathcal{X}^{(k)}$ are iterates and $\mathcal{X}^*$ is the exact unique solution

4.2 *Tensor Regression with Factorized Measurement Operator*

We now turn our attention to problems in which the measurement operator is given by the product of two tensor operators. This setting naturally occurs in some applications, such as image deblurring with two known blurring operators [31], or can be statistically motivated when the measurement matrix $\boldsymbol{A}$ has an intrinsic low dimension. In that case, an SVD can be used to replace $\boldsymbol{A}$ by its lower-dimensional representation. Reducing computational costs can also drive the use of a product of the type (13) in lieu of $\boldsymbol{A}$ [17]. For example, algorithms for low-rank matrix completion, such as alternating minimization, include a step where a low-rank $\boldsymbol{A}$ is decomposed into two matrices $\boldsymbol{U}$ and $\boldsymbol{V}$, ideally with $r < \min(m, n)$, and proceed with working with the smaller matrices instead of the original measurement matrix.

Existing Approaches to Regression with Factorized Measurement Operator For matrices, the problem can be formally stated as follows: solve the system $\boldsymbol{A}\boldsymbol{x} = \boldsymbol{b}$, where $\boldsymbol{A} \in \mathbb{R}^{m\times n}$ comes in the product form $\boldsymbol{A} = \boldsymbol{U}\boldsymbol{V}$, where $\boldsymbol{U} \in \mathbb{R}^{m\times r}$ and $\boldsymbol{V} \in \mathbb{R}^{r\times n}$. In [41], the authors consider the system $\boldsymbol{A}\boldsymbol{x} = \boldsymbol{b}$ in the form

$$\boldsymbol{U}\boldsymbol{V}\boldsymbol{x} = \boldsymbol{b}, \tag{13}$$

which they solve in multi-steps by tackling the associated subsystems

$$\boldsymbol{U}z = \boldsymbol{b}, \tag{14}$$

and

$$\boldsymbol{V}\boldsymbol{x} = z \tag{15}$$

using an algorithm that interlaces the steps of the Kaczmarz method applied to each individual system.

The authors in [41] develop the two algorithms described below, RK-RK and REK-RK, that find the optimal solution for (13) using the factors of $\boldsymbol{A}$. Both methods, which are variants of the randomized Kaczmarz method in [66], solve the systems (14) and (15) by interlacing Kaczmarz steps. That is, instead of first finding the solution of (14) iteratively and then using this solution to solve (15) in a subsequent sequence of iterations, we compute updates for both systems in each iteration of our algorithm such that the most recent update for $z^{(k)}$ is employed to evaluate the next update $\boldsymbol{x}^{(k)}$. This leads to a more efficient algorithm [41].

The variant RK-RK focuses on a consistent system $\mathbf{A}\boldsymbol{x} = \boldsymbol{b}$, while REK-RK deals with the case in which $\mathbf{A}\boldsymbol{x} = \boldsymbol{b}$ is inconsistent. Note that the latter method draws from REK, an extension of RK proposed by [81] to produce the optimal solution for any system (with linear convergence for both inconsistent systems and consistent systems) following the demonstration in [46] that RK may not yield the optimal solution for the inconsistent system—it is only guaranteed to converge to within a radius of convergence of this solution.

Algorithm 9 Matrix RK-RK [41]

procedure RK-RK($\boldsymbol{U}, \boldsymbol{V}, \boldsymbol{b}, K$)
  $\boldsymbol{z}^{(0)} = \boldsymbol{0}$
  $\boldsymbol{x}^{(0)} = \boldsymbol{0}$
  for $k = 1, \dots, K$ **do**
    Sample i_k with probability $\|\boldsymbol{U}_{i_k:}\|_2^2/\|\boldsymbol{U}\|_F^2$.
    $\boldsymbol{z}^{(k)} = \boldsymbol{z}^{(k-1)} - \boldsymbol{U}^*_{i_k:}\frac{(\boldsymbol{U}_{i_k:}\boldsymbol{z}^{(k-1)} - \boldsymbol{b}_{i_k})}{\|\boldsymbol{U}_{i_k:}\|_2^2}$
    Sample j_k with probability $\|\boldsymbol{V}_{j_k:}\|_2^2/\|\boldsymbol{V}\|_F^2$.
    $\boldsymbol{x}^{(k)} = \boldsymbol{x}^{(k-1)} - \boldsymbol{V}^*_{j_k:}\frac{(\boldsymbol{V}_{j_k:}\boldsymbol{x}^{(k-1)} - \boldsymbol{z}^{(k)}_{j_k})}{\|\boldsymbol{V}_{j_k}\|_2^2}$
  end for return $\boldsymbol{x}^{(K)}$
end procedure

Algorithm 10 Matrix REK-RK [41]

procedure REK-RK($\boldsymbol{U}, \boldsymbol{V}, \boldsymbol{b}, K$)
  $\boldsymbol{z}^{(0)} = \boldsymbol{0}$
  $\boldsymbol{x}^{(0)} = \boldsymbol{0}$
  $\boldsymbol{w}^{(0)} = \boldsymbol{b}$
  for $k = 1, \dots, K$ **do**
    Sample i_k with probability $\|\boldsymbol{U}_{i_k:}\|_2^2/\|\boldsymbol{U}\|_F^2$.
    Sample j_k with probability $\|\boldsymbol{U}_{:j_k}\|_2^2/\|\boldsymbol{U}\|_F^2$
    $\boldsymbol{w}^{(k)} = \boldsymbol{w}^{(k-1)} - \boldsymbol{U}_{:j_k}\frac{\boldsymbol{U}^*_{:j_k}\boldsymbol{w}^{(k-1)}}{\|\boldsymbol{U}_{:j_k}\|_2^2}$
    $\boldsymbol{z}^{(k)} = \boldsymbol{z}^{(k-1)} - \boldsymbol{U}^*_{i_k:}\frac{(\boldsymbol{U}_{i_k:}\boldsymbol{z}^{(k-1)} - \boldsymbol{y}_{i_k})}{\|\boldsymbol{U}_{i_k:}\|_2^2}$
    Sample j_k with probability $\|\boldsymbol{V}_{j_k:}\|_2^2/\|\boldsymbol{V}\|_F^2$.
    $\boldsymbol{x}^{(k)} = \boldsymbol{x}^{(k-1)} - \boldsymbol{V}^*_{j_k:}\frac{(\boldsymbol{V}_{j_k:}\boldsymbol{x}^{(k-1)} - \boldsymbol{z}^{(k)}_{j_k})}{\|\boldsymbol{V}_{j_k}\|_2^2}$
  end for return $\boldsymbol{x}^{(K)}$
end procedure

Extension to Tensor Settings In this section, we study methods for consistent tensor systems; hence, we will focus on generalizing RK-RK to the tensor setting. Our generalization will build upon TRK (Algorithm 1) and the factorized RK method (Algorithm 9) [41]. We propose the factorized tensor randomized Kaczmarz (FacTRK) procedure for solving a factorized system:

$$\mathcal{U}\mathcal{V}\mathcal{X} = \mathcal{B},$$

where $\mathcal{U} \in \mathbb{R}^{m\times r\times p}$, $\mathcal{V} \in \mathbb{R}^{r\times n\times p}$, $\mathcal{X} \in \mathbb{R}^{n\times l\times p}$ and $\mathcal{B} \in \mathbb{R}^{m\times l\times p}$, by alternatively applying tensor RK to the two systems $\mathcal{U}\mathcal{Z} = \mathcal{B}$, and $\mathcal{V}\mathcal{X} = \mathcal{Z}$. The pseudocode for FacTRK is provided in Algorithm 11.

Comparing FacTRK to Matrix Method and TRK We begin with a comparison of FacTRK (updates using $\mathcal{U}$ and $\mathcal{V}$) to TRK (updates using $\mathcal{A} = \mathcal{U}\mathcal{V}$) on a consistent tensor linear system. Figure 7 compares the performance of our algorithm FacTRK,

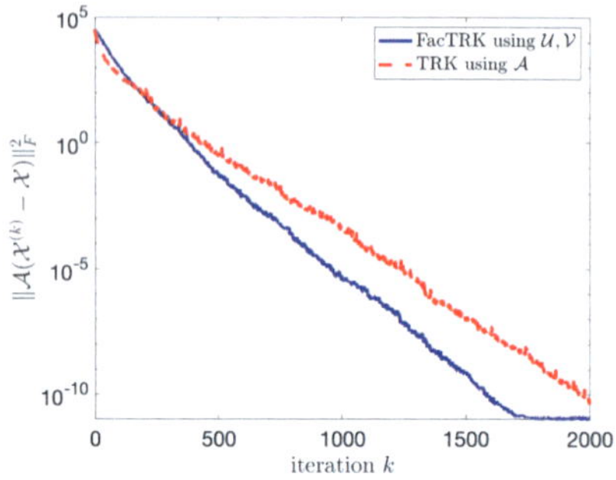

Fig. 7 Residual error against iterations for FacTRK vs. TRK

Algorithm 11 Factorized Tensor Randomized Kaczmarz (FacTRK)

1: **procedure** FACTRK($\mathcal{U}, \mathcal{V}, \mathcal{B}, K$)
2: $\mathcal{Z}^{(0)} = \mathbf{0}$
3: $\mathcal{X}^{(0)} = \mathbf{0}$
4: **for** $k = 1, \dots, K$ **do**
5: Sample $i_k \in [m]$.
6: $\mathcal{Z}^{(k)} = \mathcal{Z}^{(k-1)} - \mathcal{U}^*_{i_k::}(\mathcal{U}_{i_k::}\mathcal{U}^*_{i_k::})^{-1}(\mathcal{U}_{i_k::}\mathcal{Z}^{(k-1)} - \mathcal{B}_{i_k::})$
7: Sample $j_k \in [r]$.
8: $\mathcal{X}^{(k)} = \mathcal{X}^{(k-1)} - \mathcal{V}^*_{j_k::}(\mathcal{V}_{j_k::}\mathcal{V}^*_{j_k::})^{-1}(\mathcal{V}_{j_k::}\mathcal{X}^{(k-1)} - \mathcal{Z}^{(k)}_{j_k::})$
9: **end for return** $\mathcal{X}^{(K)}$
10: **end procedure**

Algorithm 11, with TRK, Algorithm 1, for solving the t-product linear system $\mathcal{A}\mathcal{X} = \mathcal{B}$, where $\mathcal{A} = \mathcal{U}\mathcal{V}$. Here, we generate tensors $\mathcal{U} \in \mathbb{R}^{50\times 10\times 30}, \mathcal{V} \in \mathbb{R}^{10\times 5\times 30}$ with i.i.d. Gaussian normal entries and generate $\mathcal{A} \in \mathbb{R}^{50\times 5\times 30}$ as $\mathcal{A} = \mathcal{U}\mathcal{V}$. We construct $\mathcal{X} \in \mathbb{R}^{5\times 7\times 30}$ with i.i.d. Gaussian normal entries and construct a consistent linear system by setting $\mathcal{B} = \mathcal{A}\mathcal{X}$. We perform 2000 iterations of each algorithm and measure the residual error $\|\mathcal{A}\mathcal{X}^{(k)} - \mathcal{B}\|_F^2$ in each iteration. The numerical results suggest that FacTRK gains a computational advantage by exploiting the factorization over naively applying TRK.

Next, we compare FacTRK to the matrix RK-RK algorithm applied to the equivalent matrix system, $\text{bcirc}(\mathcal{U})\text{bcirc}(\mathcal{V})\mathcal{X}_{[2]} = \mathcal{B}_{[2]}$, defined in Fact 1 for a suite of examples. The following table describes different cases for numerical experiments that compare FacTRK and matrix RK-RK (Algorithm 9) for solving t-linear system $\mathcal{U}\mathcal{V}\mathcal{X} = \mathcal{B}$ where $\mathcal{U} \in \mathbb{R}^{m\times r\times p}, \mathcal{V} \in \mathbb{R}^{r\times n\times p}$ and $\mathcal{A} = \mathcal{U}\mathcal{V} \in \mathbb{R}^{m\times n\times p}$. Throughout these experiments, the matricized error is $\|(\mathcal{X}^{(k)} - \mathcal{X}^*)_{[2]}\|/\|\mathcal{X}^*_{[2]}\|$ and the matricized residual error is $\|\text{bcirc}(\mathcal{U})\text{bcirc}(\mathcal{V})\mathcal{X}^{(k)}_{[2]} - \mathcal{B}_{[2]}\|$. We notice that in every case, FacTRK enjoys faster decreasing residual error than that of matrix RK-RK. In every case, except those where $\mathcal{U}$ is under-determined and $\mathcal{V}$ is over-determined, the residual error follows a generally decreasing trend and does not approach a nonzero (numerically) horizon. We additionally note that in the cases in which $\mathcal{U}$ and $\mathcal{A}$ are both under-determined, the relative errors of FacTRK appear to be worse than those of matrix RK-RK for late iterations. We hypothesize that tensor methods require fewer iterations and in fact the shapes of the error curves are

Table 3 Cases of the numerical experiments of FacTRK and matrix RK-RK algorithms on various sizes of tensors $\mathcal{A}$, $\mathcal{U}$, and $\mathcal{V}$. The dimensions of the $\mathcal{X}$ were $20 \times 10 \times 30$, the same for all the cases. We indicate those cases that are impossible to form with "–"

Cases	$\mathcal{U}$ under-determined $\mathcal{V}$ under-determined	$\mathcal{U}$ over-determined $\mathcal{V}$ over-determined	$\mathcal{U}$ over-determined $\mathcal{V}$ under-determined	$\mathcal{U}$ under-determined $\mathcal{V}$ over-determined
1. $\mathcal{A} \in \mathbb{R}^{15\times 20\times 30}$ under-determined	$r = 17$ Fig. 8 (left)	–	$r = 10$ Fig. 8 (center)	$r = 25$ Fig. 8 (right)
2. $\mathcal{A} \in \mathbb{R}^{10\times 20\times 30}$ under-determined	$r = 15$ Fig. 9 (left)	–	$r = 5$ Fig. 9 (center)	$r = 25$ Fig. 9 (right)
3. $\mathcal{A} \in \mathbb{R}^{40\times 20\times 30}$ over-determined	–	$r - 30$ Fig. 10 (left)	$r - 15$ Fig. 10 (center)	$r = 45$ Fig. 10 (right)
4. $\mathcal{A} \in \mathbb{R}^{30\times 20\times 30}$ over-determined	–	$r = 25$ Fig. 11 (left)	$r = 15$ Fig. 11 (center)	$r = 35$ Fig. 11 (right)
5. $\mathcal{A} \in \mathbb{R}^{20\times 20\times 30}$ over-determined	–	$r = 20$ Fig. 12	–	–

the same, and only matricized methods require more iterations to achieve the same errors as the tensor method. This is due to the fact that matrix RK-RK accesses significantly less of the problem defining data than FacTRK, since the matricization spreads the data from a single row slice of the tensor system $\mathcal{A}\mathcal{X} = \mathcal{B}$ into a block of rows in the equivalent matrix system $\text{bcirc}(\mathcal{A})\mathcal{X}_{[2]} = \mathcal{B}_{[2]}$. Outside of the four cases where $\mathcal{U}$ and $\mathcal{A}$ are both under-determined, the relative error of FacTRK decreases at least as quickly as that of matrix RK-RK.

4.3 *Tensor Regression with Adversarial Corruption*

We now consider the challenging setting in which the tensor linear system has arbitrary and even possibly adversarial corruptions and develop iterative methods that are robust to such corruptions. This setting is relevant in most modern applications where measurements must be collected, stored, and repeatedly accessed. These steps often introduce transmission or transcription corruption into the data; on any one instance, this corruption is rare, but across a large-scale tensor, corruption is likely and could be of arbitrary size. Because these corruptions could be quite large, running a standard least-squares solver will likely not give a solution anywhere near the desired solution.

Simple iterative methods like the Kaczmarz method are prime candidates for corruption-robust methods [20–22]. The information calculated within an iteration (e.g., residual entries) can often additionally provide information about the geometry of the problem, the trustworthiness of data, and the nearness and existence

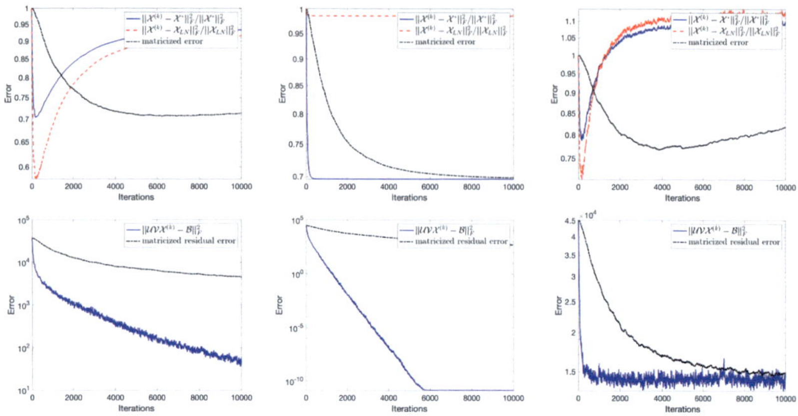

Fig. 8 Relative (top) and residual (bottom) solution error versus iterations for FacTRK and matrix RK-RK (on equivalent system $\text{bcirc}(\mathcal{U})\text{bcirc}(\mathcal{V})\mathcal{X}_{[2]} = \mathcal{B}_{[2]}$ defined in Fact 1) for systems from Case 1 in Table 3 ($\mathcal{A} \in \mathbb{R}^{15\times20\times30}$ under-determined), where $\mathcal{X}^{(k)}$ are iterates, $\mathcal{X}^*$ is the exact solution, and $\mathcal{X}_{LN}$ is the least-norm solution. (Left) $r = 17$, $\mathcal{U}$ under-determined, $\mathcal{V}$ under-determined; (Center) $r = 10$, $\mathcal{U}$ over-determined, $\mathcal{V}$ under-determined; (Right) $r = 25$, $\mathcal{U}$ under-determined, $\mathcal{V}$ over-determined

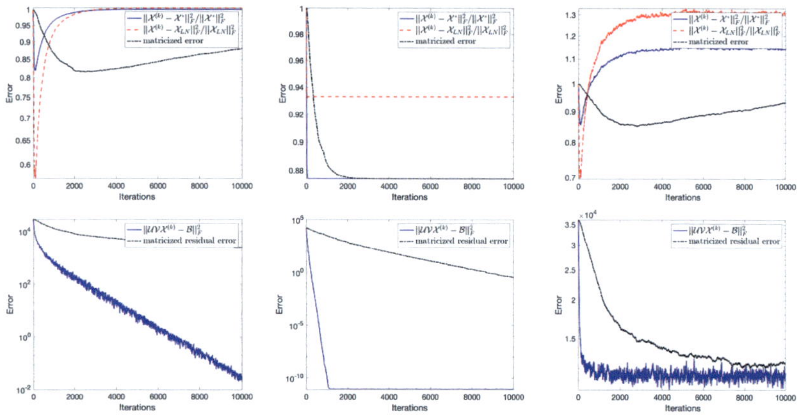

Fig. 9 Relative (top) and residual (bottom) solution error versus iterations for FacTRK and matrix RK-RK (on equivalent system $\text{bcirc}(\mathcal{U})\text{bcirc}(\mathcal{V})\mathcal{X}_{[2]} = \mathcal{B}_{[2]}$ defined in Fact 1) for systems from Case 2 in Table 3 ($\mathcal{A} \in \mathbb{R}^{10\times20\times30}$ under-determined), where $\mathcal{X}^{(k)}$ are iterates, $\mathcal{X}^*$ is the exact solution, and $\mathcal{X}_{LN}$ is the least-norm solution. (Left) $r = 15$, $\mathcal{U}$ under-determined, $\mathcal{V}$ under-determined; (Center) $r = 5$, $\mathcal{U}$ over-determined, $\mathcal{V}$ under-determined; (Right) $r = 25$, $\mathcal{U}$ under-determined, $\mathcal{V}$ over-determined

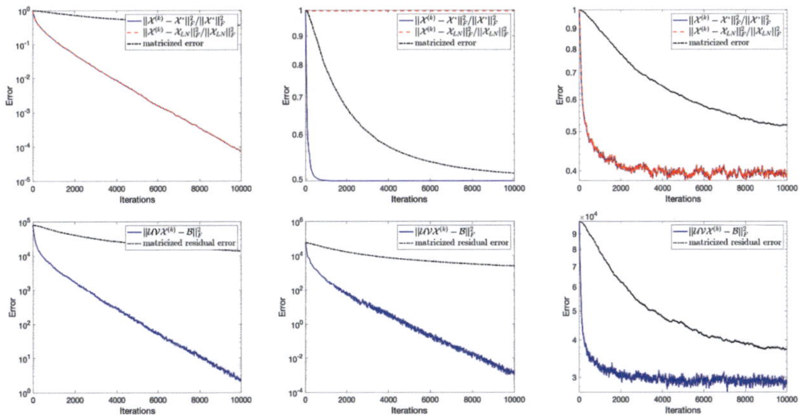

Fig. 10 Relative (top) and residual (bottom) solution error versus iterations for FacTRK and matrix RK-RK (on equivalent system $\text{bcirc}(\mathcal{U})\text{bcirc}(\mathcal{V})\mathcal{X}_{[2]} = \mathcal{B}_{[2]}$ defined in Fact 1) for systems from Case 3 in Table 3 ($\mathcal{A} \in \mathbb{R}^{40\times 20\times 30}$ over-determined), where $\mathcal{X}^{(k)}$ are iterates, $\mathcal{X}^*$ is the exact solution, and $\mathcal{X}_{LN}$ is the least-norm solution. (Left) $r = 30$, $\mathcal{U}$ over-determined, $\mathcal{V}$ over-determined; (Center) $r = 15$, $\mathcal{U}$ over-determined, $\mathcal{V}$ under-determined; (Right) $r = 45$, $\mathcal{U}$ under-determined, $\mathcal{V}$ over-determined

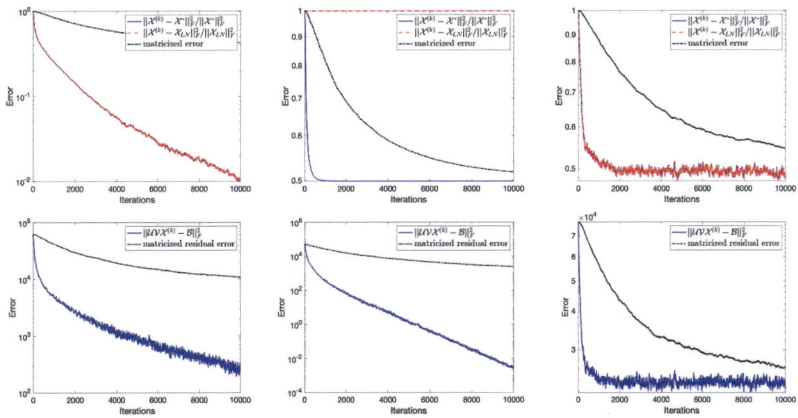

Fig. 11 Relative (top) and residual (bottom) solution error versus iterations for FacTRK and matrix RK-RK (on equivalent system $\text{bcirc}(\mathcal{U})\text{bcirc}(\mathcal{V})\mathcal{X}_{[2]} = \mathcal{B}_{[2]}$ defined in Fact 1) for systems from Case 4 in Table 3 ($\mathcal{A} \in \mathbb{R}^{30\times 20\times 30}$ over-determined), where $\mathcal{X}^{(k)}$ are iterates, $\mathcal{X}^*$ is the exact solution, and $\mathcal{X}_{LN}$ is the least-norm solution. (Left) $r = 25$, $\mathcal{U}$ over-determined, $\mathcal{V}$ over-determined; (Center) $r = 15$, $\mathcal{U}$ over-determined, $\mathcal{V}$ under-determined; (Right) $r = 35$, $\mathcal{U}$ under-determined, $\mathcal{V}$ over-determined

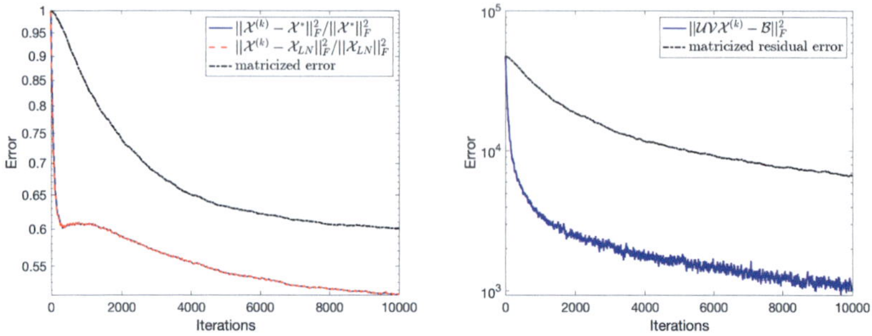

Fig. 12 Relative (left) and residual (right) solution error versus iterations for FacTRK and matrix RK-RK (on equivalent system $\text{bcirc}(\mathcal{U})\text{bcirc}(\mathcal{V})\mathcal{X}_{[2]} = \mathcal{B}_{[2]}$ defined in Fact 1) for system with $r = 20$ from Case 5 in Table 3 ($\mathcal{A} \in \mathbb{R}^{20\times 20\times 30}$, $\mathcal{U}$, and $\mathcal{V}$ over-determined), where $\mathcal{X}^{(k)}$ are iterates, $\mathcal{X}^*$ is the exact solution, and $\mathcal{X}_{LN}$ is the least-norm solution

of a solution. It has become common to aggregate information across multiple iterations to attempt to mitigate the effect of benign noise [2, 44], but variants using this information to avoid the devastating effects of adversarial corruption in the problem-defining data are newer and less well-understood [11, 22, 63, 65]. Since these problems arise in medical imaging, sensor networks, error correction, and data science, the effects of adversarial corruptions in the data could be catastrophic downstream. An ill-timed update using corrupted data can destroy the valuable information learned by many updates produced with uncorrupted data, making iterative methods challenging to employ on large-scale data.

Existing Statistical Approaches to Adversarial Corruption Since arbitrary corruptions can be quite large, simply ignoring them and hoping for convergence is not realistic. On the other hand, given that the corruptions can be large, iterative steps that encounter such a corruption should be statistically different than non-affected iterative steps. This notion is what lies behind current methods for the matrix case that tolerate large corruptions [11, 22, 63, 65]. This approach, coined *quantile randomized Kaczmarz* (QRK), utilizes the *residual error*, $\boldsymbol{A}\boldsymbol{x}^{(k)} - \boldsymbol{b}$, where $\boldsymbol{x}^{(k)}$ is the current iterate, to detect outliers corresponding to corruptions. If a residual entry has magnitude above a certain quantile of all residuals, that entry is deemed unreliable, and its corresponding hyperplane will be rejected if selected in that iteration. The QRK method shows reliable convergence to the true solution of the uncorrupted system under mild assumptions on the number of corruptions both empirically and theoretically.

Some QRK follow-up works considered the case when sparse corruptions are mixed with small noise [29, 75], gave an alternative approach for the convergence proof [65], and considered a variant of the algorithm in which the quantile is computed from only a subsample of the residual [19]. Additionally, there has been significant interest in methods for robust linear regression [5, 58, 70] due to the ubiquity of linear problems with a small number of outlier measurements. Other works relevant to this problem include *min-k loss SGD* [63], robust SGD [11, 56], and Byzantine approaches [1, 6].

Extension to Tensor Linear Systems While the QRK approach works well in the matrix setting, the tensor setting provides some significant challenges. In the matrix case, a corruption in $\boldsymbol{b}$ simply implies that (once identified) one can ignore that entry and otherwise proceed as usual in the iterative process. In the tensor setting, however, a single corruption in $\mathcal{B}$ affects an entire *slice* of the solution $\mathcal{X}$. If there are few corruptions, ignoring each slice during iteration is possible. However, with a moderate number of corruptions, ignoring each corrupted slice is not feasible as this would result in never updating the estimation and thus never converging. Therefore, a different approach is necessary. Of course, one could matricize the system and run QRK and be able to tolerate a moderate amount of corruptions. However, the downside to doing this, as shown in [39], is that the contraction rate, and thus the computational time, is much worse for the matrix version of RK than it is for TRK (Algorithm 1). In the future work, we will investigate an approach that has the benefit of a fast convergence rate like TRK with the ability to handle a moderate number of corruptions like QRK.

Building on the work in the matrix case [22], we have developed a simple quantile tensor RK (QTRK) method that utilizes the Frobenius norms of the updates to decide whether an update is unreliable (i.e., is likely to correspond to a corruption). We denote

$$P_j(\mathcal{X}) := \mathcal{A}^*_{j::}(\mathcal{A}_{j::}\mathcal{A}^*_{j::})^{-1}(\mathcal{A}^*_{j::}\mathcal{X} - \mathcal{B}_{j::}). \tag{16}$$

We write $Q_q(\boldsymbol{X}, S)$ as the qth quantile of all $\boldsymbol{X}$ over the set S. We assume $\mathcal{A} \in \mathbb{R}^{m\times n\times p}$ and $\mathcal{B} \in \mathbb{R}^{m\times l\times p}$. The pseudocode for our QTRK method is given in Algorithm 12.

Algorithm 12 Quantile Tensor Randomized Kaczmarz (QTRK)

procedure QTRK($\mathcal{A}$, $\mathcal{B}$, quantile level q, block size t, K)
  $\mathcal{X}^{(0)} = \mathbf{0}$
  for k = 1, 2, . . . , K **do**
    sample $i_1, \ldots i_t \sim$ Uniform$(1, \ldots, m)$
    sample $j \sim$ Uniform$(1, \ldots, m)$
    if $\|\mathcal{P}_j(\mathcal{X}^{(k-1)})\|_F \le Q_q(\|P_{i_l}(\mathcal{X}^{(k-1)})\|_F, \{i_l : l \in [t]\})$ **then**
      $\mathcal{X}^{(k)} = \mathcal{X}^{(k-1)} - \mathcal{P}_j(\mathcal{X}^{(k-1)})$ ▷ See definition in (16)
    else
      $\mathcal{X}^{(k)} = \mathcal{X}^{(k-1)}$
    end if
  end for
  return $\mathcal{X}^{(K)}$
end procedure

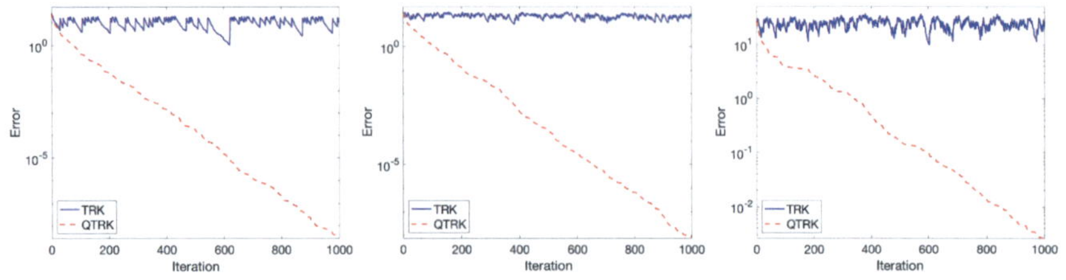

Fig. 13 Solution error versus iteration for TRK vs QTRK, for a simple example, where $\mathcal{A} \in \mathbb{R}^{12\times5\times10}$, $\mathcal{B} \in \mathbb{R}^{12\times7\times10}$, $\mathcal{X} \in \mathbb{R}^{5\times7\times10}$, and $t = 12$. (Left) One corruption, $q = 0.9$; (Center) Three corruptions, $q = 0.7$; (Right) Five corruptions, $q = 0.5$

Comparing QTRK to TRK We have found that, empirically, this method performs well for a small number of corruptions of arbitrary size. Indeed, we compare the classical TRK method to our proposed QTRK method in Fig. 13, where the error is displayed between the true and QTRK and TRK approximated solution for various levels of corruptions in the system. In this experiment, $\mathcal{A} \in \mathbb{R}^{12\times5\times10}$, $\mathcal{B} \in \mathbb{R}^{12\times7\times10}$, $\mathcal{X} \in \mathbb{R}^{5\times7\times10}$. We generate $\mathcal{A}$ and $\mathcal{X}$ to have i.i.d. Gaussian normal random entries and produce $\mathcal{B}$ by taking $\mathcal{A}\mathcal{X}$ and corrupting the indicated number of entries (one in the left plot, three in the center plot, and five in the right plot of Fig. 13). The number of iterations is $K = 1,000$, and the block size is $t = 12$, which aligns with the number of horizontal slices in tensor $\mathcal{A}$. By randomly corrupting varying numbers (1, 3, and 5) of entries in tensor $\mathcal{A}$, the performance of both methods is assessed in terms of error between the true solution and the approximated solution achieved by QTRK and TRK. Note that for a greater number of corruptions, one has to be more conservative with the choice of quantile level q. We see that, unsurprisingly, QTRK offers convergence to the solution of the underlying uncorrupted system, whereas TRK has no hope of such convergence, given that its projections are using corrupted data that pushes the iterate from the true solution.

5 Image Deconvolution Experiments

Image deconvolution or deblurring is the process of removing blurring artifacts from images and returning a sharp image from an image convolved with a known blurring operator. Image blurring may be represented in terms of a convolution where each pixel value is replaced with a weighted sum of nearby pixel values. This averaging diminishes sharp contrasts in the image. This convolution can be represented by multiplication by a given circulant matrix and thus is connected to the t-product. The equivalency between 2D convolution and multiplication by a tensor operator under the t-product has been established [30] and exploited [10].

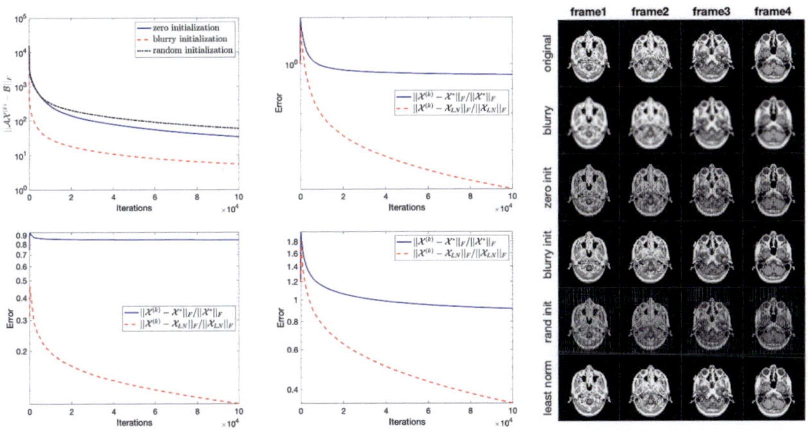

Fig. 14 Video frame deconvolution with TRGS method on consistent system $\mathcal{A}\mathcal{X} = \mathcal{B}$. (Upper left) residual error of TRGS iterates in three experiments varying initialization; (Upper center) relative errors of TRGS iterates produced with zero $\mathcal{X}^{(0)}$; (Bottom left) relative errors of TRGS iterates produced with blurry measurement $\mathcal{X}^{(0)}$; (Bottom center) relative errors of TRGS iterates produced with random $\mathcal{X}^{(0)}$; (Right) visualization of original frames (top row), blurry frames (second row), TRGS recovered frames (third- fifth row), and least-norm solution frames (bottom row)

We can consider the task of deblurring to be a tensor recovery optimization problem subject to the constraint $\mathcal{A}\mathcal{X} = \mathcal{B}$ [10, 30]. In this notation, $\mathcal{B}$ is the measurement tensor (containing the pixel information in the blurred image), $\mathcal{A}$ is the known convolutional blurring operator, and our goal is to recover $\mathcal{X}$, the sharp image tensor. In the right columns of Fig. 14, we can see the original (sharp) images in the top row and the degraded, blurry images after convolution with the known blurring operator in the second row. The goal is to recover the resolution of the original images, given only access to the blurry ones.

In this section, we illustrate the use of our proposed methods in image deconvolution. We demonstrate the promise of these methods for variants of the deblurring problem and discuss the interesting theoretical questions these experiments and the image deconvolution application raise. The experiments presented in this section were performed in MATLAB 2021b on a MacBook Pro 2019 with a 2.3 GHz 8-Core Intel Core i9 and 16GB RAM.

In these experiments, we focus on deblurring an image sequence, or video, $\bar{\mathcal{B}} \in \mathbb{R}^{m \times l \times p}$ with l frames. We assume that all frames are degraded by the same known spatial blurring operator, represented in its tensor form $\mathcal{A} \in \mathbb{R}^{m \times m \times p}$. In all given experiments, we use the MRI video data set `mri` in MATLAB. This data set contains 12 frames of size 128×128 from an MRI data scan of a human head. In these experiments, the blurry frames are generated by convolving the ground truth $\bar{\mathcal{X}}^* \in \mathbb{R}^{128 \times 12 \times 128}$ with (a) tensor(s) representing frame-wise 2D Gaussian smoothing kernel(s). We note that $\mathcal{A}\mathcal{X}^* = \mathcal{B}$; however, we cannot hope to exactly recover the original image from only these measurements, as the matrix operator and tensor operator representing the Gaussian smoothing kernel are not full-rank. For this reason, we are interested in examining the recovered images and not just measuring the relative error or residual error of the recovered images.

5.1 Deconvolution with TRGS Method

In our first experiment, we construct the degraded, blurry images by convolving the `mri` tensor $\bar{\mathcal{X}}^*$ described above with a tensor representing a Gaussian smoothing kernel of size 5×5 with standard deviation two and replicating the edge pixels twice on all edges (to construct a "tall" system). The blurring and replication may be represented by t-product multiplication with a tensor operator $\mathcal{A} \in \mathbb{R}^{132\times 128\times 132}$. We hope to recover $\mathcal{X}^*$, which is $\bar{\mathcal{X}}^*$ with the edge pixels on the left and right edge of every frame replicated twice. Indeed, the blurry and replicated image $\mathcal{B} \in \mathbb{R}^{132\times 12\times 132}$ satisfies $\mathcal{A}\mathcal{X}^* = \mathcal{B}$.

We apply the TRGS method to this consistent tensor linear system with various initial iterates. In our first experiment, we begin with initial iterate $\mathcal{X}^{(0)} \in \mathbb{R}^{128\times 12\times 132}$ with all entries zero. In our second experiment, we begin with initial iterate $\mathcal{X}^{(0)} = \mathcal{B}_{2:130,1:12,1:132}$; that is, we trim the left and right replicated pixels from every frame of the blurry video. In our third experiment, we begin with initial iterate $\mathcal{X}^{(0)} \in \mathbb{R}^{128\times 12\times 132}$ with each entry sampled i.i.d. from the unif[0, 1] distribution. In each experiment, we run 100,000 iterations of the TRGS method. For all iterations of each experiment, we measure the residual error $\|\mathcal{A}\mathcal{X}^{(k)} - \mathcal{B}\|_F$; see the upper left plot of Fig. 14. We also measure the relative error to the original frames, $\|\mathcal{X}^{(k)} - \mathcal{X}^*\|_F/\|\mathcal{X}^*\|_F$, and to the least-norm solution frames $\mathcal{X}_{\text{LN}} = \mathcal{A}^\dagger\mathcal{B}$, $\|\mathcal{X}^{(k)} - \mathcal{X}_{\text{LN}}\|_F/\|\mathcal{X}_{\text{LN}}\|_F$; see the upper middle plot of Fig. 14 for the relative errors of TRGS initialized with the zero tensor, the lower left plot of Fig. 14 for the relative errors of TRGS initialized with the blurry tensor, and the lower middle plot of Fig. 14 for the relative errors of TRGS initialized with the random tensor.

We compare the frames recovered from the TRGS method with these three initializations (rows 3–5) to the original frames (top row), the degraded blurry frames (second row), and the least-norm solution frames produced as $\mathcal{A}^\dagger\mathcal{B}$ (bottom row) in the array of frames on the right of Fig. 14. As evidenced by the visualization of the recovered frames, the residual error, and the final relative error to the least-norm solution, TRGS initialized with the blurry frame tensor provides the strongest recovery. We note that the current theory in the matrix regime does not account for or explain the improved performance of TRGS initialized with the blurry frame tensor. In the future work, we will investigate what property of the blurry frames had made them a good initialization for the TRGS method.

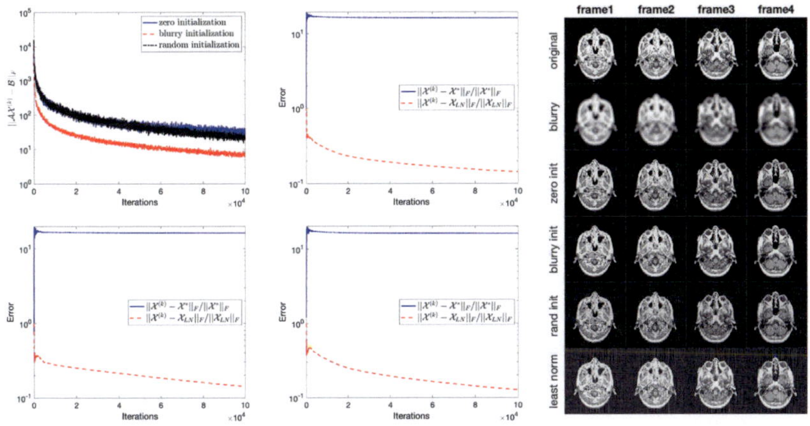

Fig. 15 Video frame deconvolution of doubly blurred frames with the FacTRK method on consistent system $\mathcal{U}\mathcal{V}\mathcal{X} = \mathcal{B}$. (Upper left) residual error of FacTRK iterates in three experiments varying initialization; (Upper center) relative errors of FacTRK iterates produced with zero $\mathcal{X}^{(0)}$; (Bottom left) relative errors of FacTRK iterates produced with blurry measurement $\mathcal{X}^{(0)}$; (Bottom center) relative errors of FacTRK iterates produced with random $\mathcal{X}^{(0)}$; (Right) visualization of original frames (top row), blurry frames (second row), FacTRK recovered frames (third–fifth row), and least-norm solution frames (bottom row)

5.2 *Deconvolution of Doubly Blurred Images with FacTRK Method*

In this experiment, we construct the degraded, blurry images by convolving the `mri` tensor $\bar{\mathcal{X}}^*$ described above with a tensor representing a Gaussian smoothing kernel of size 5×5 with standard deviation two *and* convolving this result with a tensor representing an averaging kernel of size 5×5 and replicating the edge pixels twice on all edges (to construct a "tall" system). The dual blurring and replication may be represented by t-product multiplication with a tensor operator, which can be written as the product of two tensors, $\mathcal{A} = \mathcal{U}\mathcal{V} \in \mathbb{R}^{132\times 128\times 132}$. We hope to recover $\mathcal{X}^*$, which is $\bar{\mathcal{X}}^*$ with the edge pixels on the left and right edge of every frame replicated twice. Indeed, the doubly blurry and replicated image $\mathcal{B} \in \mathbb{R}^{132\times 12\times 132}$ satisfies $\mathcal{A}\mathcal{X}^* = \mathcal{U}\mathcal{V}\mathcal{X}^* = \mathcal{B}$.

We apply the FacTRK method to this consistent tensor linear system with various initial iterates. In our first experiment, we begin with initial iterate $\mathcal{X}^{(0)} \in \mathbb{R}^{128\times 12\times 132}$ with all entries zero. In our second experiment, we begin with initial iterate $\mathcal{X}^{(0)} = \mathcal{B}_{2:130,1:12,1:132}$; that is, we trim the left and right replicated pixels from every frame of the blurry video. In our third experiment, we begin with initial iterate $\mathcal{X}^{(0)} \in \mathbb{R}^{128\times 12\times 132}$ with each entry sampled i.i.d. from the unif[0, 1] distribution. In each experiment, we run 100, 000 iterations of the FacTRK method. For all iterations of each experiment, we measure the residual error $\|\mathcal{A}\mathcal{X}^{(k)} - \mathcal{B}\|_F$; see the upper left plot of Fig. 15. We also measure the relative error to the original frames, $\|\mathcal{X}^{(k)} - \mathcal{X}^*\|_F/\|\mathcal{X}^*\|_F$, and to the least-norm solution frames $\mathcal{X}_{\mathrm{LN}} = \mathcal{A}^\dagger\mathcal{B}$, $\|\mathcal{X}^{(k)} - \mathcal{X}_{\mathrm{LN}}\|_F/\|\mathcal{X}_{\mathrm{LN}}\|_F$; see the upper middle plot of Fig. 15 for the relative errors of FacTRK initialized with the zero tensor, the lower left plot of Fig. 15 for the relative errors of FacTRK initialized with the blurry tensor, and the lower middle plot of Fig. 15 for the relative errors of FacTRK initialized with the random tensor.

We compare the frames recovered from the FacTRK method with these three initializations (rows 3–5) to the original frames (top row), the doubly degraded blurry frames (second row), and the least-norm solution frames produced as $\mathcal{A}^\dagger\mathcal{B}$ (bottom row) in the array of frames on the right of Fig. 15. We note again, as evidenced by the visualization of the recovered frames and the residual error, FacTRK initialized with the blurry frame tensor provides the strongest recovery and that, again, the current theory in the matrix regime does not account for or explain the improved performance of FacTRK initialized with the blurry frame tensor. Additionally, in this problem, the least-norm solution (right bottom row of Fig. 15) suffers from some artifacts due to an error in the numerical calculation of $\mathcal{A}^\dagger$ and that FacTRK appears to better avoid these. Finally, we note that in each experiment, the relative error to the original frames, $\|\mathcal{X}^{(k)} - \mathcal{X}^*\|_F/\|\mathcal{X}^*\|_F$, *increases* even as the residual error, $\|\mathcal{A}\mathcal{X}^{(k)} - \mathcal{B}\|_F$, *decreases*. In the future work, we will investigate the description of the element of the solution space to which FacTRK is converging.

5.3 *Deconvolution of Blurred and Corrupted Images with QTRK Method*

In this experiment, we construct the corrupted and degraded, blurry images by convolving the `mri` tensor $\bar{\mathcal{X}}^*$ described above with a tensor representing a Gaussian smoothing kernel of size 5×5 with standard deviation two and replicating the edge pixels twice on all edges (to construct a "tall" system) *and* introducing a corruption into the blurry frame tensor by setting a randomly sampled entry in the first frame slice to value 1000. The blurring and replication may be represented by t-product multiplication with a tensor operator, which can be written as the product of two tensors, $\mathcal{A} = \mathcal{U}\mathcal{V} \in \mathbb{R}^{132\times128\times132}$. We hope to recover $\mathcal{X}^*$, which is $\bar{\mathcal{X}}^*$ with the edge pixels on the left and right edge of every frame replicated twice. However, in this case, the corrupted and blurry image $\mathcal{B} \in \mathbb{R}^{132\times12\times132}$ *does not* satisfy $\mathcal{A}\mathcal{X}^* = \mathcal{U}\mathcal{V}\mathcal{X}^* = \mathcal{B}$.

We apply the QTRK method with $q = 0.99$ to this corrupted tensor linear system with various initial iterates. In our first experiment, we begin with initial iterate $\mathcal{X}^{(0)} \in \mathbb{R}^{128\times12\times132}$ with all entries zero. In our second experiment, we begin with initial iterate $\mathcal{X}^{(0)} = \mathcal{B}_{2:130,1:12,1:132}$; that is, we trim the left and right replicated pixels from every frame of the blurry video. In our third experiment, we begin with initial iterate $\mathcal{X}^{(0)} \in \mathbb{R}^{128\times12\times132}$ with each entry sampled i.i.d. from the unif[0, 1] distribution. In each experiment, we run 100, 000 iterations of the QTRK method.

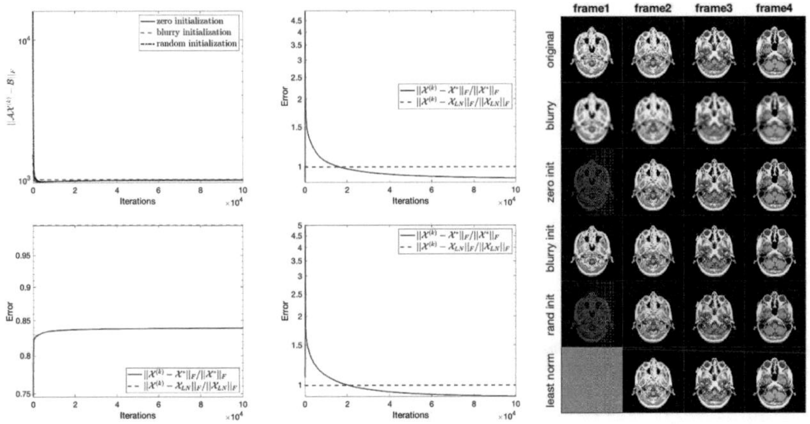

Fig. 16 Video frame deconvolution of blurred and corrupted frames with the QTRK method on the corrupted (inconsistent) system defined by $\mathcal{A}$ and $\mathcal{B}$. (Upper left) residual error of QTRK iterates in three experiments varying initialization; (Upper center) relative errors of QTRK iterates produced with zero $\mathcal{X}^{(0)}$; (Bottom left) relative errors of QTRK iterates produced with blurry measurement $\mathcal{X}^{(0)}$; (Bottom center) relative errors of QTRK iterates produced with random $\mathcal{X}^{(0)}$; (Right) visualization of original frames (top row), blurry frames (second row), QTRK recovered frames (third–fifth row), and least-norm solution frames (bottom row)

For all iterations of each experiment, we measure the residual error $\|\mathcal{A}\mathcal{X}^{(k)} - \mathcal{B}\|_F$; see the upper left plot of Fig. 16. We also measure the relative error to the original frames, $\|\mathcal{X}^{(k)} - \mathcal{X}^*\|_F/\|\mathcal{X}^*\|_F$, and to the least-norm solution frames $\mathcal{X}_{\text{LN}} = \mathcal{A}^\dagger\mathcal{B}$, $\|\mathcal{X}^{(k)} - \mathcal{X}_{\text{LN}}\|_F/\|\mathcal{X}_{\text{LN}}\|_F$; see the upper middle plot of Fig. 16 for the relative errors of QTRK initialized with the zero tensor, the lower left plot of Fig. 16 for the relative errors of QTRK initialized with the blurry tensor, and the lower middle plot of Fig. 16 for the relative errors of QTRK initialized with the random tensor.

We compare the frames recovered from the QTRK method with these three initializations (rows 3–5) to the original frames (top row), the doubly degraded blurry frames (second row), and the least-norm solution frames produced as $\mathcal{A}^\dagger\mathcal{B}$ (bottom row) in the array of frames on the right of Fig. 16. We note again that, as demonstrated by the visualization of the recovered frames, QTRK initialized with the blurry frame tensor provides the strongest recovery and that, again, the current theory in the matrix regime does not account for or explain the improved performance of QTRK initialized with the blurry frame tensor, nor the meaningful recovery of QTRK at all (as the underlying uncorrupted system is less than full rank and nearly square). Additionally, in this problem, the least-norm solution (right bottom row of Fig. 16) is entirely ruined by the presence of the corruption in the blurry slice corresponding to the first frame, yet QTRK is able to approximately solve the deconvolution problem. In the future work, we will investigate the theoretical convergence of QTRK and the matrix method QRK on structured systems that are under-determined.

6 Conclusion

In this paper, we accomplish several things. First, we summarize the current state of tensor regression in various settings, e.g., consistent or inconsistent linear systems or convex optimization over linear system constraints. Then, we develop new algorithms, which extend the Gauss-Seidel and Kaczmarz methods to new settings, namely, tensors for the Gauss-Seidel method and, for the Kaczmarz algorithms, tensor regression with factorized measurement operator and tensor regression with adversarial corruption. We test the performance of these algorithms in various critical scenarios, e.g., consistent or inconsistent cases and a deblurring problem, by empirically comparing their convergence rate and/or accuracy to matrix counterparts. Our results are promising; in all our experiments, our methods match or (significantly) surpass the empirical convergence rate and accuracy of matrix methods.

Furthermore, our numerical work prompts several follow-up questions; for example, we see from Figs. 2, 3, 4, 5, 6 that TRGS converges to an element of the solution space, but we have yet to establish the nature of this solution. Additionally, several of our experiments suggest convergence of our iterative tensor regression methods outside of the regime for which theoretical results for their matrix counterparts guarantee convergence; for instance, consider the residual convergence of TRGS and FacTRK on under-determined systems and the success of QTRK on the under-determined image deconvolution system. We hope that our future work proving convergence guarantees for these tensor system iterative methods may offer, additionally, novel insights into the matrix system setting. Additionally, in Figs. 8, 9, 10, 11, right panels, we see that when $\mathcal{U}$ is under-determined and $\mathcal{V}$ is over-determined, convergence of the relative and residual solution error differs significantly from the other cases. This suggests further theoretical work is necessary to understand why this case is different. We hypothesize that when $\mathcal{V}$ is over-determined, the solution to the inner system is overfitted, and this does not allow enough freedom for the outer system to stabilize and converge. Additional future work will consider the image deconvolution problem in more detail and treat it more fully. In particular, we are interested in understanding what properties of the image deconvolution problem may allow for useful recovery (by TRGS, FacTRK, and QTRK) on systems for which existing convergence guarantees in the matrix regime do not apply. Our ongoing efforts are focused on investigating these interesting theoretical questions and providing a comprehensive theoretical framework for convergence of TRGS, FacTRK, and QTRK.

Acknowledgments The initial research for this effort was conducted at the Research Collaboration Workshop for Women in Data Science and Mathematics (WiSDM), held in August 2023 at the Institute for Pure and Applied Mathematics (IPAM). IPAM, AWM, and DIMACS funded the workshop (NSF grant CCF1144502).

This material is based upon work supported by the National Science Foundation under Grant No. DMS-1928930, while several of the authors were in residence at the Mathematical Sciences Research Institute in Berkeley, California, during the summer of 2024.

Several of the authors also appreciate support provided to them at a SQuaRE at the American Institute of Mathematics. The authors thank AIM for providing a supportive and mathematically rich environment.

This material is based upon work supported by the National Science Foundation under Grant No. DMS-1929284 while the author was in residence at the Institute for Computational and Experimental Research in Mathematics in Providence, RI, during the "Randomized Algorithms for Tensor Problems with Factorized Operations or Data" Collaborate@ICERM.

JH was partially supported by NSF DMS #2211318. DN was partially supported by NSF DMS #2011140. KYD was partially supported by NSF #2232344.

Competing Interests The authors have no conflicts of interest to declare that are relevant to the content of this chapter.

References

1. Alistarh, D., Allen-Zhu, Z., Li, J.: Byzantine stochastic gradient descent. Adv. Neur. Inf. **31** (2018)
2. Bach, F.: Adaptivity of averaged stochastic gradient descent to local strong convexity for logistic regression. J. Mach. Learn. Res. **15**(1), 595–627 (2014)
3. Bao, W., Zhang, F., Li, W., Wang, Q., Gao, Y.: Randomized average Kaczmarz algorithm for tensor linear systems. Mathematics **10**(23) (2022). https://doi.org/10.3390/math10234594. URL https://www.mdpi.com/2227-7390/10/23/4594
4. Beck, A., Ben-Tal, A.: A global solution for the structured total least squares problem with block circulant matrices. SIAM J. Matrix Anal. Appl. **27**, 238–255 (2005). https://doi.org/10.1137/040612233
5. Bhatia, K., Jain, P., Kamalaruban, P., Kar, P.: Consistent robust regression. In: Adv. Neur. Inf., vol. 30 (2017)
6. Blanchard, P., El Mhamdi, E.M., Guerraoui, R., Stainer, J.: Machine learning with adversaries: Byzantine tolerant gradient descent. Adv. Neur. Inf. **30** (2017)
7. Boyd, S., Vandenberghe, L.: Convex Optimization. Cambridge Univ. Press, Cambridge (2004)
8. Carroll, J.D., Chang, J.J.: Analysis of individual differences in multidimensional scaling via an N-way generalization of "eckart-young" decomposition. Psychometrika **35**(3), 283–319 (1970)
9. Chen, X., Powell, A.: Almost sure convergence of the Kaczmarz algorithm with random measurements. J. Fourier Anal. Appl. pp. 1–20 (2012). http://doi.org/10.1007/s00041-012-9237-2
10. Chen, X., Qin, J.: Regularized Kaczmarz algorithms for tensor recovery. SIAM J. Imag. Sci. **14**(4), 1439–1471 (2021)
11. Diakonikolas, I., Kamath, G., Kane, D., Li, J., Steinhardt, J., Stewart, A.: Sever: A robust meta-algorithm for stochastic optimization. In: Int. Conf. Machine Learning, pp. 1596–1606. PMLR (2019)
12. Du, K., Sun, X.H.: Randomized regularized extended Kaczmarz algorithms for tensor recovery. arXiv preprint arXiv:2112.08566 (2021)
13. Dumitrescu, B.: On the relation between the randomized extended Kaczmarz algorithm and coordinate descent. BIT Num. Math. pp. 1–11 (2014)
14. Eldar, Y.C., Needell, D.: Acceleration of randomized Kaczmarz method via the Johnson-Lindenstrauss lemma. Num. Algorithms **58**(2), 163–177 (2011). https://doi.org/10.1007/s11075-011-9451-z
15. Frommer, A., Szyld, D.B.: On the convergence of randomized and greedy relaxation schemes for solving nonsingular linear systems of equations. Num. Algorithms **92**(1), 639–664 (2023)

16. Glusa, C., Boman, E.G., Chow, E., Rajamanickam, S., Szyld, D.B.: Scalable asynchronous domain decomposition solvers. SIAM J. Sci. Comput. **42**(6), C384–C409 (2020)
17. Goes, J., Zhang, T., Arora, R., Lerman, G.: Robust stochastic principal component analysis. In: AISTATS, pp. 266–274 (2014)
18. Gordon, R., Bender, R., Herman, G.T.: Algebraic reconstruction techniques (ART) for three-dimensional electron microscopy and X-ray photography. J. Theoret. Biol. **29**, 471–481 (1970)
19. Haddock, J., Ma, A., Rebrova, E.: On subsampled quantile randomized Kaczmarz. In: Proc. Allerton Conf. Communication, Control, and Computing (2023)
20. Haddock, J., Needell, D.: Randomized projections for corrupted linear systems. In: AIP Conf. Proc., vol. 1978, p. 470071. AIP Publishing LLC (2018)
21. Haddock, J., Needell, D.: Randomized projection methods for linear systems with arbitrarily large sparse corruptions. SIAM J. Sci. Comput. **41**(5), S19–S36 (2019)
22. Haddock, J., Needell, D., Rebrova, E., Swartworth, W.: Quantile-based iterative methods for corrupted systems of linear equations. SIAM J. Matrix Anal. Appl. **43**(3), 605–637 (2022)
23. Hao, N., Kilmer, M.E., Braman, K., Hoover, R.C.: Facial recognition using tensor-tensor decompositions. SIAM J. Imag. Sci. **6**(1), 437–463 (2013)
24. Harshman, R.A.: Foundations of the parafac procedure: Models and conditions for an "explanatory" multimodal factor analysis. UCLA Working Papers in Phonetics (1970)
25. Hefny, A., Needell, D., Ramdas, A.: Rows versus columns: Randomized kaczmarz or gauss-seidel for ridge regression. SIAM J. Sci. Comput. **39**(5), S528–S542 (2017)
26. Herman, G., Meyer, L.: Algebraic reconstruction techniques can be made computationally efficient. IEEE T. Med. Imag. **12**(3), 600–609 (1993)
27. Hillar, C.J., Lim, L.H.: Most tensor problems are NP-hard. J. ACM **60**(6), 1–39 (2013)
28. Huang, G.X., Zhong, S.Y.: Tensor randomized extended Kaczmarz methods for large inconsistent tensor linear equations with t-product. Num. Algorithms, pp. 1–24 (2023)
29. Jarman, B., Needell, D.: QuantileRK: Solving large-scale linear systems with corrupted, noisy data. In: Asilomar Conf. Sig. Sys. Comput., pp. 1312–1316. IEEE (2021)
30. Kernfeld, E., Kilmer, M.E., Aeron, S.: Tensor–tensor products with invertible linear transforms. Linear Algebra Appl. **485**, 545–570 (2015)
31. Kilmer, M.E., Braman, K., Hao, N., Hoover, R.C.: Third-order tensors as operators on matrices: A theoretical and computational framework with applications in imaging. SIAM J. Matrix Anal. Appl. **34**(1), 148–172 (2013)
32. Kilmer, M.E., Martin, C.D.: Factorization strategies for third-order tensors. Linear Algebra Appl. **435**(3), 641–658 (2011)
33. Kolda, T.G., Bader, B.W.: Tensor decompositions and applications. SIAM Rev. **51**(3), 455–500 (2009)
34. Leventhal, D., Lewis, A.S.: Randomized methods for linear constraints: convergence rates and conditioning. Math. Oper. Res. **35**(3), 641–654 (2010)
35. Liu, J., Wright, S.J., Sridhar, S.: An asynchronous parallel randomized Kaczmarz algorithm. arXiv preprint arXiv:1401.4780 (2014)
36. Liu, Y., Liu, J., Long, Z., Zhu, C.: Tensor Regression. Springer (2022)
37. Lock, E.F.: Tensor-on-tensor regression. J. Comput. Graph. Stat. **27**(3), 638–647 (2018)
38. Lu, C., Feng, J., Chen, Y., Liu, W., Lin, Z., Yan, S.: Tensor robust principal component analysis: Exact recovery of corrupted low-rank tensors via convex optimization. In: Proceedings of the IEEE Conference on Computer Vision and Pattern Recognition, pp. 5249–5257 (2016)
39. Ma, A., Molitor, D.: Randomized Kaczmarz for tensor linear systems. BIT Num. Math. **62**(1), 171–194 (2022)
40. Ma, A., Needell, D., Ramdas, A.: Convergence properties of the randomized extended Gauss–Seidel and Kaczmarz methods. SIAM J. Matrix Anal. Appl. **36**(4), 1590–1604 (2015)
41. Ma, A., Needell, D., Ramdas, A.: Iterative methods for solving factorized linear systems. SIAM J. Matrix Anal. Appl. **39**(1), 104–122 (2018)
42. Magoules, F., Szyld, D.B., Venet, C.: Asynchronous optimized Schwarz methods with and without overlap. Num. Math. **137**(1), 199–227 (2017)
43. Mallat, S.: A Wavelet Tour of Signal Proces. Elsevier, Amsterdam (1999)

44. Moorman, J.D., Tu, T.K., Molitor, D., Needell, D.: Randomized Kaczmarz with averaging. BIT Num. Math. **61**, 337–359 (2020)
45. Morshed, M.S., Islam, M.S., Noor-E-Alam, M.: Accelerated sampling Kaczmarz Motzkin algorithm for the linear feasibility problem. J. Global Optim., 1–22 (2019)
46. Needell, D.: Randomized Kaczmarz solver for noisy linear systems. BIT Num. Math. **50**(2), 395–403 (2010). https://doi.org/10.1007/s10543-010-0265-5
47. Needell, D., Srebro, N., Ward, R.: Stochastic gradient descent and the randomized Kaczmarz algorithm. Math. Program. A **155**(1), 549–573 (2016)
48. Needell, D., Tropp, J.A.: Paved with good intentions: Analysis of a randomized block Kaczmarz method. Linear Algebra Appl. **441**, 199–221 (2013)
49. Newman, E., Horesh, L., Avron, H., Kilmer, M.: Stable tensor neural networks for rapid deep learning. arXiv preprint arXiv:1811.06569 (2018)
50. Newman, E., Kilmer, M.E.: Nonnegative tensor patch dictionary approaches for image compression and deblurring applications. SIAM J. Imag. Sci. **13**(3), 1084–1112 (2020)
51. Olson, B., Shaw, S., Shi, C., Pierre, C., Parker, R.: Circulant matrices and their application to vibration analysis. Appl. Mech. Rev. **66**, 040803 (2014). https://doi.org/10.1115/1.4027722
52. Pollock, S.: Circulant matrices and time-series analysis. Int. J. Math. Educ. Sci. Technol. **33**, 213–230 (2002). https://doi.org/10.1080/00207390110118953
53. Popa, C.: A fast Kaczmarz-Kovarik algorithm for consistent least-squares problems. Korean J. Comput. Appl. Math. **8**(1), 9–26 (2001)
54. Popa, C.: A Kaczmarz-Kovarik algorithm for symmetric ill-conditioned matrices. An. Ştiinţ. Univ. Ovidius Constanţa Ser. Mat. **12**(2), 135–146 (2004)
55. Popa, C., Preclik, T., Köstler, H., Rüde, U.: On Kaczmarz's projection iteration as a direct solver for linear least squares problems. Linear Algebra Appl. **436**(2), 389–404 (2012)
56. Prasad, A., Suggala, A.S., Balakrishnan, S., Ravikumar, P., et al.: Robust estimation via robust gradient estimation. J. R. Stat. Soc. B **82**(3), 601–627 (2020)
57. Richtárik, P., Takáč, M.: Parallel coordinate descent methods for big data optimization. Math. Program. **156**, 433–484 (2016)
58. Rousseeuw, P.J.: Least median of squares regression. J. Am. Stat. Assoc. **79**(388), 871–880 (1984)
59. Rüde, U.: Mathematical and Computational Techniques for Multilevel Adaptive Methods. SIAM, Philadelphia (1993)
60. Saad, Y., Van Der Vorst, H.A.: Iterative solution of linear systems in the 20th century. J. Comput. Appl. Math. **123**(1-2), 1–33 (2000)
61. Savvides, A., Han, C.C., Strivastava, M.B.: Dynamic fine-grained localization in ad-hoc networks of sensors. In: Proc. 7th Ann. Int. Conf. Mobile Computing and Networking, pp. 166–179 (2001)
62. Semerci, O., Hao, N., Kilmer, M.E., Miller, E.L.: Tensor-based formulation and nuclear norm regularization for multienergy computed tomography. IEEE T. Image Proces. **23**(4), 1678–1693 (2014)
63. Shah, V., Wu, X., Sanghavi, S.: Choosing the sample with lowest loss makes SGD robust. In: Proc. Int. Conf. Artif. Intel. Stat., pp. 2120–2130 (2020)
64. Soltani, S., Kilmer, M.E., Hansen, P.C.: A tensor-based dictionary learning approach to tomographic image reconstruction. BIT Num. Math. **56**, 1425–1454 (2016)
65. Steinerberger, S.: Quantile-based random Kaczmarz for corrupted linear systems of equations. Inform. Infer. **12**(1), 448–465 (2023)
66. Strohmer, T., Vershynin, R.: A randomized Kaczmarz algorithm with exponential convergence. J. Fourier Anal. Appl. **15**(2), 262 (2009)
67. Tang, L., Yu, Y., Zhang, Y., Li, H.: Sketch-and-project methods for tensor linear systems. Num. Linear Algebra **30**(2), e2470 (2023)
68. Tang, L., Zhang, Y., Li, H.: On sketch-and-project methods for solving tensor equations. arXiv preprint arXiv:2210.08241 (2022)
69. Trottenberg, U., Oosterlee, C.W., Schuller, A.: Multigrid. Elsevier, Amsterdam (2000)
70. Víšek, J.Á.: The least trimmed squares. part I: Consistency. Kybernetika **42**(1), 1–36 (2006)

71. Wang, X., Che, M., Wei, Y.: Tensor neural network models for tensor singular value decompositions. Comput. Optim. Appl. **75**, 753–777 (2020)
72. Wolfson-Pou, J., Chow, E.: Distributed Southwell: an iterative method with low communication costs. In: Proc. Int. Conf. High Perform. Comput. Network. Stor. Anal., pp. 1–13 (2017)
73. Yu, R., Liu, Y.: Learning from multiway data: Simple and efficient tensor regression. In: Int. Conf. Machine Learning, pp. 373–381. PMLR (2016)
74. Zhang, J., Saibaba, A.K., Kilmer, M.E., Aeron, S.: A randomized tensor singular value decomposition based on the t-product. Num. Linear Algebra Appl. **25**(5), e2179 (2018)
75. Zhang, L., Wang, H., Zhang, H.: Quantile-based random sparse Kaczmarz for corrupted, noisy linear inverse systems. arXiv preprint arXiv:2206.07356 (2022)
76. Zhang, Z., Aeron, S.: Denoising and completion of 3D data via multidimensional dictionary learning. arXiv preprint arXiv:1512.09227 (2015)
77. Zhang, Z., Aeron, S.: Exact tensor completion using t-SVD. IEEE T. Signal Proces. **65**(6), 1511–1526 (2016)
78. Zhang, Z., Ely, G., Aeron, S., Hao, N., Kilmer, M.E.: Novel methods for multilinear data completion and de-noising based on tensor-SVD. In: Proc. IEEE Conf. Computer Vision and Pattern Recognition, pp. 3842–3849 (2014)
79. Zhou, H., Li, L., Zhu, H.: Tensor regression with applications in neuroimaging data analysis. J. Amer. Stat. Ass. **108**(502), 540–552 (2013)
80. Zhou, P., Lu, C., Lin, Z., Zhang, C.: Tensor factorization for low-rank tensor completion. IEEE T. Image Proces. **27**(3), 1152–1163 (2017)
81. Zouzias, A., Freris, N.M.: Randomized extended Kaczmarz for solving least squares. SIAM J. Matrix Anal. Appl. **34**(2), 773–793 (2013)

Matrix Exponentials: Lie–Trotter–Suzuki Fractal Decomposition, Gauss Runge–Kutta Polynomial Formulation, and Compressible Features

Rachel E. Emrick, Emily H. Huang, Yidan Mei, Joaquin E. Drut, Jingfang Huang, and Yifei Lou

1 Introduction

In classical linear algebra analysis, a matrix exponential e^{At} can be intuitively defined by its Taylor expansion:

$$e^{At} = I + At + \frac{1}{2!}A^2t^2 + \frac{1}{3!}A^3t^3 + \cdots, \tag{1}$$

where A is an $n \times n$ matrix, I is the Identity matrix, and t is a scalar variable to show the connections between matrix exponentials and time-dependent problems. When matrix A can be diagonalized using its eigensystems in the form $A = BDB^{-1}$, where

$$D = \begin{pmatrix} \lambda_1 & 0 & \cdots & 0 \\ 0 & \lambda_2 & \cdots & 0 \\ \vdots & \vdots & \ddots & \vdots \\ 0 & 0 & \cdots & \lambda_n \end{pmatrix} \tag{2}$$

R. E. Emrick · E. H. Huang · Y. Mei · J. Huang · Y. Lou (✉)
Department of Mathematics, The University of North Carolina at Chapel Hill, Chapel Hill, NC, USA
e-mail: remrick@alumni.unc.edu; huanghe@email.unc.edu; ymei@unc.edu; huang@email.unc.edu; yflou@email.unc.edu

J. E. Drut
Department of Physics and Astronomy, The University of North Carolina at Chapel Hill, Chapel Hill, NC, USA
e-mail: drut@email.unc.edu

C. Garcia-Cardona, H. Lee (eds.), *Advances in Data Science*, Association for Women in Mathematics Series 37, https://doi.org/10.1007/978-3-031-87804-6_2

is a diagonal matrix containing all the eigenvalues λ_i on the diagonal and the column vectors of B are the corresponding eigenvectors, then the matrix exponential is given by

$$e^{At} = Be^{Dt}B^{-1},$$

where

$$e^{Dt} = \begin{pmatrix} e^{\lambda_1 t} & 0 & \cdots & 0 \\ 0 & e^{\lambda_2 t} & \cdots & 0 \\ \vdots & \vdots & \ddots & \vdots \\ 0 & 0 & \cdots & e^{\lambda_n t} \end{pmatrix}. \tag{3}$$

The eigen-decomposition-based matrix exponential definition can be generalized to Jordan forms when the matrix is not diagonalizable, and we skip the details.

This paper presents the work of three female undergraduate students (RE, EH, and YM) at UNC-Chapel Hill. Each student focused on a particular research topic and collaborated to generate interesting analysis and numerical results. The students contributed equally to this work. The *main contributions* of their research include the explorations of two techniques for efficient calculations of the matrix exponentials using the operator splitting techniques, efficient numerical extractions of different compressible features in a matrix, and a study of the interactions between diagonal (or banded) structures with low-rank structures when time t evolves in a matrix exponential. In this paper, we focus on Hermitian matrices A that are always diagonalizable. When the size of the matrix, denoted by n, is large, the well-defined matrix exponential using the straightforward eigen-decomposition becomes computationally intractable due to the $O(n^3)$ operations and $O(n^2)$ storage required to find the eigensystem. Instead of general Hermitian matrices, this paper considers special cases of the matrix exponentials when matrix A can be decomposed as the sum of two (or more) simple structured matrices, i.e., $A = T + V$, where matrices T and V can be diagonal or banded in the physical or frequency domains or have low-rank structures. Note that compressible matrix structures commonly exist in real-world applications, e.g., the Laplace operator is diagonal in the Fourier/frequency domain; (nonlinear) reactions in physical or biological systems are diagonal in the physical domain; discretized Laplace differential operator is banded (tridiagonal) in 1D setting; a banded matrix is the sum of a diagonal matrix and hierarchically low-rank off-diagonal matrix blocks; and the submatrix blocks representing the "well-separated" far-field electrostatic, hydrodynamic, and other types of interactions (also related to the inverse of the Laplacian and other elliptic operators) are always low-rank. We present two examples with different compressible features. The first example comes from quantum statistical mechanics. As the wave function ϕ satisfies the Schrödinger's equation,

$$i\hbar\frac{\partial \phi}{\partial t} = H\phi = \left(-\frac{\hbar^2}{2m}\nabla^2 + U\right)\phi. \tag{4}$$

Using matrix A to represent the discretized Hamiltonian H that characterizes the total energy of a quantum physical system, matrix A can be split into two matrices as

$$A = T + V, \tag{5}$$

where T (representing discretized $-\dfrac{\hbar^2}{2m}\nabla^2$) is a diagonal matrix when expressed in the momentum space, also known as Fourier space, and V (representing the potential U from the environment in which the quantum system exists) often contains low-rank structures when considered either in the physical or frequency domain. In the second example, we consider a reaction–diffusion equation model from biological or physical applications given by

$$u_t(\mathbf{x}, t) = \Delta u(\mathbf{x}, t) + f(u(\mathbf{x}, t)),$$

where u_t is the time derivative, Δ is the spatial Laplace operator describing the diffusion process, and $f(u)$ is the reaction term. Note that the Laplace operator is diagonal in the Fourier frequency domain and the reaction term is diagonal in the physical domain. Therefore, the operator can be split into a diagonal operator and a convolution operator that is diagonal in the frequency domain. For systems that can be decomposed into the summation of simple structured blocks, "operator splitting" is a commonly used technique to take advantage of simple structures and accelerate numerical simulations. For example, a time-splitting spectral approximation scheme [1] is applied to the general nonlinear Schrödinger equations (NLS) in the semi-classical regimes, where the original operator is decomposed into the summation of a diagonal operator in the frequency domain (Laplace operator) and a different diagonal operator in the physical domain. Then, the frequency domain diagonal system can be accurately and efficiently solved using a spectral method, and the physical domain diagonal system can be solved separately using high-order ordinary differential equation (ODE) initial value problem solvers for each decoupled single-variable scalar differential equation.

In this paper, we focus on a class of applications when the original operator (matrix A) is the sum of a diagonal operator (matrix T) and a low-rank operator (matrix V). We first study the efficient computation of the matrix exponential e^{tA} as t evolves. We present two numerical approaches for one time step from 0 to Δt when the time stepsize Δt is small. In the first approach that is widely used in the quantum statistics community [2, 4, 11, 19, 20], the local fractal Lie–Trotter–Suzuki (LTS) decompositions approximate $e^{A\Delta t}$ using the products of terms in the form $e^{\alpha_k T\Delta t}$

and $e^{\beta_k V \Delta t}$ with scalar values α_k, β_k. Examples of the LTS decompositions include the Lie product formula [15], also widely called the Trotter decomposition [21]:

$$e^{A\Delta t} = e^{T\Delta t} e^{V\Delta t} + O(\Delta t)^2 = e^{V\Delta t} e^{T\Delta t} + O(\Delta t)^2,$$

and the second-order Strang splitting [18]:

$$e^{A\Delta t} \approx e^{T\frac{\Delta t}{2}} e^{V\Delta t} e^{T\frac{\Delta t}{2}} \approx e^{V\frac{\Delta t}{2}} e^{T\Delta t} e^{V\frac{\Delta t}{2}}.$$

In the second approach, note that $Y(t) = e^{At}$ satisfies the ordinary different equations:

$$\begin{cases} Y'(t) = A \cdot Y(t) \\ Y(0) = I_{n\times n}. \end{cases} \tag{6}$$

As there generally exists no polynomial matrix with a bounded degree that exactly satisfies the differential equation in one time marching step $[0, \Delta t]$, one searches for a degree p polynomial matrix that satisfies a pseudo-spectral (collocation) formulation by requiring that the differential equation is exactly satisfied at p collocation points $\{0 \leq t_1, t_2, \cdots, t_p \leq \Delta t\}$. Following the terminology in mathematical analysis, a spectral method is a technique where one studies the function expansion coefficients (frequency domain) instead of the function values (physical domain) where the basis can be any orthogonal eigensystems from the Sturm–Liouville theory, e.g., the Fourier series or orthogonal polynomial basis functions. A pseudo-spectral method refers to the case when one implicitly studies the expansion using the function values at a special set of sample points that are often related to the zeros of the eigenfunctions. When the Gaussian quadrature nodes (zeros of the Legendre polynomials) are used, the resulting Gauss Runge–Kutta (GRK) method (also called the Gauss collocation or pseudo-spectral formulations) has order $2p$ and is A-stable, B-stable, symmetric, and symplectic [10]. To avoid the numerically unstable differentiation operations, we consider the equivalent Picard integral equation reformulation:

$$\begin{cases} Y(x) = Y(0) + \int_0^x A \cdot Y(t)dt \\ Y(0) = I_{n\times n}, \end{cases} \tag{7}$$

and solve the resulting discretized GRK formulation efficiently using the spectral deferred correction (SDC) or Krylov deferred correction (KDC) methods, where a low-order method (e.g., a low-order LTS method) is applied as a preconditioner and the preconditioned system is solved either by the geometric series expansion (fixed-point iterations in SDC) [6] or by the least squares-based Krylov subspace methods (in KDC) [12]. We show how both the LTS decompositions and SDC/KDC accelerated GRK methods can compute the matrix exponentials efficiently when the $n \times n$ matrix $A = T + V$, T is diagonal, and V is rank-$k \ll n$. Note that the exponential of a diagonal matrix is diagonal and the exponential of a rank-k matrix is the Identity matrix plus a rank-k matrix. Therefore, the required storage and number of operations in a related matrix–vector multiplication are both $O(n)$ for T and $e^{\alpha T}$, and $O(kn)$ for V and $e^{\beta V}$. The matrix–matrix multiplications for T, $e^{\alpha T}$, V, and $e^{\beta V}$ are either $O(n)$ (diagonal times diagonal) or $O(kn)$ (otherwise). When solving a differential equation system with a given initial value vector, both the LTS decompositions and the SDC/KDC-GRK algorithms can utilize the compressible features in the split matrices T and V, reducing each matrix–vector multiplication cost from the direct $O(n^2)$ to the asymptotically optimal $O(kn)$. Therefore, the storage and number of operations are bounded by $O(kn)$ in each time marching step.

For $A = T + V$ where T is diagonal and V is low rank, we have theoretically and numerically explored the interesting question on how the two special structures in the matrix exponential e^{At} evolve as time t increases. Is e^{At} still the sum of a diagonal matrix and a low-rank matrix? If so, how does the numerical rank of the low-rank matrix change? Is the rank bounded? And are there any low-rank structures in the submatrix blocks representing the far-field relations between different well-separated subsystems? Understanding the hidden compressible features allows more efficient computations of matrix exponentials. By combining randomized rank-revealing and low-rank decomposition algorithms, we present some preliminary results on the study of the "interactions" of the low-rank and diagonal structures in the matrix exponential e^{At}.

We organize this paper as follows. Section 2 focuses on different Lie–Trotter–Suzuki decompositions. In Sect. 3, we discuss the polynomial expansion-based Gauss Runge–Kutta formulation and how the discretized system can be solved efficiently using the spectral or Krylov deferred correction methods. In Sect. 4, we describe several compressible structures that can be utilized to accelerate the involved algebraic computations, and discuss the interactions of low-rank and diagonal structures in a matrix exponential. In Sect. 5, we present numerical results to demonstrate the performance of various methods under different choices of algorithmic parameters and how the low-rank structures change in matrix exponentials as time t evolves. Finally in Sect. 6, we summarize our results and discuss our future work.

2 Lie–Trotter–Suzuki Decompositions

The Lie–Trotter–Suzuki (LTS) decompositions are a class of varying order methods commonly used in the study of quantum mechanical systems for numerically approximating the matrix e^A using easier to compute factors of e^T and e^V, where $A = T + V$. In this section, we present several decompositions we have studied.

2.1 S1: Order 1 Trotter Decomposition

The simplest decomposition is the Trotter approximation, denoted as S1, given by

$$e^{\Delta t(T+V)} = e^{\Delta t T} e^{\Delta t V} + O(\Delta t^2) = S1(\Delta t) + O(\Delta t^2). \tag{8}$$

The order of S1 can be derived by the Taylor expansion of matrix exponentials where the matrix–matrix multiplications are noncommutative. Following the work in [11], we compare the following Taylor expansions of matrix exponentials:

$$\begin{aligned} e^{\Delta t(T+V)} &= I + \Delta t(T+V) + \frac{\Delta t^2}{2}(T+V)^2 + O(\Delta t^3) \\ &= I + \Delta t(T+V) + \frac{\Delta t^2}{2}(T^2 + TV + VT + V^2) + O(\Delta t^3), \end{aligned} \tag{9}$$

and

$$e^{\Delta t T} e^{\Delta t V} = I + \Delta t(T+V) + \frac{\Delta t^2}{2}(T^2 + 2TV + V^2) + O(\Delta t^3). \tag{10}$$

This analysis shows that the leading order in local truncation error is the second order given by $\frac{\Delta t^2}{2}(VT - TV)$; hence, the global error in a time marching scheme is first-order $O(\Delta t)$.

2.2 S2 and P2: Symmetric Order 2 Decompositions

A famous second-order decomposition is the Strang splitting [18, 19], referred to as S2 in this paper, given by the formulas

$$e^{\Delta t(T+V)} = e^{\Delta t T/2} e^{\Delta t V} e^{\Delta t T/2} + O(\Delta t^3) = S2(\Delta t) + O(\Delta t^3), \tag{11}$$

or equivalently,

$$e^{\Delta t(T+V)} = e^{\Delta t V/2} e^{\Delta t T} e^{\Delta t V/2} + O(\Delta t^3) = S2(\Delta t) + O(\Delta t^3). \tag{12}$$

The "symmetric structure" in the decomposition makes the Strang splitting a globally second-order method when solving time-dependent differential equations.

When expanded in terms of nested commutators (Hall bases) and minimizing the 1-norm of the coefficients, the optimal second-order expansion, referred to as P2, is given by the following formula:

$$e^{\Delta t(T+V)} \approx P2(\Delta t) = e^{a_1 \Delta t T} e^{b_1 \Delta t V} e^{a_2 \Delta t T} e^{b_1 \Delta t V} e^{a_1 \Delta t T}, \tag{13}$$

where $a_1 = \frac{1}{6}(3 - \sqrt{3})$, $a_2 = 1 - 2a_1$, and $b_1 = \frac{1}{2}$ are the parameters that optimize the 1-norm of the coefficients [2, 19].

2.3 S3, Q3, and Q4: Higher-Order Decompositions

We also study higher-order LTS decompositions. By properly combining the decomposition S2, a new decomposition S3 is given by the formula

$$
\begin{aligned}
e^{\Delta t(T+V)} \approx S3(\Delta t) &= S2(s\Delta t)S2((1-2s)\Delta t)S2(s\Delta t) \\
&= e^{\frac{s}{2}\Delta t T} e^{s\Delta t V} e^{\frac{1-s}{2}\Delta t T} e^{(1-2s)\Delta t V} e^{\frac{1-s}{2}\Delta t T} e^{s\Delta t V} e^{\frac{s}{2}\Delta t T},
\end{aligned}
\tag{14}
$$

where $s = \frac{1}{2-\sqrt[3]{2}}$. Because of the special choice of s, S3 is an order 4 method [19].

In addition to real-valued coefficients, complex coefficients can be used for potentially improved accuracy and stability for special physical systems. We present two recursively defined decompositions in this category [19]. The decomposition Q3 is given by

$$
Q_3(\Delta t) = S_2(p_3\Delta t)S_2(\bar{p_3}\Delta t), \tag{15}
$$

where $p_3 = \frac{1}{6}(3 + \sqrt{3}i)$ and $\bar{p_3}$ is the complex conjugate of p_3. Using Q3, we can define $Q4$ as

$$
Q_4(\Delta t) = Q_3(p_4\Delta t)Q_3(\bar{p_4}\Delta t) \tag{16}
$$

where $p_4 = \left(1 + e^{i\pi/4}\right)^{-1}$. For general complex differential equations or complex matrix $A = T + V$, Q3 is the third order and Q4 is the fourth order. An interesting feature of Q3 is that when it is applied to a system with all real numbers, as its leading order local truncation error is purely imaginary, it effectively becomes a fourth-order method, which will be numerically shown in Sect. 5.

We refer interested readers to [2, 7, 11, 19, 20] for detailed discussions of existing LTS decompositions. In Sect. 5, we present preliminary numerical experiments to demonstrate the performance of the aforementioned decompositions.

3 Gauss Runge–Kutta Formulation and Its Accelerated Solutions

In addition to the LTS decompositions, we rely on classical numerical analysis for ODE initial value problems to handle the matrix exponential. As the matrix exponential $Y(t) = e^{At}$ for an $n \times n$ matrix A satisfies the ODE

$$\begin{cases} Y'(t) = A \cdot Y(t) \\ Y(0) = I_{n\times n}, \end{cases} \tag{17}$$

we present a polynomial-based approach to approximate $Y(t) = e^{At} \approx P_p(t)$, where $P_p(t)$ is a polynomial matrix of degree p, i.e., each entry of $P_p(t)$ is a polynomial in t of degree no more than p.

3.1 *Pseudo-spectral Differentiation Formulation*

In general, there exists no polynomial matrix with a bounded degree that exactly satisfies the differential equation; therefore, one can search for a degree p polynomial matrix P that satisfies a pseudo-spectral (collocation) formulation by requiring that the differential equation is exactly satisfied at p collocation points $\{t_j\}$. We define the pseudo-spectral differentiation formulation for the ODE initial value problem in Eq. (17) as follows:

Definition 1 (Pseudo-spectral Differentiation Formulation) For a given set of collocation points $\{t_1, t_2, \cdots, t_p\}$, the pseudo-spectral formulation finds a polynomial matrix $P_p(t)$, which satisfies

$$\begin{cases} P_p'(t_j) = A \cdot P_p(t_j) \\ P_p(0) = I_{n\times n}. \end{cases} \tag{18}$$

Comment on the Choice of $\{t_j\}$ Clearly, the choice of the nodes $\{t_j\}$ has impacts on the numerical properties of the spectral differentiation formulation. We briefly discuss the following two important classes of node choices.

(i) Gauss–Legendre Nodes When the zeros of a Legendre polynomial are used, the resulting pseudo-spectral formulation is often referred to as the Gauss collocation or Gauss Runge–Kutta (GRK) formulation. The numerical algorithm for Eq. (18) has the following nice properties:

Theorem 1 *When p Gauss–Legendre nodes are used, the Gauss Runge–Kutta formulation in Eq. (18) for approximating the solution of Eq. (17) is order $2p$, A-stable, B-stable, symmetric, and symplectic.*

Interested readers are referred to [10] to understand why the Gauss collocation formulation can be considered as a special case of the implicit Runge–Kutta methods and numerical properties of the resulting GRK formulations.

(ii) Gauss–Chebyshev Nodes One can also use the zeros of the Chebyshev polynomial to take advantage of the near-minimax properties of the Chebyshev polynomial approximation and the efficiency of the fast Fourier transform to compute the Chebyshev expansion coefficients from the function values at all the node points. We refer to the resulting formulation as the Gauss–Chebyshev collocation formulation.

Note that the node choices have significant impacts on numerical integration and differentiation operations. It is well documented in classical numerical analysis literature that interpolations based on the zeros of orthogonal polynomials (Gauss-type quadrature nodes) have improved accuracy and stability properties. In this paper, we focus on the Gauss quadrature nodes. We have also considered evenly spaced points when the desired order of the method is no more than 8, to better "recycle" previous computations for improved algorithm efficiency (we skip the details).

3.2 Spectral Integration and Picard Integral Equation Reformulation

Given the unknown values $P_p(t_j)$, where $0 \le t_j \le \Delta t$ are the (scaled) Gauss quadrature points in $[0, \Delta t]$, to solve the pseudo-spectral differentiation formulation in Eq. (18) in one time step $[0, \Delta t]$, a spectral differentiation operator is needed.

Definition 2 ((Pseudo-)Spectral Differentiation Matrix) Given the function values $f(t_j)$ at $t_0 = 0$ and scaled Gauss quadrature points $\{t_1, t_2, \cdots, t_p\}$ in the interval $[0, \Delta t]$, one can construct an interpolating polynomial of degree p denoted by $P(t)$. If one differentiates the polynomial and evaluates the derivative at the same set of points, one can construct a linear mapping from the function values $\{f(t_j)\}_{j=1,\cdots,p}$ to the derivative values $\{f'(t_j) \approx P'(t_j)\}_{j=1,\cdots,p}$. The transformation matrix D is defined as the spectral differentiation matrix and we have

$$\begin{bmatrix} f'(t_1) \\ f'(t_2) \\ \vdots \\ f'(t_p) \end{bmatrix} = \Delta t D_{p\times p} \begin{bmatrix} f(t_1) \\ f(t_2) \\ \vdots \\ f(t_p) \end{bmatrix}.$$

Similarly, a spectral integration matrix can be defined as follows:

Definition 3 ((Pseudo-)Spectral Integration Matrix) Given the function values $f(t_j)$ at $t_0 = 0$ and scaled Gauss quadrature points $\{t_1, t_2, \cdots, t_p\}$ in the interval $[0, \Delta t]$, one can construct an interpolating polynomial of degree p denoted by $P(t)$. If one defines $F(t_j) = \int_0^{t_j} f(\tau)d\tau$, which can be approximated by evaluating the polynomial $\int_0^{t_j} P(\tau)d\tau$ at the same set of points, one can construct a linear mapping from the function values $\{f(t_j)\}_{j=1,\cdots,p}$ to the integral values $\{F(t_j)\}_{j=1,\cdots,p}$. The transformation matrix S is defined as the spectral integration matrix and we have

$$\begin{bmatrix} F(t_1) \\ F(t_2) \\ \vdots \\ F(t_p) \end{bmatrix} = \Delta t S_{p\times p} \begin{bmatrix} f(t_1) \\ f(t_2) \\ \vdots \\ f(t_p) \end{bmatrix}.$$

It is interesting to compare the spectral differentiation with spectral integration matrix. It was shown in [9] that the spectral integration matrix is almost a tri-diagonal matrix in the frequency domain (when one considers the relation between expansion coefficients rather than the function values). One can use the fast Fourier/Cosine transform (for Chebyshev nodes) or fast Legendre transform (for Legendre nodes) to go back and forth efficiently between the physical domain (function values) and frequency domain (expansion coefficients). Such algorithm accelerations are particularly helpful when the number of nodes p is large. A more important difference is the condition numbers of the spectral differentiation and integration matrices. In [9], it was shown that the condition number of the spectral differentiation matrix is approximately $O(p^2)$. The condition number of the spectral integration matrix, on the other hand, is bounded by a constant (for the Chebyshev polynomial, the constant is approximately 2.4; see Eq. (21) in [9]).

We numerically demonstrate the condition numbers of the spectral integration and differentiation matrices for the Gauss quadrature nodes. For the function $f(t) = \sin(t)$, we compute its derivative and integral values using spectral differentiation and integration matrices, respectively, and compare the numerical results with the analytical derivatives and integrals. We present the numerical results in Fig. 1, showing how errors decay when the number of Gauss nodes increases for each matrix. The numerical results match the theoretical analysis, and spectral integration is numerically more stable than spectral differentiation.

To avoid the numerical instability associated with the differential operator in Eq. (17), we consider an equivalent Picard integral equation reformulation:

$$\begin{cases} Y(t) = Y(0) + \int_0^t A \cdot Y(\tau) d\tau \\ Y(0) = I_{n\times n} \end{cases}, \tag{19}$$

and search for a polynomial matrix $P_p(t)$ that satisfies the discretized pseudo-spectral integral equation formulation defined as follows:

Definition 4 (Pseudo-spectral Integral Equation Reformulation) For the equivalent Picard integral equation reformulation of the ODE initial value problem presented in Eq. (19) and a given set of scaled Gauss quadrature-type node points $\{t_1, t_2, \cdots, t_p\}$ in one marching step from $[0, \Delta t]$, the pseudo-spectral integral equation reformulation finds a polynomial matrix $P_p(t)$ that satisfies

$$\begin{cases} P_p(t_j) = P_p(0) + \int_0^{t_j} A \cdot P_p(\tau) d\tau \\ P_p(0) = I_{n\times n}. \end{cases} \tag{20}$$

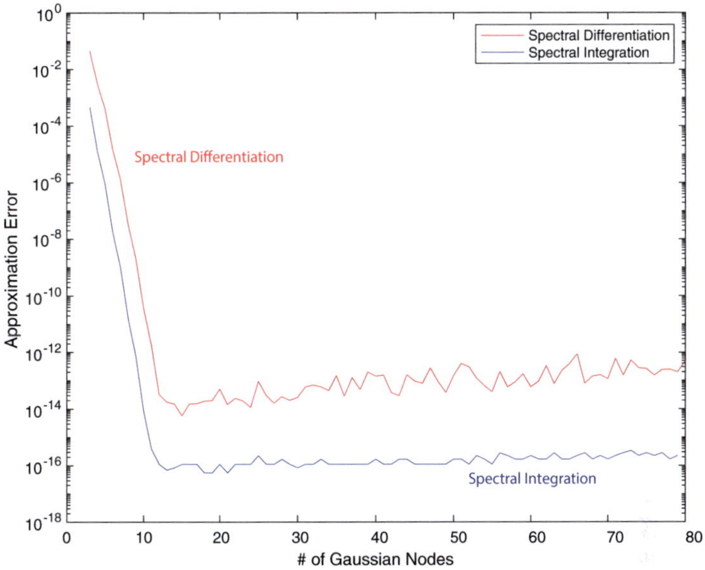

Fig. 1 A comparison of spectral integration versus spectral differentiation

The pseudo-spectral integral equation reformulation is capable of achieving machine precision accuracy when it is solved accurately.

3.3 Spectral Deferred Correction Methods

To solve the pseudo-spectral integral equation reformulation in Eq. (20), we apply the spectral deferred correction (SDC) method first introduced in [6] to improve the efficiency of the algorithm. The SDC steps are listed next.

Step 1: Find a Low-Order Approximate Solution The first step of a deferred correction method is to find an approximate matrix polynomial solution $\tilde{P}(t)$ using a low-order method. To demonstrate the idea, we simply apply the first-order Trotter decomposition and compute $\tilde{P}(t)$ as follows:

$$\tilde{P}(0) = I_{n\times n},$$

$$\tilde{P}(t_{j+1}) = e^{T(t_{j+1}-t_j)} e^{V(t_{j+1}-t_j)} \tilde{P}(t_j).$$

Instead of a first-order approximation, one can apply higher-order approximations discussed in Sect. 2. An interesting question is how different "low-order" predictors impact algorithm efficiency, which will be studied in the future.

Step 2: Compute the Residue Once the approximate solution $\tilde{P}(t)$ is available, we define $P(t) = \tilde{P}(t) + \delta(t)$ and plug it into the Picard integral equation:

$$\begin{cases} \tilde{P}(t) + \delta(t) = Y(0) + \int_0^t A \cdot (\tilde{P}(\tau) + \delta(\tau))d\tau \\ \delta(0) = 0_{n\times n}. \end{cases} \tag{21}$$

One can derive a new set of equations for the error (also called defect) $\delta(t)$:

$$\begin{cases} \delta(t) = \int_0^t A \cdot \delta(\tau)d\tau + \left(Y(0) + \int_0^t A \cdot \tilde{P}(\tau)d\tau - \tilde{P}(t)\right) \\ \delta(0) = 0_{n\times n}. \end{cases} \tag{22}$$

Define the residue as

$$\epsilon(t) = Y(0) + \int_0^t A \cdot \tilde{P}(\tau)d\tau - \tilde{P}(t),$$

and then the error's equation becomes a new (inhomogeneous) Picard integral equation:

$$\begin{cases} \delta(t) = \int_0^t A \cdot \delta(\tau)d\tau + \epsilon(t) \\ \delta(0) = 0_{m\times m}. \end{cases} \tag{23}$$

Note that the approximate solution $\tilde{P}(t)$ is known; therefore, the integral $\int_0^t A \cdot \tilde{P}(\tau)d\tau$ can be accurately evaluated using a high-order (and stable) pseudo-spectral integration matrix that integrates the interpolating polynomial of $\tilde{P}(t)$ exactly.

Step 3: Apply a Low-Order Method to the Error's Equation The third step of an SDC method is to apply a low-order method to get a low-order estimate $\tilde{\delta}(t)$ of the analytical error $\delta(t)$. There are many low-order approaches; among which we demonstrate a second-order algorithm based on the trapezoidal rule.

We consider the differential equation form of the Picard integral equation:

$$\begin{cases} \delta'(t) = A \cdot \delta(t) + \epsilon'(t) \\ \delta(0) = 0_{n\times n}. \end{cases} \tag{24}$$

For given $\delta(t_j)$, the analytical solution at t_{j+1} is given by

$$\delta(t_{j+1}) = e^{A(t_{j+1}-t_j)}\delta(t_j) + \int_{t_j}^{t_{j+1}} e^{A(t_{j+1}-\tau)}\epsilon'(\tau)d\tau.$$

Applying integration by parts, we have

$$\delta(t_{j+1}) = e^{Ah_{j+1}}\delta(t_j) + e^{A(t_{j+1}-\tau)}\epsilon(\tau)\big|_{\tau=t_j}^{t_{j+1}} + \int_{t_j}^{t_{j+1}} Ae^{A(t_{j+1}-\tau)}\epsilon(\tau)d\tau$$

where $h_{j+1} = t_{j+1} - t_j$. Therefore, we obtain

$$\delta(t_{j+1}) = e^{Ah_{j+1}}\delta(t_j) + \epsilon(t_{j+1}) - e^{Ah_{j+1}}\epsilon(t_j) + \int_{t_j}^{t_{j+1}} Ae^{A(t_{j+1}-\tau)}\epsilon(\tau)d\tau.$$

Again, $e^{Ah_{j+1}}$ can be efficiently applied using the Lie–Trotter–Suzuki operator splitting-based low-order methods. We apply the second-order trapezoidal rule to evaluate $\int_{t_j}^{t_{j+1}} Ae^{A(t-\tau)}\epsilon(\tau)d\tau$; hence, the updating formula becomes

$$\tilde{\delta}(t_{j+1}) = e^{Ah_{j+1}}(\delta(t_j) - \epsilon(t_j)) + \epsilon(t_{j+1}) + \frac{h_{j+1}}{2}(Ae^{Ah_{j+1}}\epsilon(t_j) + A\epsilon(t_{j+1})), \tag{25}$$

where $\tilde{\delta}(t)$ is the low-order approximation of the exact error $\delta(t)$. We can define Steps 2 and 3 as a function $\tilde{\delta}(t) = Fun(\tilde{P}(t))$ where the input is the given approximate solution $\tilde{P}(t)$ and the output is the low-order approximation of the error $\tilde{\delta}(t)$. Each such (implicit) function evaluation is considered as one SDC correction.

Step 4: Repeat Steps 2–4, or Stop, or Reset and Restart If the low-order estimate of the error $\tilde{\delta}(t)$ is within a prescribed error tolerance, then the approximate solution is considered accurate enough and one can output it as the converged solution. Otherwise, one can use the low-order estimate of the error to improve the approximate solution simply by using

$$\tilde{P}_{new} = \tilde{P}_{old} + \tilde{\delta},$$

and then returning to Step 2 until the iterations are convergent or a maximum number of iterations is reached. In the latter case when the method is not convergent, one reduces the time stepsize and restarts SDC from Step 1. As the low-order method becomes more accurate for smaller time stepsizes, the method is guaranteed to converge when a sufficiently small stepsize is used.

Comment The SDC method is equivalent to a sequence of fixed-point (stationary) iterations representing a particular Neumann series expansion for a low-order-method-preconditioned formulation. As the spectral integration matrix is applied in the final converged collocation formulation, the resulting algorithm is referred to as the spectral deferred correction (SDC) method in existing literature [6]. Instead of a naive Neumann series expansion, one can use the terms in the Neumann series to construct a Krylov subspace and search for the optimal least squares solution in the Krylov subspace to further accelerate the convergence. The resulting algorithm is referred to as the Krylov deferred correction method (KDC) [12]. For general nonlinear ODE initial value problems, the implementation of KDC is a simple application of an existing Jacobian-free Newton–Krylov (JFNK) solver [13, 14] to find the zero of the low-order-method-preconditioned function $\tilde{\delta}(t) = Fun(\tilde{P})$.

4 Compressible Matrix Structures and Accelerated Calculations

By representing the original matrix A as the sum of two (or more) simple structured matrices T and V, many of the involved algebraic operations can be accelerated. In this section, we study the diagonal and low-rank compressible features, their interactions in the matrix exponentials, and accelerated computation techniques.

First, using the Taylor expansion definition of a matrix exponential, it is straightforward to show that the exponential of a diagonal matrix T is also diagonal. The required storage for an $n \times n$ diagonal matrix is $O(n)$. When a diagonal matrix is applied to any vector, the amount of required operations is $O(n)$. The multiplication of two diagonal matrices requires asymptotically optimal $O(n)$ operations.

Next, we consider a real symmetric low-rank (rank $k \ll n$) matrix and consider its "complete" eigen-decomposition $V_{n\times n} = U_{n\times n}\Sigma_{n\times n}U^T_{n\times n}$ where U is an orthogonal matrix containing all the orthonormal eigenvectors and Σ is diagonal with only the first k diagonal entries nonzero. As $e^V = Ue^{\Sigma}U^T$ and only the first k diagonal entries of $e^{\Sigma} \neq 1$, we find that e^V is the sum of the Identity matrix and a low-rank matrix $U\left(e^{\Sigma} - I\right)U^T$ with the same rank k. Assuming the "compact" low-rank decomposition $V_{n\times n} = U_{n\times k}\Sigma_{k\times k}U^T_{k\times n}$ is already available, the storage of this decomposed form is $O(k \cdot n)$. When V or e^V is applied to a given vector using the decomposed form, the number of operations is only $O(k \cdot n)$. Furthermore, it only requires $O(k \cdot n)$ operations to compute the product of a low-rank matrix with a diagonal matrix, or a low-rank matrix with another low-rank matrix, or the sum of a diagonal matrix and low-rank matrix with the sum of another diagonal matrix and low-rank matrix. These products are basic building blocks in the LTS decompositions.

When studying the solution of ODE initial value problem using $e^{tA}\mathbf{y}_0$ where the vector $\mathbf{y_0}$ contains the initial conditions and $A = T + V$ with T and V either diagonal or low-rank, both the LTS decompositions and SDC accelerated GRK (SDC-GRK) methods are efficient for large-size matrices as the time marching scheme only requires the storage of vectors, special structured matrices, and matrix–vector multiplications of T, e^T, V, and e^V with given vectors. The required storage and number of operations for each time marching step are both asymptotically optimal $O(k \cdot n)$.

For applications where the matrix exponential e^{tA} is required, due to the interactions of different compressible features in matrix A, the corresponding compressible features in the matrix exponential e^{tA} may become complicated when t increases. This can be demonstrated using the following example: Assume D_1 and D_2 are two diagonal matrices and L_1 and L_2 are two low-rank matrices with rank k. Then, the matrix product (products of different terms in the LTS decompositions) is

$$(D_1 + L_1)(D_2 + L_2) = D_1D_2 + (D_1L_2 + L_1(D_2 + L_2)) = D_3 + L_3.$$

Although we conclude with another sum of a diagonal matrix $D_3 = D_1 D_2$ and a low-rank matrix $L_3 = (D_1 + L_1)L_2 + L_1 D_2$, the rank of L_3 may become larger than k and increase to as large as $2k$. As such products appear many times in the LTS decompositions and SDC-GRK algorithms, in the worst-case scenario, both the storage and number of operations in a related matrix–matrix multiplication increase as t increases due to the change in the rank k of the low-rank matrix, and the numerical computation may soon become impossible for large-size matrices. One motivation for our research is to understand how the compressible features (e.g., diagonal, banded, and low-rank structures in the physical and frequency domains) of matrix exponentials evolve as t increases. Understanding and identifying the hidden compressible features allow more efficient computations of matrix exponentials.

We have explored the interactions of the diagonal structure with the low-rank structure in the matrix exponential e^{tA} when $A = T + V$ is the sum of a diagonal matrix and a rank $k \ll n$ matrix. We have implemented two numerical approaches to efficiently find the numerical rank of the matrix L_3, its compressed SVD representation, and the diagonal matrix D_3 after each matrix–matrix product $(D_1+L_1)(D_2+L_2) = D_3+L_3$. In the first approach, as we know the rank of L_3 is no more than $2k$ and the matrices D_1, D_2, L_1, and L_2 have compressed structures, the matrix–vector multiplication $L_3\mathbf{v}$ can be efficiently evaluated for any given vector $\mathbf{v}$. Therefore, we apply the randomized algorithm from [17] where $2k + q$ vectors $\{\mathbf{v}_i\}$, $i = 1, \cdots, 2k + q$ are randomly generated and the compressed representation of L_3 is derived by analyzing the matrix–vector multiplications $\{L_3\mathbf{v}_i\}$. We set $q = 8$ and refer interested readers to [17] for how to choose q to minimize the probability of large approximation errors and details of the randomized algorithm. In the second approach, we combine ideas from the CUR matrix decompositions [16] and randomized CUR decomposition algorithms presented in [5]. By studying a properly sampled "skeleton" of the off-diagonal entries of $(D_1+L_1)(D_2+L_2)$, we find the numerical rank of the off-diagonals of L_3 (defined as the minimal rank of $L_3 - D$ for all possible diagonal matrices D). Once the compressed representations of the off-diagonals of L_3 are available, we find the diagonal matrix D_3 by simply computing the differences between the diagonals of $(D_1 + L_1)(D_2 + L_2)$ and those from the compressed representations of the off-diagonals of L_3. We skip the algorithmic details and refer interested readers to [5]. Note that the first approach finds the numerical rank r_1 of L_3, while the second approach studies the rank r_2 of the off-diagonals of L_3. Clearly, $r_1 \geq r_2$. For most L_3 matrices, we have observed numerically that $r_1 = r_2$.

We have applied our randomized rank-revealing algorithms to decompose the matrix exponential e^{tA} (derived using either the LTS decompositions or SDC accelerated GRK techniques) as the sum of a diagonal matrix D_3 and a low-rank matrix L_3. We present some of our theoretical findings. First, when $A = \lambda_0 I + V$ where V is rank k, $e^{tA} = \lambda_1 I + L_3$ where the rank of L_3 is no more than k for any time t. Next, when A is symmetric negative definite, as all the eigenvalues λ are real and negative, and $e^{\lambda t} \to 0$ exponentially fast as $t \to \infty$, the numerical rank of L_3 eventually decays to zero as $t \to \infty$. For finite time t, we can group the eigenvalues of tA into two groups, those less than a threshold $-\sigma$ and those in the interval

$[-\sigma, 0]$. If σ is a large positive number and $e^{-\sigma}$ is less than the error tolerance, then all the eigenvalues less than $-\sigma$ can be neglected when computing the numerical rank of L_3. For the other group, as the function e^x can be approximated by a finite degree polynomial for $x \in [-\sigma, 0]$, the rank of L_3 is bounded as there are only a finite number of $(D_1 + L_1)(D_2 + L_2)$ type matrix–matrix multiplications when computing the matrix exponential e^{tA} using its polynomial approximation. For more general matrix A, the rank of L_3 depends on the eigenvalue distributions of matrix A, which can be studied numerically in a time marching scheme. We present some preliminary numerical results in Sect. 5.

We end this section by citing two relevant results along this research direction. It was found in [8] that the eigenvector trajectory generated by smooth changes (e.g., in time) of the Hamiltonian matrix can be well approximated by a low-dimensional manifold; therefore, one can "learn" the eigenvector trajectory using data where the eigenvectors are computable. In addition, the eigensystems of a matrix X_n after a low-rank perturbation are studied in [3]. It was shown that adding some randomness to the eigenspaces permits further progress in analysis and a phase transition phenomenon was discovered after exact answers (interpreted in a probabilistic sense) are derived.

5 Preliminary Numerical Results

In this section, we present numerical results for selected matrix A examples. The LTS decompositions and SDC-GRK formulations are applied to the same set of problems to understand their accuracy and stability properties. As these approaches are based on different mathematical ideas and a lot of algorithm parameters still need fine-tuning, we find that a fair comparison is often hard and the method of choice is highly dependent on the accuracy requirements, problem settings, algorithm parameter selections, and compressible features of the split matrices and their exponentials.

In the first example, we consider a simplified two-body interacting quantum system in three dimensions modeled by $A = T + V$ where $T_{i,j} = -\delta_{i,j}\frac{p_x^2+p_y^2+p_z^2}{2m}$ representing the kinetic energy (or the noninteracting) part of A. It is a diagonal matrix and its diagonal entries are determined by a vector $\mathbf{p}$ with three entries (p_x, p_y, p_z), each representing momentum in one of three dimensions. $V_{i,j} = -g$ is the potential energy (or interacting) part of A; thus, V is a constant matrix where g is a constant scalar representing the coupling strength of the two bodies. Such a contact (zero-range) interaction is often used to describe ultracold atomic gases in dilute regimes; it is also often used to model dilute neutron matter in the crust of neutron stars (see, e.g., [22]).

We first demonstrate the "order" behaviors of the LTS decompositions for Example 1. In Fig. 2, we present the local truncation errors from different LTS decompositions. For an $n \times n$ real matrix A and its numerical approximation $\tilde{A}$, we

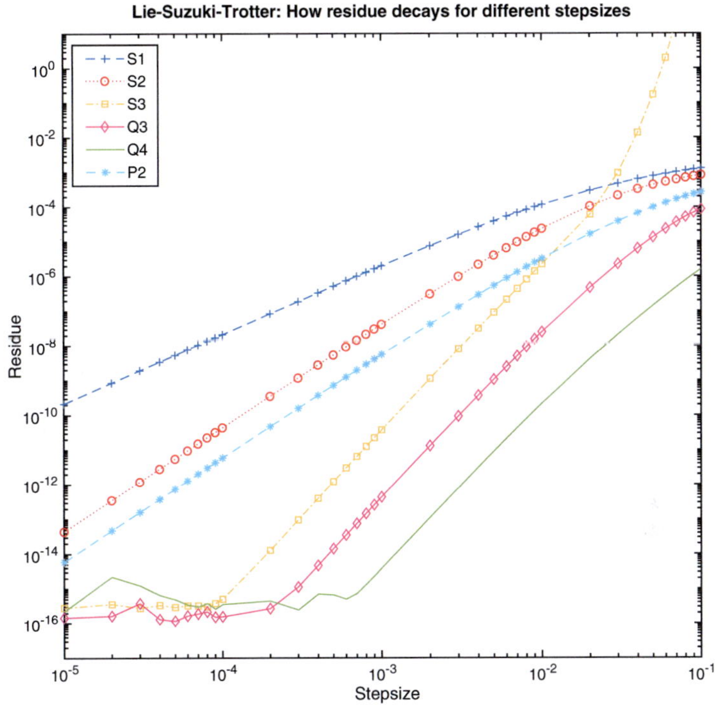

Fig. 2 Example 1: Local truncation error analysis for LTS decompositions

measure the error (residue) using the scaled Frobenius norm (also called Euclidean norm) defined as $\frac{1}{n}\sqrt{\sum_{i,j}\left((a_{i,j}-\tilde{a}_{i,j})^2\right)}$, where $a_{i,j}$ and $\tilde{a}_{i,j}$ are the entries in matrices A and $\tilde{A}$, respectively. When the decomposition Q3 or Q4 is used, we only consider the real part of the approximating matrix. Our numerical results show that the local truncation errors of S1, S2, S3, P2, and Q4 decay to zero with the expected order when the stepsize converges to zero. As discussed in Sect. 2, our numerical results also confirmed that the real part of the local truncation error of the method Q3 has order 5; therefore, its global order becomes the same as those of the decompositions S3 and Q4.

The order analysis of the SDC-GRK algorithm becomes more complicated. It depends on the type and number of node points used in the Picard integral equation formulation and on the number of SDC corrections. In this paper, because of its optimal numerical properties as shown in Theorem 1, we focus on the Gauss–Legendre collocation nodes in the GRK formulation. We compare the algorithm performance for different numbers of nodes and SDC corrections. In Fig. 3, we show the convergence properties of the SDC accelerated GRK algorithms with $p = 4$ and $p = 10$ Gauss nodes for different stepsize choices after two, three, and four SDC corrections. Instead of the scaled Frobenius norm, we use the element-wise

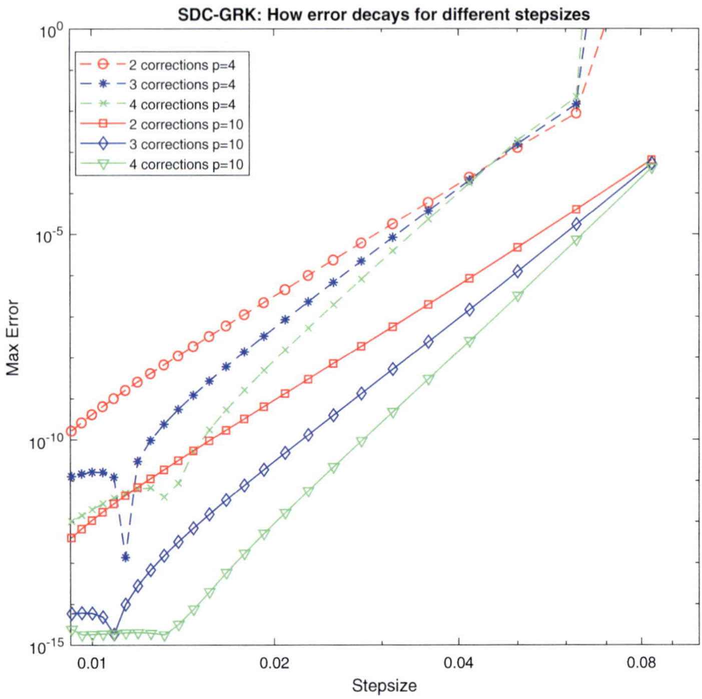

Fig. 3 Example 1: Local truncation error analysis for SDC-GRK formulations

max norm defined as $\max_{i,j} |a_{i,j} - \tilde{a}_{i,j}|$ to measure the error. We first observe that when the stepsize is too large (e.g., stepsize > 0.1), the SDC method becomes divergent and the error increases as the number of corrections increases. The reason is that the low-order method preconditioned system is not yet close to the Identity matrix and has "bad" eigenvalues that cause the error in the fixed-point iterations to grow exponentially as a function of the number of corrections. Second, when the SDC method is convergent, the order of the method heavily depends on the number of corrections, e.g., after two corrections, the slope of the algorithm with $p = 4$ is almost the same as that from $p = 10$. Similar results can be seen after three and four corrections. We want to mention that after reaching the intrinsic order $2p$ of the GRK formulation, further SDC corrections can no longer improve the order of the algorithm. Finally, when the stepsize is approximately 0.01 (10 Gauss points are used in the interval $[0, 0.01]$), the numerical results from the SDC-GRK method after four corrections provide better accuracy than all of the tested LTS decompositions with stepsize 0.001.

In order to better compare the performance of the LTS decompositions with that of the SDC-GRK formulations, we also show the achieved accuracy as a function of the number of matrix–matrix multiplications (we assume matrix exponential e^{At} is required). The results for the LTS decompositions are presented in Fig. 4.

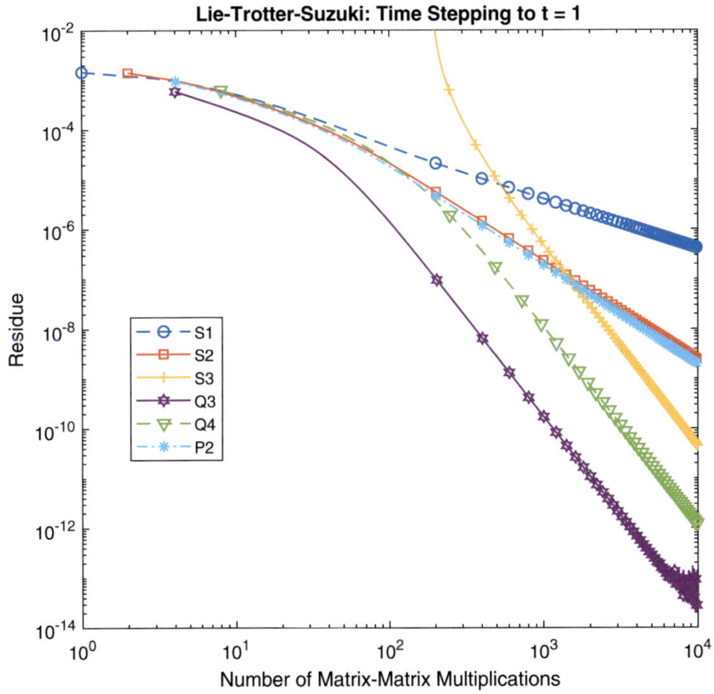

Fig. 4 Example 1: Achieved accuracy as a function of the number of matrix–matrix multiplications for LTS decompositions

For the same stepsize, Q4 provides better accuracy than Q3, but it requires more matrix–matrix multiplications as each Q4 is the product of two Q3s. When the performance is measured by the number of matrix–matrix multiplications for the same prescribed accuracy requirement, Q3 becomes a clear winner. In Fig. 5, we present the performance results for the SDC-GRK methods. The numerical results show that lower-order SDC-GRK method may perform better for low-accuracy requirements and higher-order SDC-GRK methods are preferred for high-accuracy requirements. As the current SDC-GRK implementation uses a first- or second-order time marching scheme as the preconditioner for the GRK formulation, the resulting algorithm outperforms the low-order LTS decompositions (e.g., S1, S2, and P2) for high-accuracy requirements. However, such first- or second-order time marching scheme-based SDC-GRK algorithm cannot yet compete with Q3 for this example. We find that using Q3 as the low-order method in Step 1 of the SDC approach can significantly improve the performance; however, it is still unclear how to use Q3 or other higher-order methods in the correction step (Step 3) of the SDC-GRK algorithms. We are continuing our research along this direction and results will be reported in the future.

In our second numerical example, we consider a 125×125 matrix T that contains two differently scaled block Identity matrices of sizes 63×63 and 62×62,

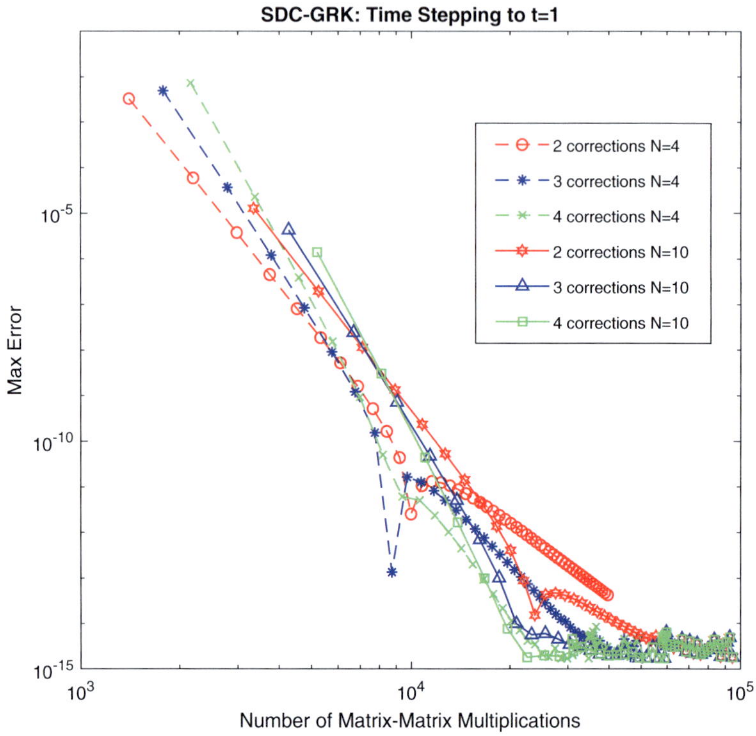

Fig. 5 Example 1: Achieved accuracy as a function of the number of matrix–matrix multiplications for SDC-GRK formulations

Table 1 Example 2: Four different cases with different c_1, c_2, and σ_{max} settings

Case	c_1	c_2	σ_{max}
1	5	4.999	5
2	5	0.005	0.005
3	5	4.999	100
4	5	0.005	5

respectively:

$$T = \begin{pmatrix} c_1 I & \mathbf{0} \\ \mathbf{0} & c_2 I \end{pmatrix}, \tag{26}$$

where c_1 and c_2 are different constant scalars. We choose V to be a rank-5 negative semi-definite matrix. The five nonzero singular values of V are randomly sampled from a uniform distribution in the interval $[-\sigma_{max}, 0]$, and its singular vectors are randomly generated. The two differently scaled Identity submatrices can be considered as two different physical systems. For this example, we consider four cases listed in Table 1, representing different interaction patterns of the two systems

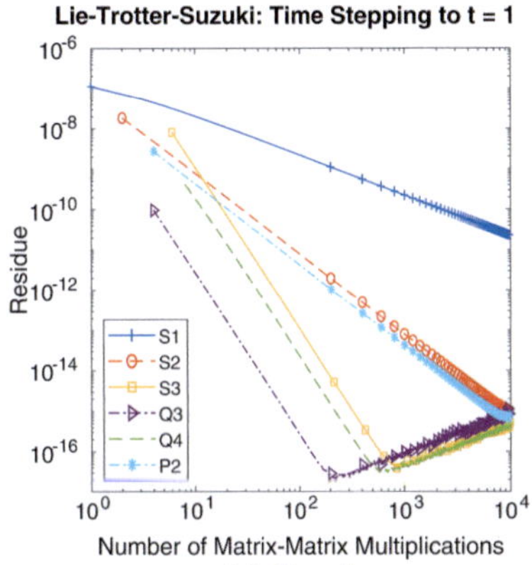

(a) Case 1

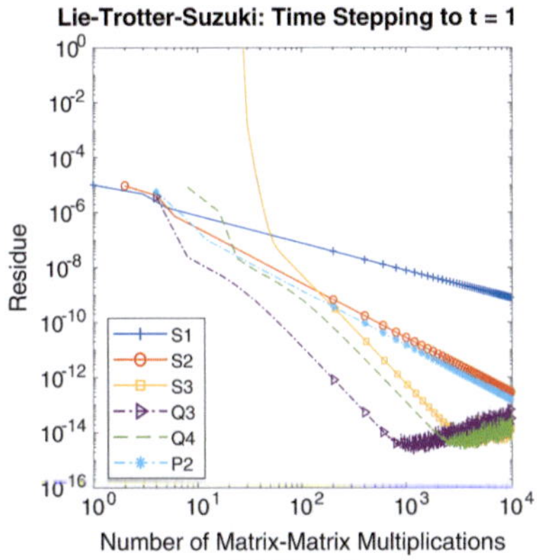

(b) Case 2

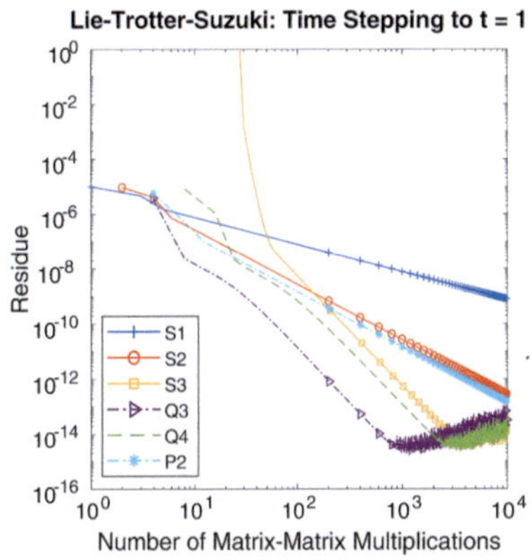

(c) Case 3

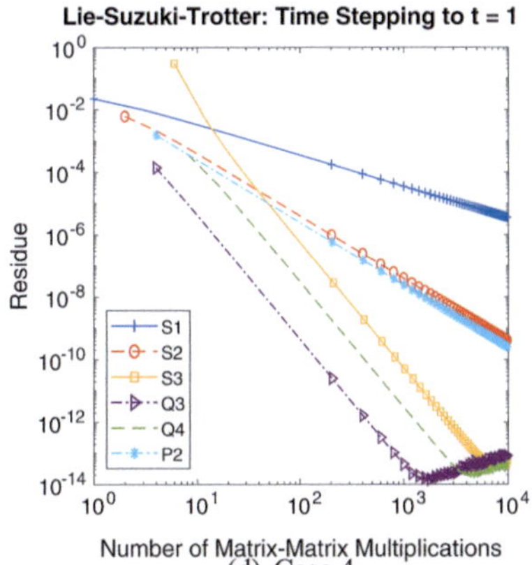

(d) Case 4

Fig. 6 Example 2: LTS decompositions for four different cases with $c_1 = 5$ and different c_2 and σ_{max} values

(c_1 and c_2) and potential energy roughly controlled by the environment parameter σ_{max}. In Fig. 6, we show how the accuracy depends on the number of matrix–matrix multiplications for different LTS decompositions for Cases 1–4. From the numerical results, we see that (i) a higher-order method is always preferred for high accuracy requirements. For low-accuracy requirements, sometimes a low-order method may perform better. (ii) For all four cases, method Q3 is a clear winner when compared with the other higher-order methods. When the time stepsize is large, there are convergence issues with the decomposition S3 (e.g., see Case 3). Similar issues with S3 are also observed in Example 1; see Fig. 4. (iii) After achieving the best possible accuracy, using smaller time stepsizes and marching more steps (more matrix–matrix multiplications) will increase the error, e.g., see Case 1, when the number of matrix–matrix multiplications is ≥ 200, the error from Q3 starts to increase. It is therefore important to choose the optimal algorithm parameters (e.g.,

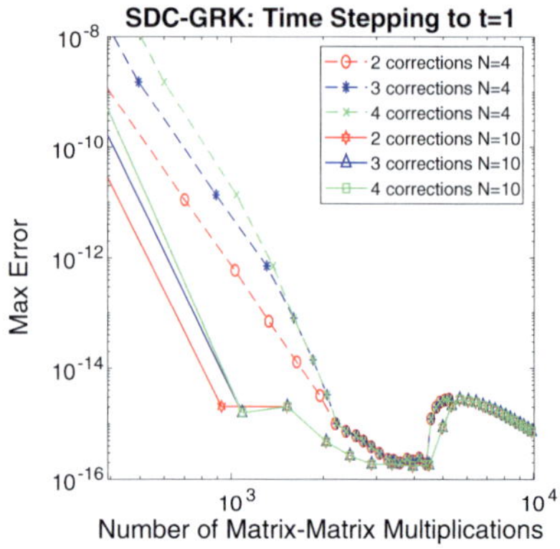

(a) Case1

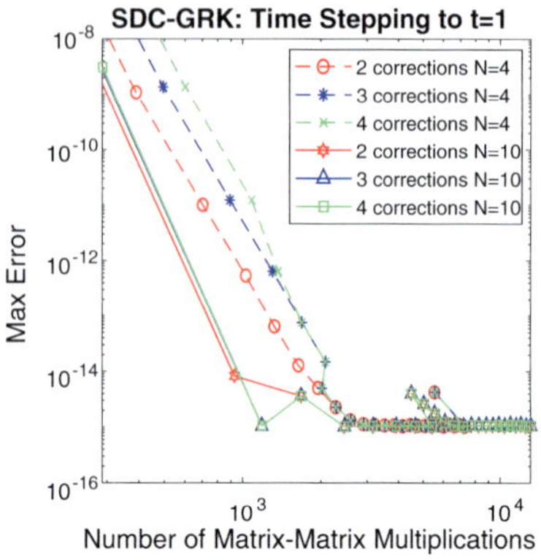

(b) Case2

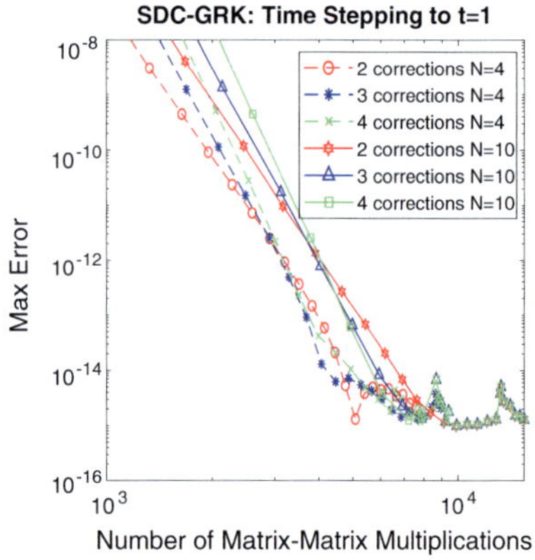

(c) Case3

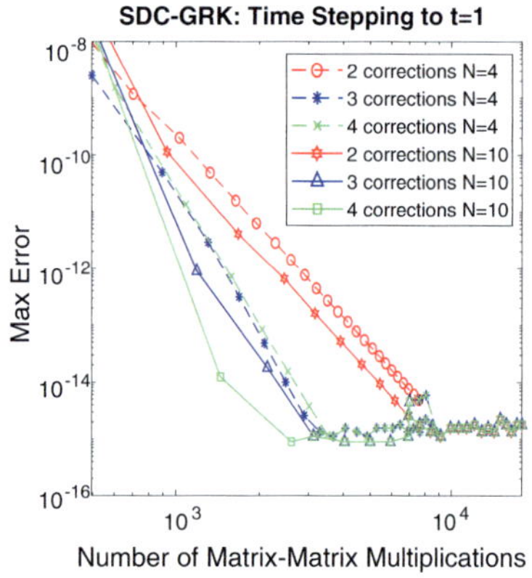

(d) Case4

Fig. 7 Example 2: SDC-GRK for four different cases with $c_1 = 5$ and different c_2 and σ_{max} values

stepsize) in the LTS decompositions. (iv) The settings c_1, c_2, and σ_{max} did change the performance of the LTS decompositions. When method Q3 is used, it requires approximately 200 matrix–matrix multiplications to achieve machine precision for Case 1. The numbers become approximately 300, 700, and > 1000 for Cases 2, 3, and 4, respectively.

Similar experiments are performed for the SDC-GRK algorithm and results are presented in Fig. 7. For Cases 1 and 2, the SDC-GRK algorithm with $p = 10$ and two SDC corrections outperforms other SDC-GRK methods. For Case 3, the SDC-GRK algorithm with $p = 4$ and two SDC corrections is the winner for lower-accuracy requirements, and the algorithm with $p = 4$ and three SDC corrections becomes the method of choice for higher-accuracy requirements. For Case 4 and higher accuracy requirements, the SDC-GRK algorithm with $p = 10$ and four SDC corrections becomes the winner. Finally, we note that unlike the LTS

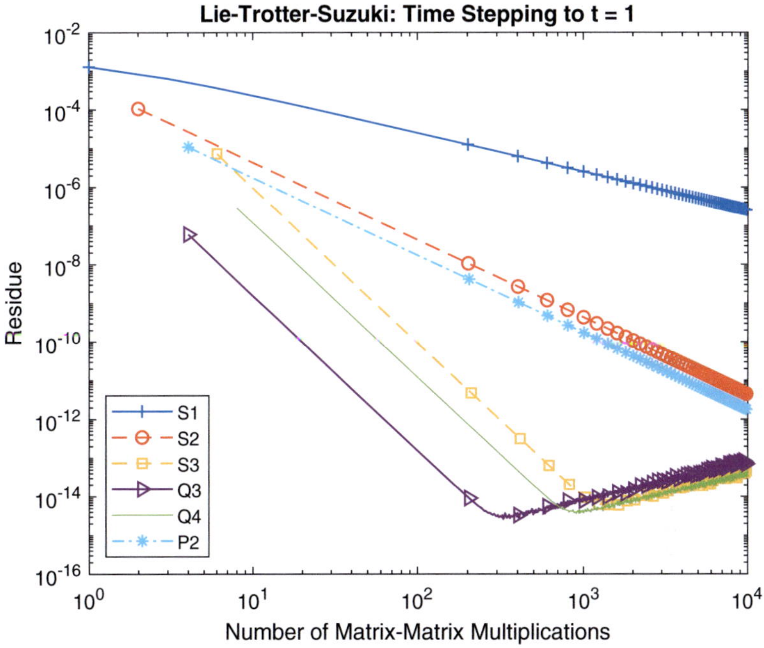

Fig. 8 Example 3: LTS decompositions, residue versus number of matrix–matrix multiplications

decompositions, the errors from the SDC-GRK algorithms for this example remain at approximately machine precision and don't seem to increase as significantly when using smaller time stepsizes (more matrix–matrix multiplications).

We have considered several additional examples including adding perturbations to the Identity matrix and increasing the rank k of the low-rank V matrix. We show the performance of different algorithms for a more general but representative setting where T is a general diagonal matrix with diagonal entries randomly sampled from the uniform probability density distribution $U[-1, 0]$ and V is a rank $k = 5$ negative semi-definite matrix constructed using $V = Q_{n\times k}\Sigma_{k\times k}Q^T_{k\times n}$ where Q is derived from the singular value decomposition of a random matrix and the k nonzero diagonals of Σ are randomly sampled from $U[-\sigma_{max}, 0]$. In Fig. 8, we set $\sigma_{max} = 1$ and plot the achieved accuracy for different numbers of matrix–matrix multiplications for different LTS decompositions. In Fig. 9, we plot the achieved accuracy as a function of the number of matrix–matrix multiplications for the SDC-GRK methods. For this example, as the collocation formulation with 10 points in the interval $[0, 1]$ approximately resolves the solution to machine precision, and the SDC iterations approximately converge to the collocation formulation in four iterations, the SDC method with ten Gauss nodes and four iterations (high-order) clearly outperforms the other SDC-GRK methods. We provide guidelines for selecting an appropriate algorithm and parameters for a given problem. For low-accuracy requirements, we find that the LTS-based Q3 outperforms most

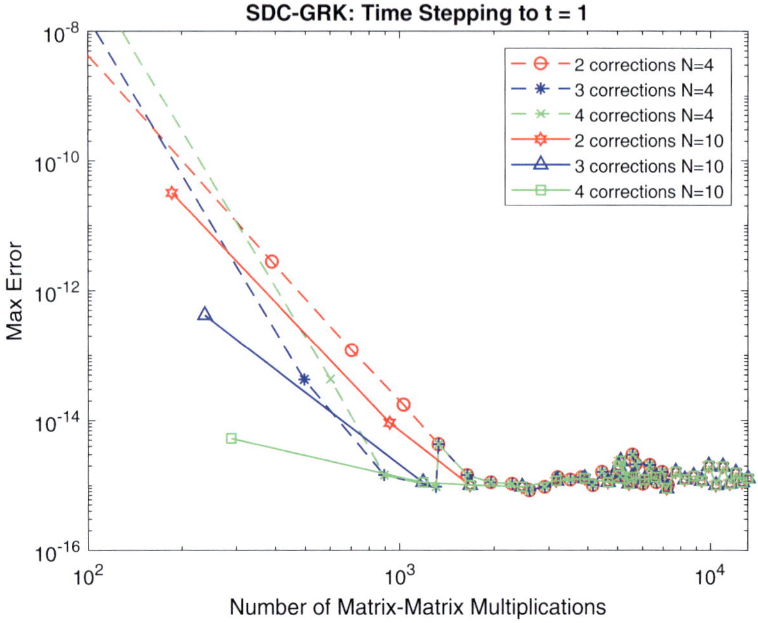

Fig. 9 Example 3: SDC-GRK methods, error vs number of matrix–matrix multiplications

of the other methods we have studied. In this regime, most current SDC-GRK implementations outperform the low-order LTS decompositions S1, S2, and P2. However, as introducing higher-order methods in the Correction Step (Step 3) of the current SDC-GRK implementations remains a challenging research topic, existing second-order preconditioner-based SDC-GRK schemes are not yet as competitive as Q3 and Q4. For high-accuracy requirements, as the SDC-GRK technique allows for much larger time stepsizes, it becomes the method of choice when proper preconditioners and acceleration techniques are introduced. The "optimal" implementations of the SDC-GRK and KDC-GRK are still actively being studied by our research community.

This paper is motivated by the research efforts to understand how the compressible features in matrix A interact with each other as t evolves in the matrix exponential e^{tA}. We present some preliminary numerical results to demonstrate how the low-rank and diagonal structures in A may impact the properties of the matrix exponentials, e.g., the numerical rank of the off-diagonals of e^{tA} defined as the minimal rank of $e^{tA} - D$ for all possible diagonal matrices D. The existence and identification of the compressible features in the matrix exponentials are important for understanding the physical systems and for accelerating the numerical simulations for large-scale problems. We consider the settings in our second example where T is given by Eq. (26). Note that when t is small and $O(t^2)$ along with higher-order terms can be neglected in Eq. (1), the numerical rank of the off-diagonals of e^{tA} is therefore determined by the numerical rank of

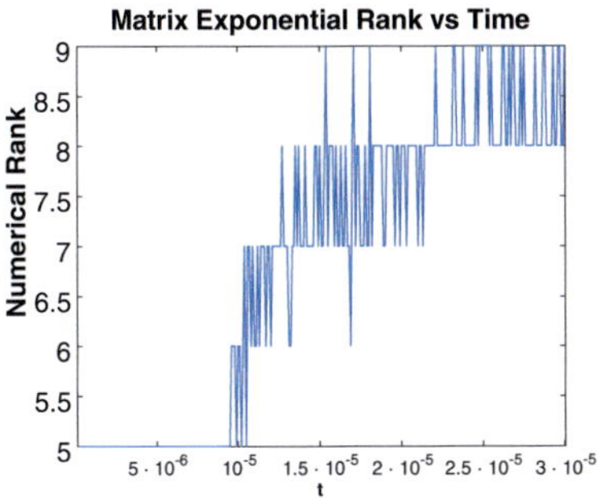

(a) Case 1: stepsize = 10^{-7}

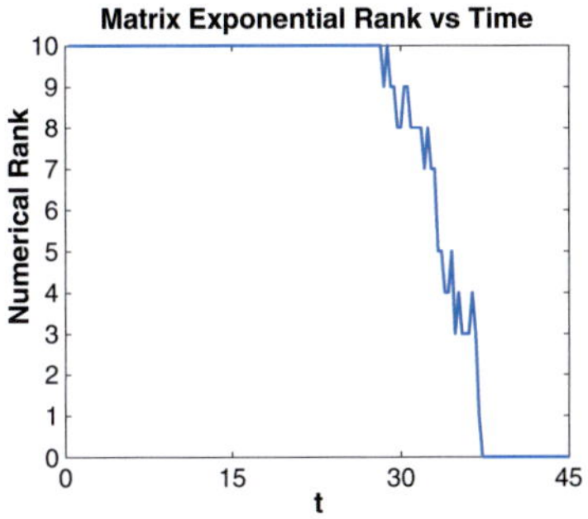

(b) Case 2: stepsize = 0.3

Fig. 10 Numerical rank of off-diagonals of e^{tA}, where $A = T + V$ and T is given by Eq. (26)

the off-diagonals of matrix tA, which is approximately 5. This rank changes as t increases. In the left of Fig. 10, we show how the numerical rank increases when $t \in [0, 3e{-}5]$. In the right of Fig. 10, we show that the solution (and rank) eventually decreases to zero when t approximately equals to 45. Therefore, the numerical rank of the off-diagonals is bounded by a constant (approximately 10) for all $t > 0$. See Sect. 4 for discussions of the off-diagonals' low-rank feature extraction algorithm and theoretical explanations of the changes in the low-rank properties in time.

Finally, to show the complexity of the relations between the compressible features in the matrix exponential e^{tA} and those in the matrix A, we consider a discretized convolution matrix where $A_{i,j} = \frac{1}{|x_i - x_j|}$, $i \neq j$, and x_i's are ordered and evenly spaced in the interval [0, 1]. To make the matrix negative definite, each diagonal $A_{i,i}$ is chosen to be the negative sum of the other matrix entries in the same row (or column). Note that such a matrix is no longer the summation of a diagonal matrix and a low-rank matrix. In many physical models, the spatially "well-separated" interactions are often low-rank. When matrix A describes the interactions of three physical systems, the first system is located in $[0, \frac{1}{3}]$, the second system is in $(\frac{1}{3}, \frac{2}{3}]$, and the last system is in $(\frac{2}{3}, 1]$. Systems 1 and 3 are thus spatially "well-separated" as they are at least "one box size" apart. The interactions of systems 1 and 3 form an off-diagonal matrix block in both A and e^{tA}. The rank of the corresponding well-separated matrix block in A is always bounded by approximately 10 using the SVD analysis, independent of the dimension of the submatrix block and locations of x_i. Our numerical results reveal that for fixed matrix size, the rank of the well-separated submatrix block will be bounded as t increases. However, when the dimension of the matrix increases, the maximum rank (defined as the maximum rank for all t values) of the corresponding well-separated submatrix block of e^{tA} increases approximately linearly as a function of the dimension of the well-separated submatrix block. This is clearly a challenge for large-scale simulations. In Fig. 11, we show how the maximum ranks (in t) of well-separated submatrices increase when the number of evenly spaced points increases. We assume the submatrix has dimension $n \times n$. We are studying this "loss

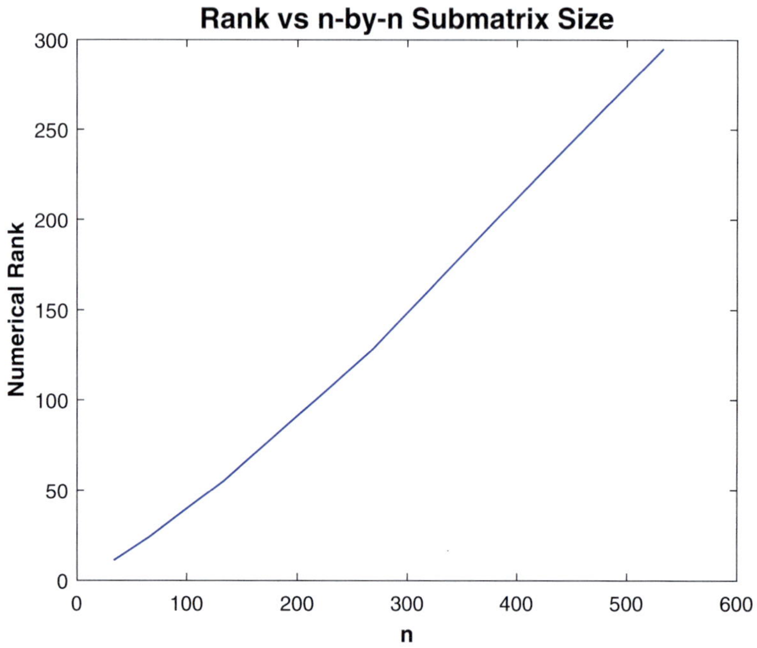

Fig. 11 Linear growth of the maximum numerical rank of well-separated submatrices of e^{At} as a function of submatrix dimension $n \times n$

of compressible features" phenomenon from both the modeling and linear algebra perspectives, and results will be reported in the future.

6 Summary and Future Work

The efficient computation of matrix exponentials is a fundamental building block in the numerical simulation of time-dependent science and engineering problems. In this paper, assuming the matrix can be split into the sum of simple structured matrices, including diagonal and low-rank matrices, we present two approaches to accelerate the computation of matrix exponentials and related matrix–matrix and matrix–vector multiplications. The first approach is based on the Lie–Trotter–Suzuki decompositions where the matrix exponential is approximated by the product of the exponentials of simple structured matrices. In the second approach, a polynomial matrix is computed to approximate the solution of the differential equation for the matrix exponentials. Compared with existing general-purpose solvers, preliminary numerical experiments show that both methods can improve simulation efficiency and provide satisfactory results in accuracy and stability.

In order to further improve the efficiency of the presented algorithms, it is important to understand and extract the compressible features in the matrix exponentials. A particularly interesting question is how the eigenvalues, eigenvectors, and rank of the off-diagonals and submatrix blocks change when the physical system evolves in time. We have implemented randomized rank-revealing algorithms to extract the low-rank structures and create compressed representations for the low-rank structures. The algorithms have been applied to several systems to better understand the complicated interactions of different compressible features. We are currently working on the rigorous analysis of the preliminary numerical experiments and results will be presented in the future.

Competing Interests and Acknowledgements The work of RE, YM, and JH was partially supported by NSF under grant No. DMS-2152289. The work of JD was funded in part by NSF under Grant No. PHY-2013078, and YL was partially supported by NSF CAREER 2414705. We also thank the anonymous reviewers for their valuable comments and suggestions.

References

1. Bao, W., Jin, S., Markowich, P.A.: On time-splitting spectral approximations for the Schrödinger equation in the semiclassical regime. J. Comput. Phys. **175**(2), 487–524 (2002)
2. Barthel, T., Zhang, Y.: Optimized Lie-Trotter-Suzuki decompositions for two and three non-commuting terms. Ann. Phys. **418**, 168165 (2020)
3. Benaych-Georges, F., Nadakuditi, R.R.: The eigenvalues and eigenvectors of finite, low rank perturbations of large random matrices. Adv. Math. **227**(1), 494–521 (2011)
4. Czejdo, A.J., Drut, J.E., Hou, Y., Morrell, K.J.: Toward an automated-algebra framework for high orders in the virial expansion of quantum matter. Condens. Matter **7**(1), 13 (2022)
5. Dong, Y., Martinsson, P.G.: Simpler is better: A comparative study of randomized algorithms for computing the cur decomposition. arXiv preprint arXiv:2104.05877 (2021)
6. Dutt, A., Greengard, L., Rokhlin, V.: Spectral deferred correction methods for ordinary differential equations. BIT Numer. Math. **40**, 241–266 (2000)
7. Emrick, R.: Rachel's undergraduate thesis (2024). UNC Chapel Hill Undergraduate Honors Thesis
8. Frame, D., He, R., Ipsen, I., Lee, D., Lee, D., Rrapaj, E.: Eigenvector continuation with subspace learning. Phys. Rev. Lett. **121**(3), 032501 (2018)
9. Greengard, L.: Spectral integration and two-point boundary value problems. SIAM J. Numer. Anal. **28**(4), 1071–1080 (1991)
10. Hairer, E., Lubich, C., Wanner, G.: Geometric Numerical Integration. Springer, Berlin (2011)
11. Hatano, N., Suzuki, M.: Finding exponential product formulas of higher orders. In: Quantum Annealing and Other Optimization Methods, pp. 37–68. Springer (2005)
12. Huang, J., Jia, J., Minion, M.: Arbitrary order Krylov deferred correction methods for differential algebraic equations. J. Comput. Phys. **221**(2), 739–760 (2007)
13. Kelley, C.: Solving Nonlinear Equations with Iterative Methods: Solvers and Examples in Julia. SIAM, Philadelphia (2022)
14. Knoll, D.A., Keyes, D.E.: Jacobian-free Newton-Krylov methods: a survey of approaches and applications. J. Comput. Phys. **193**(2), 357–397 (2004)
15. Lie, S., Engel, F.: Theorie der transformationsgruppen, vol. 3. Teubner, Leipzig (1893)
16. Mahoney, M.W., Drineas, P.: CUR matrix decompositions for improved data analysis. Proc. Natl. Acad. Sci. **106**(3), 697–702 (2009)

17. Martinsson, P.G., Rokhlin, V., Tygert, M.: A randomized algorithm for the decomposition of matrices. Appl. Comput. Harmonic Anal. **30**(1), 47–68 (2011)
18. Strang, G.: On the construction and comparison of difference schemes. SIAM J. Numer. Anal. **5**(3), 506–517 (1968)
19. Suzuki, M.: Fractal decomposition of exponential operators with applications to many-body theories and Monte Carlo simulations. Phys. Lett. A **146**(6), 319–323 (1990)
20. Suzuki, M.: Improved Trotter-like formula. Phys. Lett. A **180**(3), 232–234 (1993)
21. Trotter, H.F.: On the product of semi-groups of operators. Proc. Am. Math. Soc. **10**(4), 545–551 (1959)
22. Zwerger, W.: The BCS-BEC Crossover and the Unitary Fermi Gas. Springer, Berlin, Heidelberg (2012)

Part II
Graph Algorithms

An Exploration of Graph Distances, Graph Curvature, and Applications to Network Analysis

Kasia Jankiewicz, Manasa Kesapragada, Anna Konstorum, Kathryn Leonard, Nazia Riasat, and Michelle Snider

1 Introduction

Various notions of curvature in geometry measure how much, and possibly in which directions, the space differs from a flat Euclidean space, e.g., how much a curve differs from a straight line and how much a surface differs from a flat plane. In Riemannian geometry, the concepts of *scalar curvature*, which is an assignment of a number to each point in a manifold, or *Riemann* or *Ricci* curvature tensors, which assign a tensor to each point in a manifold, are considered as an *intrinsic* property of a geometric object, i.e., they are independent to the embedding of the object in an ambient space [17].

The classic examples of constant curvature are the surface of a sphere, which has constant positive curvature; the Euclidean plane, whose curvature is zero everywhere; and the hyperbolic plane, which has constant negative curvature. More generally, the curvature at a given point is defined for more general Riemannian

K. Jankiewicz
Department of Mathematical Sciences, University of California-Santa Cruz, Santa Cruz, CA, USA

M. Kesapragada
Department of Applied Mathematics, Baskin School of Engineering, University of California-Santa Cruz, Santa Cruz, CA, USA

A. Konstorum (✉) · M. Snider
Center for Computing Sciences, Institute for Defense Analyses, Bowie, MD, USA
e-mail: akonsto@super.org

K. Leonard
Department of Computer Science, Occidental College, Los Angeles, CA, USA

N. Riasat
Department of Statistics, North Dakota State University, Fargo, ND, USA

C. Garcia-Cardona, H. Lee (eds.), *Advances in Data Science*, Association for Women in Mathematics Series 37, https://doi.org/10.1007/978-3-031-87804-6_3

manifolds, and it may vary as the point varies, so there might be points of positive curvature where the manifold locally resembles a sphere; points with zero curvature, where the manifold locally looks flat; and points of negative curvature. There also exist notions of curvature of higher-dimensional manifolds. The curvature captures various geometric properties, e.g., the area of the disk of a fixed radius, or the isoperimetry, i.e., the relation between the circumference of a closed curve and the area enclosed by it.

There is a natural interest in defining a concept of curvature that captures some of those aspects in discrete spaces, such as graphs, to help understand the connections and structure therein. In this paper, we focus on the notions of curvature that are defined for graphs, as opposed to continuous spaces classically considered in differential geometry. We are also interested in intrinsic definitions of curvature. Among the first studied graph curvature notions were the Ollivier-Ricci curvature [23], defined in terms of optimal transport, and the Forman-Ricci curvature [10], in terms of the discrete Laplacian. More recently, the concept of graph curvature, defined in terms of shortest-path and resistance distance, has been proposed and studied by Devriendt-Lambiotte [7] Steinerberger [24], and Devriendt-Ottolini-Steinerberger [8].

Classically, nodes in graphs have often been analyzed in terms of their centrality, according to various types of centrality measures, and how that is related to the connections and structure of the graph. These measures can be thought of as how "important" a node is to some defined information flow across a network [5]. Thus, when considering the curvature of a graph, it is natural to ask how the curvature at each node is related to the centrality or if they give us different or complementary information.

In this work, we focus on studying associated properties of *centrality* and *curvature* at the node level of a graph. Definitions for different kinds of curvature and centrality depend heavily on the choice of distance metric on the graph. As such, we consider curvatures and centralities using two different distances on graphs: *shortest-path distance* and *resistance distance*. Unlike the standard shortest-path distance, the resistance distance takes into account not only the shortest path between two vertices but the lengths of all paths. As such, it provides much more information about how information can flow across a graph.

A connection between the centrality measures and other notions of discrete curvature has been recently investigated by other authors. In [19], variants of Ricci curvature (Baker-Émery, Forman, and Ollivier) on graphs and their relations to centrality measures are explored. In [18], a bound on the average shortest-path distance in terms of the average vertex degree and the average Ollivier-Ricci curvature "weighted" by the betweenness centrality is established.

In this paper, in Sect. 2, we cover the necessary definitions for the Laplacian, two graph distances, centrality measures, and graph curvature. In Sect. 3, we discuss mathematical interpretations of graph distance and curvature. We then investigate the relationships between these metrics and the intuition behind them, in particular between node-level curvature and node-centrality measures on a set of synthetic and real-world graphs. The studied graphs are described in Sect. 4, and our computations are presented in Sect. 5. Further discussion is included in Sect. 6.

2 Background

Let $G = (V, E, w)$ be an undirected, weighted graph with vertices $v \in V = [1, n]$, edges $e_{i,j} \in E$, for $i, j \in [1, n]$ between vertices i and j with edge weights in $\mathbb{R}_{\geq 0}$ given as $a_{ij} \geq 0$. Define the adjacency matrix $A = (a_{ij})$ and the diagonal degree matrix $D = \sum_j A_{ij}$, where the ith diagonal entry is the generalized degree of vertex i. Let $\mathbf{1} = (1, \dots, 1)$, the vector of all ones, with the length determined by context.

2.1 The Laplacian

Definition 1 For a graph G, we define the Laplacian matrix as

$$L := D - A.$$

This matrix is symmetric positive semi-definite; therefore, all eigenvalues are nonnegative. By definition, $L\mathbf{1} = 0$, so 0 is an eigenvalue with normalized eigenvector $1/\sqrt{n}\mathbf{1}$. Let us assume that G is a connected graph, in which case all other eigenvalues will be positive. That is, $0 < \lambda_2 \leq \lambda_3 \ldots \leq \lambda_n$ with eigenvectors $[\frac{1}{\sqrt{n}}\mathbf{1}, v_2, \dots, v_n]$. The singular value decomposition of L can then be written as

$$L = \left[\tfrac{1}{\sqrt{n}}\mathbf{1}\ v_2 \ldots v_n\right] \begin{bmatrix} 0 & & & \\ & \lambda_2 & & \\ & & \ldots & \\ & & & \lambda_n \end{bmatrix} \left[\tfrac{1}{\sqrt{n}}\mathbf{1}\ v_2 \ldots v_n\right]^T$$

$$:= V \begin{bmatrix} 0 & & & \\ & \lambda_2 & & \\ & & \ldots & \\ & & & \lambda_n \end{bmatrix} V^T.$$

The Laplacian matrix is not invertible, but we describe two pseudo-inverses. The Moore-Penrose, or generalized, inverse of the Laplacian is defined as

$$L^{\dagger} = V \begin{bmatrix} 0 & & & \\ & 1/\lambda_2 & & \\ & & \ldots & \\ & & & 1/\lambda_n \end{bmatrix} V^T. \tag{1}$$

This matrix is also symmetric, positive definite, and has zero row and column sums. In fact,

$$L^\dagger L = LL^\dagger = \Pi_n$$

where $\Pi_n = I - \frac{1}{n}\mathbf{1}\mathbf{1}^T$ is the projector onto the subspace orthogonal to the kernel of L, the space spanned by $\mathbf{1}$. From this, we can verify that

$$L^\dagger = (L + \frac{1}{n}\mathbf{1}\mathbf{1}^T)^{-1} - \frac{1}{n}\mathbf{1}\mathbf{1}^T$$

as in Ghosh et al. [12]. That is, we add a perturbation to the Laplacian matrix to make it invertible and then subtract the perturbation off after we have inverted it.

Second, we define the regularized Laplacian Γ as

$$\Gamma := L + \frac{\beta}{n}\mathbf{1}\mathbf{1}^T$$

for some choice of $\beta > 0$. This matrix has the same eigenvectors as L, but the 0 eigenvalue is shifted so that Γ is non-singular and as such we can find

$$\begin{aligned}\Gamma^{-1} &:= (L + \frac{\beta}{n}\mathbf{1}\mathbf{1}^T)^{-1} \\ &= V \begin{bmatrix} 1/\beta & & & \\ & 1/\lambda_2 & & \\ & & \cdots & \\ & & & 1/\lambda_n \end{bmatrix} V^T. \end{aligned} \tag{2}$$

That is, for $\beta = 1$, we get the following relationship between the two pseudo-inverses:

$$\Gamma^{-1} = L^\dagger + \frac{1}{n}\mathbf{1}\mathbf{1}^T$$

A note on overloaded notation: we use D with no subscripts to indicate the degree matrix and D_* with subscripts to represent the two types of distance matrices we use to compute other metrics.

2.2 Graph Distances

We define a path $p_{i,j} = \{v_i, v_{i+1}, .., v_j\}$ as a sequence of vertices on G such that for each $v_k \in p_{i,j}$, $\exists\ e_k \in E$ that connects $\{v_k, v_{k+1}\}$ and $k \neq l\ \forall\ v_k, v_l \in p_{i,j}$. We consider two distance metrics: the shortest-path distance is the length of the shortest path between two points, while the resistance distance depends on all parallel paths between them, as detailed below.

Definition 2 The shortest-path distance between nodes $v_i, v_j \in V$ is taken to be

$$D_{S,ij} = \min_{p_{i,j}} \left(\sum_{k=i}^{j-1} w(e_k) \right),$$

where $\{e_i, e_{i+1}, ..., e_{j-1}\}$ correspond to edges of a path $p_{i,j} = \{v_i, v_{i+1}, ..., v_j\}$.

The resistance distance on graphs, also known as the *effective resistance*, was introduced by Klein and Randic [15] and inspired by electrical network theory. Informally, all edges $e_{i,j} \in E$ for a graph $G = (V, E, w)$ are conceptualized as resistors with the weight of the edge proportional to the conductance, and for nodes $v_i, v_j \in E$, the resistance distance is calculated by considering a unit of current entering the network at v_i and leaving at v_j, and calculating the potential difference $D_{R,ij}$ between the two using Kirchhoff's circuit laws:

1. The sum of all currents at a node is 0.
2. The sum of all voltages around a closed loop is 0.

Definition 3 The resistance distance $D_{R,ij}$ between nodes $v_i, v_j \in V$ is taken to be

$$D_{R,ij} = (\Gamma^{-1})_{ii} + (\Gamma^{-1})_{jj} - 2(\Gamma^{-1})_{ij}, \tag{3}$$

where the inverse of the regularized Laplacian is as in Eq. 2, as defined in [6, 9].

Note that we could have equivalently defined this as

$$D_{R,ij} = L^\dagger_{ii} + L^\dagger_{jj} - 2L^\dagger_{ij}, \tag{4}$$

by Definition 2, where $L^\dagger$ is the Moore-Penrose pseudo-inverse of the Laplacian as defined in Eq. 1. We note this since some authors [24] use Eq. 4, while others [8] use Eq. 3.

The resistance distance is a useful metric for assessing the "connectedness" of two nodes in a graph, in that it incorporates not just the single shortest path but also the lengths of all the paths connecting those two nodes. As a simple example, consider a graph G with just two nodes and a single edge of weight 1, versus a graph G' with two nodes but two edges of weight 1 each. The shortest-path distance between the two nodes is 1 in both graphs, but the resistance distance in G is 1, while in G', it is $\frac{1}{2}$, accounting for the two possible paths. This can be useful in graphs that represent information flow, as it captures in some sense how easy it is for information to get from one node to another.

Bozzo et al. [6] highlight that in an acyclic graph, the shortest-path and resistance distances are equivalent, as there is only one path for current to flow from any starting point to any ending point. As graphs become denser, there are more paths between nodes, creating more paths for the current to split its flow, and, thus, the resistance distance becomes smaller than the shortest-path distance.

2.3 Centrality Measures

Network centrality refers to node properties capturing their "importance," or centrality to a network. The most basic measure is *degree centrality*, which is simply the degree of a node scaled by the number of nodes in the graph, $|V|$.

Definition 4 *Degree centrality* $Deg_{cent,i}$ of a node i is the degree of the node divided by $(|V| - 1)$.

Several other centrality measures are parameterized by the distance on the nodes. We look at two such measures, the *closeness centrality* and *betweenness centrality*, which were first defined in [4] and [11], respectively, and can be considered with respect to any distance on a graph. We provide details of their formulation using both the shortest-path distance D_S (Definition 2) and resistance distance D_R (Eq. 4).

Closeness centrality of a node is defined as the mean distance between that node and all other nodes in a network for a particular distance metric. The closeness centrality represents a relative measure of how long it will take for information to spread to and from each node. Betweenness centrality measures the proportion of paths that a node lies in between all pairs of nodes and, as such, captures how often random walks in the graph pass through a particular node. This can be thought of as how important a node is for information flow across a network. The mathematical definitions for these centrality metrics parametrized by the two different distances are provided below.

Definition 5 *Shortest-path closeness centrality*, $C_{S,i}$ of a node i is taken to be the average of the shortest-path distance from that node to all the others:

$$C_{S,i} = \frac{n}{\sum_{j=1,j\neq i}^{n} D_{S,ij}},$$

where $n = |V|$.

We will refer to $C_{S,i}$ as defined above to be the closeness centrality parameterized by the shortest distance metric. Alternately, we will consider the closeness centrality parameterized by the resistance distance as in [6], sometimes also called the *current-flow closeness centrality* due to its interpretation of how current flows between nodes.

Definition 6 *Resistance closeness centrality* $C_{R,i}$ of a node i is defined as

$$C_{R,i} = \frac{n}{\sum_{j=1,j\neq i}^{n} D_{R,ij}},$$

where the distance metric used is the resistance distance as in Eq. 3.

Definition 7 *Shortest-path betweenness centrality*, $B_{S,i}$, for a node i is the proportion of shortest paths that node i lies in between all nodes:

$$B_{S,i} = \sum_{a \neq b} \frac{\eta(a, i, b)}{\eta(a, b)},$$

where $\eta(a, i, b)$ is the number of shortest paths connecting vertices a and b that pass through node i and $\eta(a, b)$ is the total number of shortest paths between a and b.

Definition 8 *Resistance betweenness centrality* (or *current-flow betweenness centrality*), $B_{R,i}$, is calculated as the average of current flow through a node i when a unit of current is injected in a source and removed from a target node across all source-target pairs. It can be shown [6] that $B_{R,i}$ can be calculated as the following sum over source a and target b pairs:

$$B_{R,i} = \frac{1}{(1/2)n(n-1)} \sum_{(a,b),a<b} F_i^{(a,b)},$$

where $n = |V|$ and $F_i^{(a,b)}$ is the current flow through node i from source a to target b, defined as

$$F_i^{(a,b)} = \frac{1}{4} \sum_j A_{ij} |D_{R,ia} - D_{R,ja} + D_{R,jb} - D_{R,ib}|,$$

where A is the weighted adjacency matrix of G and $D_{R,**}$ is the resistance distance as calculated in Eq. 4.

Note that the ordering within the pair of source a and target b is irrelevant, as $F_i^{(a,b)} = F_i^{(b,a)}$. If we think of the graph as a network over which information flows, betweenness centrality at a node gives us information about how much each node has control over information flowing through it.

2.4 Graph Curvature

While the reader may have some intuition about the curvature of surfaces, this does not directly translate to a notion of curvature on a graph. One can think of curvature as a limiting factor on the size of a graph—just as a highly positively curved surface is limited in size, so is a highly positively curved graph [8]. Another interpretation of graph curvature is a measure of the average distance between two random walks that start at a particular vertex and take independent random paths [7].

Definition 9 The curvature $\kappa_i \in \mathbb{R}$ in the vertex $v_i \in V$ is defined such that the vector $\kappa = (\kappa_1, \ldots, \kappa_n) \in \mathbb{R}^n$ solves the linear system of equations:

$$D_* \kappa = n \cdot \mathbf{1} = (n, \ldots, n)$$

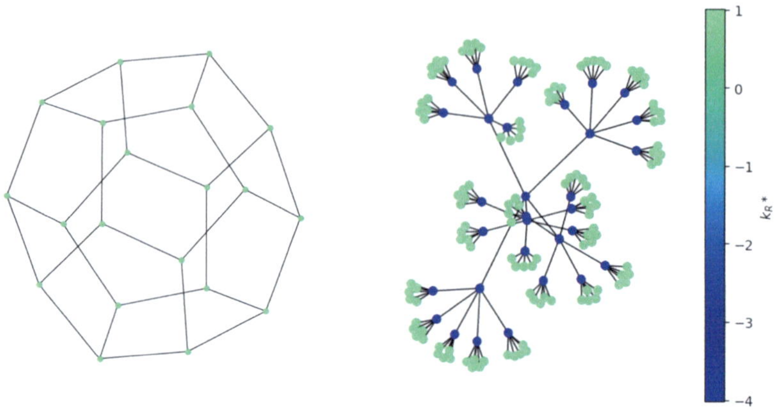

Fig. 1 (Left) The dodecahedron has constant positive resistance curvature ($\kappa_R = 1.0949$), just as the sphere has positive curvature. (Right) A balanced tree with five branches per node and a depth of 3 is an example of the following: every finite tree has positive resistance curvature at all leaf nodes and negative resistance curvature at interior nodes

as in [24], where it is noted that it is "exceedingly rare" that this system does not have a solution. We denote curvature calculated with the shortest-path distance as $\kappa_{S,i}$ and curvature calculated with the resistance distance as $\kappa_{R,i}$.

To provide some intuition, in Fig. 1, we show examples of positive and negative resistance curvature κ_R.

We note that the literature includes several slightly different definitions of graph curvature. First, in their 2022 work, Devriendt and Lambiotte [7] define it as the solution to

$$\kappa = \frac{D_*^{-1}\mathbf{1}}{\langle \mathbf{1}, D_*^{-1}\mathbf{1}\rangle}. \tag{5}$$

The denominator normalization factor has the effect that every graph with constant curvature has curvature $\kappa_i = 1/n$.

Second, in their 2024 paper, Devriendt et al. [8] make a slight change to [24] and define it as the solution to

$$D_*\kappa = \mathbf{1} = (1, \ldots, 1). \tag{6}$$

The authors note that this change allows for the computation of useful bounds relating to hitting and commute times [8].

3 Mathematical Interpretations

To help motivate the intuition behind graph distances for newcomers to the field, we provide mathematical interpretation in the language of linear algebra. One interpretation of graph distances comes from matrix theory and the link between symmetric positive definite matrices and Euclidean distance matrices, as in [14] whose framing we now summarize. Let Ω_n be the cone of symmetric positive semi-definite matrices of order n, contained in S_n, the subspace of symmetric matrices of order n. Let Λ_n be the cone of Euclidean distance matrices of order n. For $\circ$ representing the Hadamard product, and $\mathbf{1}$ representing the n-vector of ones, we may define a mapping from $\Omega_n \to \Lambda_n$:

$$\begin{aligned} K(A) &= (A \circ I)\,\mathbf{1}\mathbf{1}^T + \mathbf{1}\mathbf{1}^T\,(A \circ I) - 2A \\ &= \text{diag}(A)\mathbf{1}^T + \mathbf{1}\,\text{diag}(A)^T - 2A. \end{aligned}$$

We note that $K(A)_{ij} = a_{ii} + a_{jj} - 2a_{ij}$. Since K is linear, a quick count of dimensions shows that $\dim(\ker(K)) = n$. If we assume that $Ax = 0$ for some $x \in \mathbb{R}^n$ satisfying $x^T\mathbf{1} = 1$, we can uniquely determine an A mapping to any particular distance matrix for our choice of x. Given a graph distance matrix, we may then search for $A \in \Omega_n$ for our choice of x. Similarly, we may generate new distance matrices by applying K to matrices $A \in \Omega_n$.

In the specific case of the resistance distance, we have a matrix $\Gamma = D - A + \frac{1}{n}\mathbf{1}$, which is an element of Ω_n together with its inverse Γ^{-1}. The term $\frac{1}{n}\mathbf{1}$ reflects the choice of x determining $\ker(A)$. Applying the mapping K, we obtain a matrix $R = K(\Gamma^{-1})$ with ijth element $R_{ij} = (\Gamma^{-1})_{ii} + (\Gamma^{-1})_{jj} - 2(\Gamma^{-1})_{ij}$. In other words, the resistance distance matrix is the image of the inverse of the regularized graph Laplacian matrix (with a particular choice of kernel) under the mapping K. This provides a mathematical context for interpreting resistance distance in particular graphs, as we do in some of the experiments that follow.

4 Methods

In this work, we are interested in the relationships between notions of centrality (closeness, betweenness, and degree) and curvature. We will consider each measure computed with both the standard shortest-path distance measure and the resistance distance. We compare these measures on the vertices of a set of graphs, some synthetic and some based on real-world data. For completeness, we list the seven measures on a vertex v_i here:

- $Deg_{\text{cent}\,i}$: degree centrality (Definition 4)
- $C_{S,i}$: shortest-path closeness centrality (Definition 5)
- $C_{R,i}$: resistance closeness centrality (Definition 6)

- $B_{S,i}$: shortest-path betweenness centrality (Definition 7)
- $B_{R,i}$: resistance betweenness centrality (Definition 8)
- $\kappa_{S,i}$: shortest-path curvature (Definition 9)
- $\kappa_{R,i}$: resistance curvature (Definition 9)

4.1 Graphs Tested

The graphs considered include a mixture of synthetic and real-world graphs, chosen to investigate the relationships between the measures listed above.

4.1.1 Synthetic Graphs Under Perturbations

To develop our intuition about how these measurements interrelate, we begin with experiments on small graphs. Looking at graphs with the same number of nodes but different graph structures helps us to understand how graph distances change as graph connectivity changes. The graphs we consider, each with eight nodes, are the binomial tree, the Sedgewick maze graph, the lollipop graph, the ladder graph, and the star graph (Fig. 7).

Another dimension of intuition comes from seeing how values of graph curvature and centrality change for a fixed graph structure as distances grow. Given a distance matrix D_*, we may apply the techniques in Sect. 3 to increase the size of the dominant eigenvalue of D. More precisely, given a distance matrix D, we compute the corresponding $A = K^{-1}(D_*)$, which is a real, symmetric matrix. We then multiply the dominant eigenvalue of A by either 10 or 1000 to obtain a perturbed A_p for $p = 10, 1000$ and then recover a perturbed $D_{*,p} = K(A_p)$. A short calculation shows that the nondominant eigenvalues of D_* are largely stable under this perturbation. We then compute graph measures for D_*, $D_{*,10}$, and $D_{*,1000}$ and see how those graph measures vary.

4.1.2 Synthetic Graphs for Measure Comparison

For the centrality and curvature correlation experiments, we consider one constructed graph and two generated graphs: the Krackhardt kite graph, a lobster graph, and a Barabási-Albert graph (Fig. 2). These graphs have been chosen because their structures highlight differences in the metrics under consideration.

The **Krackhardt kite graph** [16] is a simple graph with ten nodes, designed to distinguish different concepts of centrality computed with the shortest-path distance: the vertex with maximum degree $Deg_{\text{cent}\,i}$, the vertex with maximum betweenness centrality $B_{S,i}$, and the two vertices with maximum closeness centrality $C_{S,i}$ are all different.

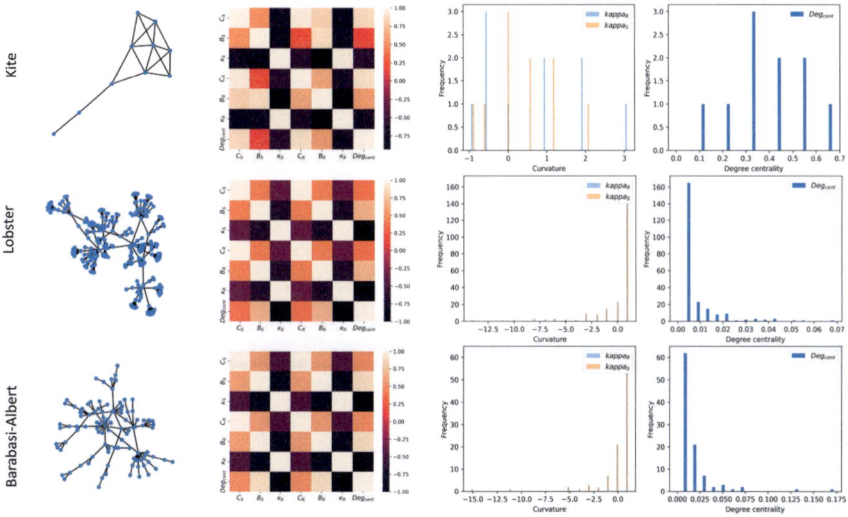

Fig. 2 Synthetic graphs and their properties: (Left column) Spring layout of the synthetic graph; (middle column) Pearson correlations between centrality measures and curvatures of the tested graphs. Axes from top-to-bottom and left-to-right are C_S, B_S, κ_S, C_R, B_R, κ_R, and Deg_{cent}; (right columns) distribution of centrality metrics

The **lobster graph** is randomly generated, parameterized by the number of nodes in the backbone n, the probability of adding an edge to the backbone p_1, and the probability of adding an edge one level beyond the backbone p_2. In our example, we used $n = 8$, $p_1 = 0.8$, $p_2 = 0.7$. This parameter selection creates a balance between high-curvature leaf nodes and low-curvature backbone nodes.

The **Barabási-Albert graph** was generated using the Barabási-Albert network growth model [3]. It generates networks with scale-free properties and a power-law distribution of node degrees. In this model, nodes are incrementally added one at a time, and each newly added node forms connections with existing nodes chosen based on their degrees. The Barabási-Albert model in this study was configured with 100 nodes ($n = 100$), where each newly introduced node forms a single connection ($m = 1$) with an existing node in the network. This parameter selection highlights the basic preferential attachment mechanism, illustrating how hubs form as the network grows, which is considered an essential feature of real-world networks [3].

4.1.3 Real-World Graphs for Measure Comparison

For more realistic centrality and curvature correlation experiments, we consider the following three graphs based on real data: the Zachary karate club graph, the Davis Southern women graph, and the co-authorship network (Fig. 3).

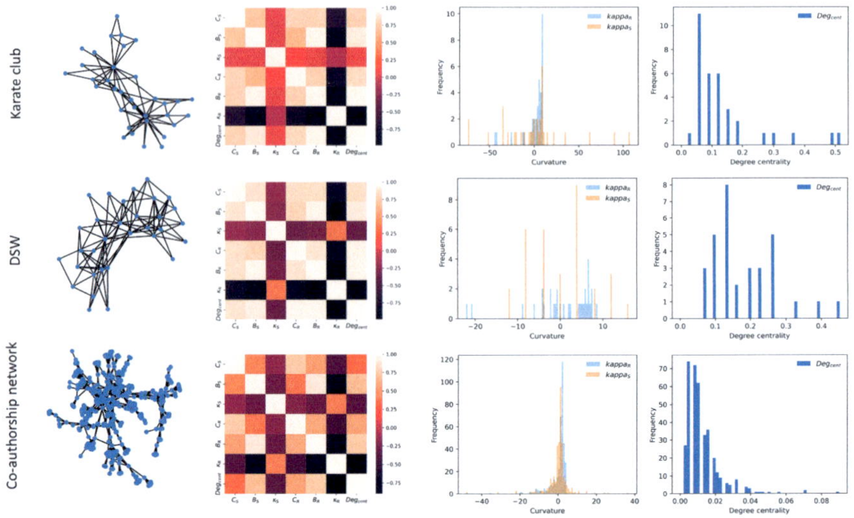

Fig. 3 Real-world graphs and their properties: (Left column) Spring layout of the real-world graph; (middle column) Pearson correlations between centrality measures and curvatures of the tested graphs. Axes from top-to-bottom and left-to-right are C_S, B_S, κ_S, C_R, B_R, κ_R, and Deg_{cent}; (right columns) distribution of centrality metrics

Zachary's karate club graph is a social network graph based on 34 members of a karate club that split in two after a conflict [26]. The structure of the graph can be used to predict which members (nodes) joined each of the two resulting groups.

The **Davis Southern Women graph (DSW)** is a bipartite graph generated from data collected by Davis et al. in the 1930s representing observed attendance at 14 social events by 18 Southern women [1].

The **co-authorship network** is based on publication co-authorship among a group of 1589 scientists working on the topic of networks in the early 2000s [20]. For this work, we only consider the largest connected component of this graph.

4.1.4 Metric Behavior on Sample Graph

In order to provide some intuition on the various metrics we consider in this paper, we describe those metrics in a constructed example. The kite graph (Fig. 4) was constructed such that the following vertices are all different under the shortest-path distance: the vertex with maximum degree $Deg_{\text{cent}\ i}$ (vertex 3), the vertex with maximum $B_{S,i}$ (vertex 7), the two vertices with maximum $C_{S,i}$ (vertices 5 and 6), and the vertex with maximum κ_S (vertex 9). When we consider the resistance distance, the vertices with the maximum $B_{R,i}$, $C_{R,i}$, and κ_R all match their shortest distance counterparts. If we consider the vertices with the lowest values of each metric, the closeness centrality is the lowest under both metrics in vertex 9. While

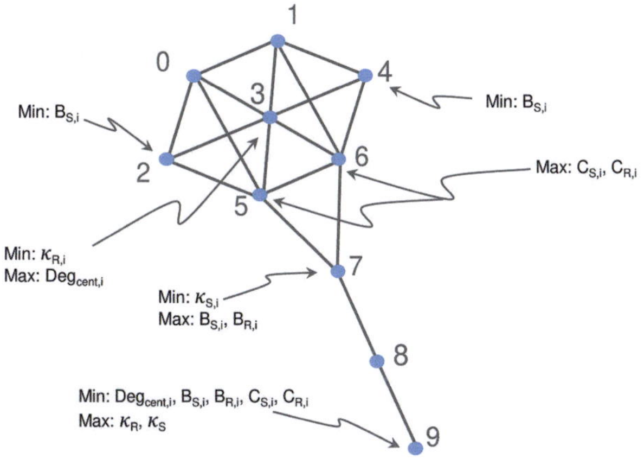

Fig. 4 Krackhardt kite graph with labeled vertices, highlighting the negative correlation between degree and curvatures. Note that vertices 2, 4, and 9 are tied for the minimum $B_{S,i}$

the betweenness centralities $B_{S,i}$ and $B_{R,i}$ are minimized at vertex 9 under both metrics, vertices 2 and 4 also tie for the minimum under only the shortest-path betweenness. For curvature, the minimum κ_S occurs at vertex 7, but the minimum κ_R is at vertex 3 (where the degree centrality is highest).

5 Results

As expected, we note a positive correlation between all centrality metrics, irrespective of the graphs tested (Figs. 2 and 3). We also observe a strong negative correlation between κ_R, and to a lesser extent κ_S, and all centrality measures across all graphs.

5.1 Distribution of Metrics as a Function of Graph Structure

We observe that a portion of the graphs are heavy-tailed in their degree distribution: the lobster and Barabási-Albert graphs from the synthetic graphs and the co-authorship and, to a lesser degree, the karate club graph from the real-world graphs. A heavy-tailed degree distribution indicates that the graphs have a "hub-like" structure, with a few nodes that are highly connected relative to the others [13]. Although similar with respect to their degree distribution, the two synthetic graphs have a tree configuration, whereas the two real-world graphs have cycles. We will show that these different structural properties give rise to different relationships between the graphs' respective node metrics.

We start with a discussion of the behavior of the two different distance metrics on tree structures. Recall that the resistance distance between two nodes depends on all the possible paths between them. First, consider a pair of nodes connected by a single edge: the shortest-path distance and resistance distance between them are equal. This will extend to a pair of nodes that only allow a single path between them, which is the case for two nodes on a single tree branch. For a tree graph with no cycles, most of the nodes will have a low degree, as we observe in the degree centrality distribution for the lobster and Barabási-Albert graphs (Fig. 2). All pairs of nodes will have only one path between them, as noted earlier and in [6], meaning that the resistance distance will be equal to the shortest-path distance for all node pairs. We can observe this relationship in the metrics, which we compute using the distance measures: in the lobster and Barabási-Albert graphs, C_S and C_R, B_S and B_R, and κ_S and κ_R, are all perfectly correlated, whereas in the real-world graphs, which display cycles, there are less consistently high correlations across these pairs of metrics.

When parameterizing both centralities and curvature with the resistance distance (C_R, B_R, and κ_R, respectively), we can refer back to Fig. 1 to gain additional intuition on the behavior of both metrics vis-a-vis the graph structure. In a tree graph, local resistance distances will be large, as current trying to flow through will have limited paths. Thus, the resistance distance centrality measure will be high, and we observe negative curvature at all nodes that aren't leaves, and for the same reason, a high C_R and B_R. Conversely, in a more dense graph that is not acyclic, a node that has a neighborhood that has many cycles connected to it will have lower local resistance distances, as the current flowing through will be able to split across multiple paths. This means the C_R and B_R will be lower, and a lower κ_R will be observed.

These relationships are well illustrated in the random lobster graph (Fig. 5, top row). The inverse correlation of resistance betweenness (B_R) and resistance curvature (κ_R) is highlighted both in the leaf nodes (where both are close to 0) and backbone nodes (where B_R is positive, and κ_R is negative). We see a very similar structure for the Barabási-Albert graph (Fig. 2).

We now consider the degree distribution of the different graphs in Figs. 2 and 3. We note that the degree distribution tends to be more heavy-tailed in the lobster and Barabási-Albert graphs (which are trees) and the karate and co-authorship graphs (which have cycles). For intuition, note that the two nodes with the highest Deg_{cent} in the karate club graph correspond to the leaders of the two factions after the original karate club split (Figs. 3 and 5). As discussed above, in the trees, the shortest-path and resistance distance matrices will be identical; therefore, κ_S and κ_R will be perfectly correlated and highly correlated with the degree centrality, mirroring the observed heavy-tailed distribution (Fig. 2). Conversely, in the karate club and co-authorship networks, the correlation between κ_S and κ_R is relatively lower due to the differences in the shortest-path and resistance distances of the nodes. Notably, κ_R maintains its high correlation to Deg_{cent}, whereas κ_S does not.

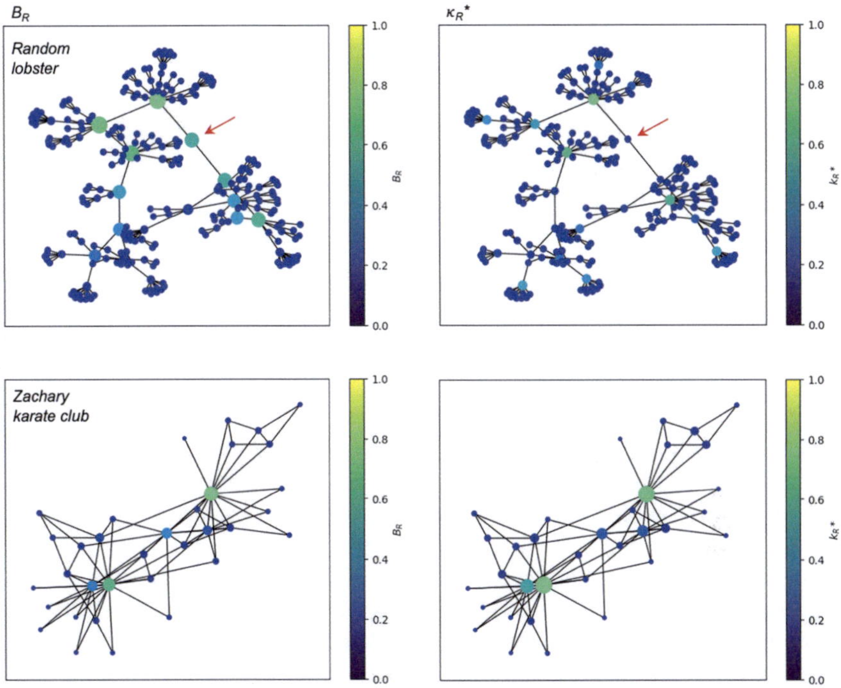

Fig. 5 Comparing centrality and curvature in (top row) the random lobster graph and (bottom row) the karate club graph. The graphs are colored by resistance betweenness (B_R) in the first column and resistance curvature (κ_R), negated and shifted, in the second column. For the lobster graph, the red arrows point to a node that has differing relative B_R and κ_r

This is due to the fact that the resistance distance captures information related to all paths between two nodes, which will be more closely correlated to the node degree than the shortest-path distance. Therefore, the hub-like nature, which is evidenced by the heavy-tailed degree distribution, is also reflected by κ_R, but not κ_S. Of note, κ_R and Deg_{cent} are also more highly correlated in the DSW graph, which does not have a heavy-tailed degree distribution but has cycles, indicating that the strong correlation observed between κ_R and Deg_{cent} does not depend on the distribution of Deg_{cent}.

5.2 *Discrepancies Between Centrality, Curvature, and Degree*

Since the resistance distance has been argued as a "natural" distance to parameterize graph curvature [24], we focus on understanding potential similarities/differences between the betweenness centrality as parameterized by the resistance distance (B_R) and graph curvature as parameterized by the resistance distance (κ_R). We visualize

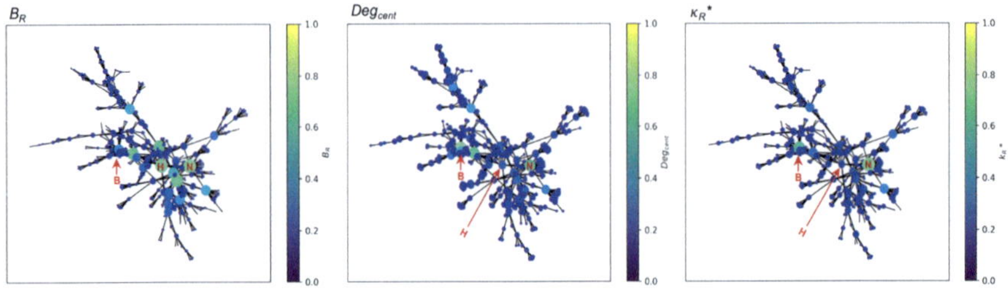

Fig. 6 The co-authorship network, with nodes colored by (left) resistance betweenness (B_R), (middle) degree centrality (Deg_{cent}), and (left) resistance curvature (κ_R), negated and shifted. In these plots, the node size is also proportional to each measure. The node labels **N, B, H** correspond to authors M. Newman, A. Barabási, and P. Holme, respectively

the co-authorship network since it has the lowest correlation between B_R and κ_R (Fig. 3). Plotting B_R, Deg_{cent}, and (the negative of) κ_R, respectively, for the co-authorship network shows that the highest-scoring nodes are not identical (Fig. 6). Indeed, we observe some very interesting patterns emerge: author M. Newman has the highest B_R and -κ_R values, whereas author A. Barabási has the highest Deg_{cent} and second highest -κ_R. Author P. Holme has the second highest B_R and does not make the top five highest scores for either Deg_{cent} or -κ_R. We observe then that A. Barabási has the highest degree but relatively lower B_R, showing that he is not positioned in a part of the network with high information flow. Nevertheless, he still achieves the second highest -κ_R, indicating that the resistance curvature is able to identify nodes with both high degree and information flow, which neither B_R nor Deg_{cent} can pick up individually. We observe a similar phenomenon in the random lobster graph (Fig. 5), where the node indicated by the red arrow has a relatively higher B_R than κ_R^*. In this case, we observe that nodes connecting two highly connected regions of a graph, which themselves are not highly connected, are more effectively resolved using the B_R metric.

We can see this phenomenon dynamically in the small graphs described in Sect. 4.1.1 under perturbations of the dominant eigenvalue of the distance matrix (Fig. 7), as discussed in Sect. 3. We recall the interpretation of resistance betweenness as a measure of how many paths through the graph a particular node is on and resistance curvature as a measure of how locally distant a particular node is (how "flattenable" it is). We can see in Fig. 7 that, as distances from a node increase with the increase of the dominant eigenvalue of the distance matrix, the nodes that have low resistance curvature and high resistance betweenness are those nodes that are distant from neighbors but are a bottleneck for paths through the graph. The ladder graph relationships do not change since no node is a bottleneck. In the lollipop graph, on the other hand, the nodes with growing distance lose their betweenness while maintaining low curvature since paths can move through the other side of the graph, where the nodes are increasing in-betweenness.

6 Discussion

The relationship between a newly defined form of discrete curvature on graphs parameterized by shortest-path or resistance distance (κ_S and κ_R, respectively) offers a new approach to graph and network analysis. In this paper, we have explored the relationship between graph curvature and established node-centrality metrics such as closeness and betweenness parameterized by two distances and degree centrality. We have observed a strong dependence of the metric relationships on graph structure: graphs that are more tree-like will have a strong correlation between κ_R and κ_S due to the similarity of the respective distance metrics and, conversely, a lower correlation of κ_R and κ_S in a graph with larger interconnectedness. A strong negative correlation exists between curvature and the centrality metrics across both distance metrics for all graphs studied, but more strongly for resistance distance. In general, local tree-like structures have higher centrality and low curvature, whereas local areas with many cycles have lower centrality but high curvature, giving rise to high correlations irrespective of structure.

Yet, we observed that this correlation is not 100%. In the co-authorship network, we observed that κ_R can provide a balance between emphasizing a high-degree node and a node with high information flow. Conversely, κ_R is less capable than B_R to capture nodes that connect two highly connected regions of the graph. Understanding what structural features κ_R is more and less likely to capture can help researchers determine whether this metric is appropriate for their respective applications. Another structural feature of networks where κ_R may find application is in graph community detection and analysis. Indeed, Ni et al. used a discrete version of the Ricci flow, which is an evolutionary metric that is curvature-dependent, for network community detection [7, 22]. In more recent work, Tian et al. considered several notions of graph curvature to implement the Ricci flow for community detection [25]. In addition to flow-based community detection, we could also consider using the resistance curvature to evaluate highly important nodes within communities: while we would expect generally higher curvature for nodes in communities due to their high interconnectivity, influential nodes (those with especially high degree and information flow) would have relatively lower curvature within communities and could thus be detected via the κ_R.

Both the resistance betweenness centrality and resistance curvature can also be interpreted in terms of random walks on the graphs. As shown in [21], the betweenness centrality with resistance distance (B_R) at vertex v_i is equivalent to a measure of how often a random walk between source node a and target node b passes through node v_i, averaged across all source-target pairs. Devriendt et al. show that the resistance curvature, based on their definition in Eq. 5, at a vertex v_i is related to the average distance between nodes in its neighborhood, or equivalently to the average distance between two walkers who start at vertex v_i and take independent random paths, as defined by the diffusion equation (Appendix B.5. in [7]). With this interpretation, we can see that nodes along this bridge structure have relatively higher resistance betweenness since they are on a high probability path to get between the two clusters, given that there are very few other ways to get between the clusters. However, they also have relatively high resistance curvature values despite not being in clusters themselves, as two random walkers who start on one of the bridge nodes can easily make it to one of the end clusters, at which point their paths will quickly diverge. This phenomenon is observed in our experiments in the lobster graph, as previously described.

It may also be possible to mathematically explain the relationship between resistance betweenness and resistance curvature more closely. For inverting the resistance distance matrix D_R, work has been done by [2]. For a vertex $v_i \in V$, let $adj(v_i)$ denote the set of vertices adjacent to v_i. Then, for $i = 1, \ldots, n$, let τ be the column vector with components given by

$$\tau_i = 2 - \sum_{j \in adj(v_i)} \frac{D_{R,ij}}{a_{ij}}$$

Then, one can derive an equation for the inverse of the resistance matrix $(D_R)^{-1}$ as

$$(D_R)^{-1} = -\frac{1}{2}L + \frac{1}{\tau^T D_R \tau}\tau\tau^T. \tag{7}$$

Further work could analyze the formulas for B_R (Definition 8) and κ_R (Definition 9 with Eq. 7) for insights.

In conclusion, we have performed an exploration of the relationship between a new notion of graph curvature and established centrality metrics and found several promising avenues for both application and future research on the utility of graph curvature for network analysis.

Acknowledgments The authors would like to thank IPAM for its generous hospitality during the workshop that incubated this work, AWM for its ongoing support of inclusive research collaboration events, and Anna Gilbert for proposing this project. Kathryn Leonard acknowledges funding support from NSF award DMS-1953052. Kasia Jankiewicz acknowledges funding support from NSF awards DMS-2203307, DMS-2238198, and DMS-1926686. The authors thank the anonymous referees for their comments and suggestions and Florentin Münch for bringing the work on graph Ricci curvatures and centrality to their attention.

Competing Interests The authors have no conflicts of interest to declare relevant to this chapter's content.

Appendix

Perturbation graphs as described in Sects. 4.1.1 and 5.2 (Fig. 7).

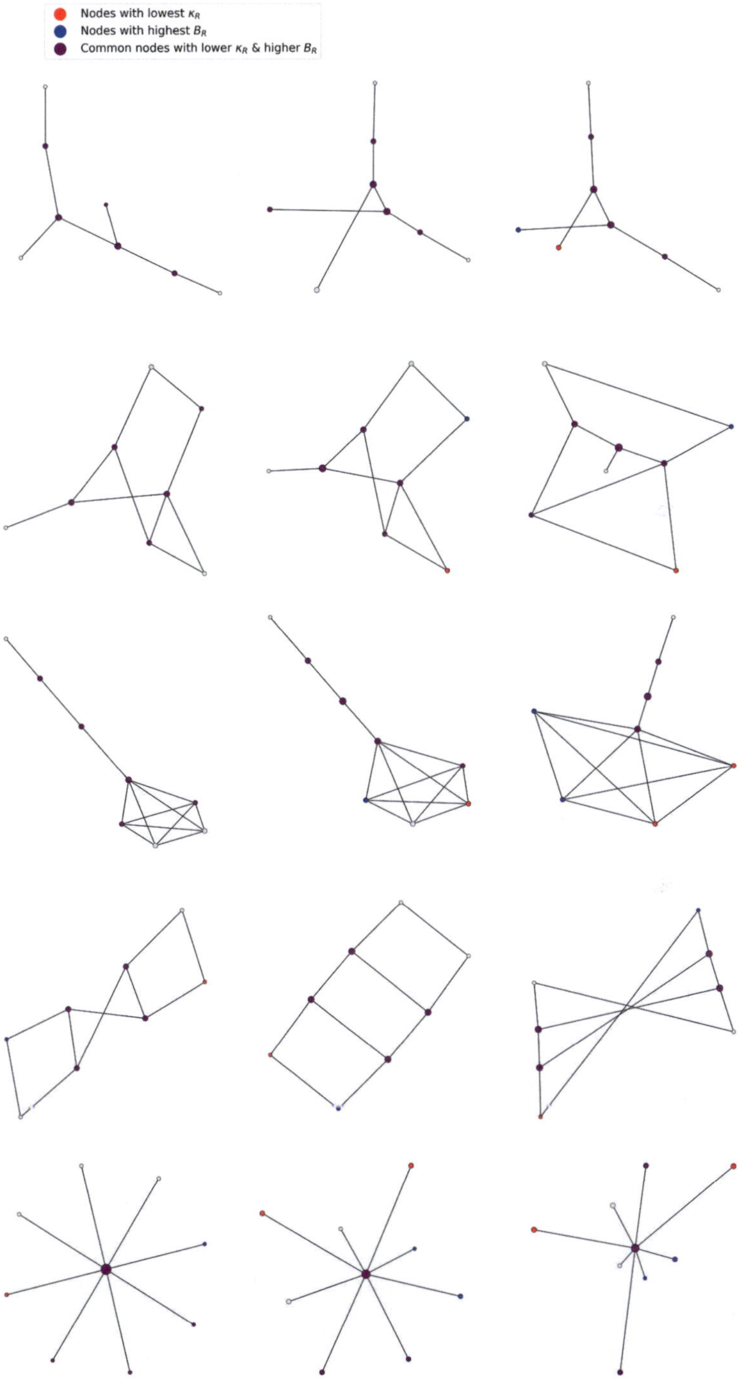

Fig. 7 Resistance betweenness (B_R) and resistance curvature (κ_R), in graphs with eight nodes. Left column: original graph. Middle column: graph with dominant distance eigenvalue increased tenfold. Right column: graph with dominant distance eigenvalue increased thousandfold

References

1. Allison Davis Burleigh Bradford Gardner, M.R.G.: Deep South: A Social Anthropological Study of Caste and Class. University of Chicago Press (1941)
2. Bapat, R.: Resistance matrix of a weighted graph. Commun. Math. Comput. Chem. (50), 73–82 (2004)
3. Barabási, A.L., Albert, R.: Emergence of scaling in random networks. Science **286**(5439), 509–512 (1999). https://doi.org/10.1126/science.286.5439.509. https://www.science.org/doi/abs/10.1126/science.286.5439.509
4. Bavelas, A.: Communication patterns in task-oriented groups. J. Acoust. Soc. Am **22**(6), 725–730 (1950)
5. Borgatti, S.P.: Centrality and network flow. Soc. Networks **27**(1), 55–71 (2005)
6. Bozzo, E., Franceschet, M.: Resistance distance, closeness, and betweenness. Soc. Networks **35**(3), 460–469 (2013). https://doi.org/10.1016/j.socnet.2013.05.003. https://www.sciencedirect.com/science/article/pii/S0378873313000488
7. Devriendt, K., Lambiotte, R.: Discrete curvature on graphs from the effective resistance*. Journal of Physics: Complexity **3**(2), 025008 (2022). https://doi.org/10.1088/2632-072X/ac730d. Publisher: IOP Publishing
8. Devriendt, K., Ottolini, A., Steinerberger, S.: Graph curvature via resistance distance (2024). https://doi.org/10.1016/j.dam.2024.01.012. https://www.sciencedirect.com/science/article/pii/S0166218X24000179
9. Dörfler, F., Simpson-Porco, J.W., Bullo, F.: Electrical networks and algebraic graph theory: Models, properties, and applications. Proc. IEEE **106**(5), 977–1005 (2018). https://doi.org/10.1109/JPROC.2018.2821924. https://ieeexplore.ieee.org/document/8347206. Conference Name: Proceedings of the IEEE
10. Forman, R.: Bochner's method for cell complexes and combinatorial Ricci curvature. Discrete Comput. Geom. **29**(3), 323–374 (2003). https://doi.org/10.1007/s00454-002-0743-x
11. Freeman, L.C.: A set of measures of centrality based upon betweenness. Sociometry **40**(1), 35–41 (1977)
12. Ghosh, A., Boyd, S., Saberi, A.: Minimizing effective resistance of a graph. SIAM Rev. **50**(1), 37–66 (2008). https://doi.org/10.1137/050645452. https://web.stanford.edu/~boyd/papers/pdf/eff_res.pdf
13. Guillaume, J.L., Latapy, M.: Bipartite structure of all complex networks. Inf. Process. Lett. **90**(5), 215–221 (2004)
14. Johnson, C., Tarazaga, P.: Connections between the real positive semidefinite and distance matrix completion problems. Linear Algebra Appl. **6**(223/224), 375–391 (1995). https://doi.org/10.1016/0024-3795(95)00096-A. https://www.sciencedirect.com/science/article/pii/002437959500096A
15. Klein, D.J., Randić, M.: Resistance distance. J. Math. Chem. **12**(1), 81–95 (1993). https://doi.org/10.1007/BF01164627
16. Krackhardt, D.: Assessing the pollivieitical landscape: Structure, cognition, and power in organizations. Admin. Sci. Quart. **35**(2), 342–369 (1990). http://www.jstor.org/stable/2393394
17. Lee, J.M.: Riemannian manifolds, Graduate Texts in Mathematics, vol. 176. Springer, New York (1997). https://doi.org/10.1007/b98852. An introduction to curvature
18. Münch, F.: Ollivier curvature, betweenness centrality and average distance (2022). https://arxiv.org/abs/2209.15564
19. Mondal, M., Samal, A., Münch, F., Jost, J.: Bakry–Émery–Ricci curvature: an alternative network geometry measure in the expanding toolbox of graph Ricci curvatures. J. Complex Networks **12**(3), cnae019 (2024). https://doi.org/10.1093/comnet/cnae019
20. Newman, M.E.J.: Finding community structure in networks using the eigenvectors of matrices. Phys. Rev. E **74**, 036104 (2006). https://doi.org/10.1103/PhysRevE.74.036104. https://link.aps.org/doi/10.1103/PhysRevE.74.036104

21. Newman, M.J.: A measure of betweenness centrality based on random walks. Soc. Networks **27**(1), 39–54 (2005). https://doi.org/10.1016/j.socnet.2004.11.009. https://www.sciencedirect.com/science/article/pii/S0378873304000681
22. Ni, C.C., Lin, Y.Y., Luo, F., Gao, J.: Community detection on networks with ricci flow. Sci. Rep. **9**(1), 9984 (2019)
23. Ollivier, Y.: Ricci curvature of metric spaces. C. R. Math. Acad. Sci. Paris **345**(11), 643–646 (2007). https://doi.org/10.1016/j.crma.2007.10.041
24. Steinerberger, S.: Curvature on graphs via equilibrium measures. J. Graph Theory **103**(3), 415–436 (2023). https://doi.org/10.1002/jgt.22925. https://onlinelibrary.wiley.com/doi/abs/10.1002/jgt.22925. eprint: https://onlinelibrary.wiley.com/doi/pdf/10.1002/jgt.22925
25. Tian, Y., Lubberts, Z., Weber, M.: Curvature-based clustering on graphs. arXiv preprint arXiv:2307.10155 (2023)
26. Zachary, W.W.: An information flow model for conflict and fission in small groups. J. Anthropol. Res. **33**(4), 452–473 (1977). http://www.jstor.org/stable/3629752

Time-Varying Graph Signal Recovery Using High-Order Smoothness and Adaptive Low-Rankness

Weihong Guo, Yifei Lou, Jing Qin, and Ming Yan

1 Introduction

Many real-world datasets are represented in the form of graphs, such as sea surface temperatures, Covid-19 cases at regional or global levels, and PM 2.5 levels in the atmosphere. Graphs play a crucial role in data science, facilitating the mathematical modeling of intricate relationships among data points. Typically composed of vertices with either undirected or directed edges, graphs regard each data point as a vertex and use edges to represent pairwise connections in terms of distances or similarities. A graph signal is a collection of values defined on the vertex set. The graph structure can be either provided by specific applications or learned from partial or complete datasets.

As an extension of (discrete) signal processing, graph signal processing [29] has become an emerging field in data science and attracted tremendous attention

W. Guo
Department of Mathematics, Applied Mathematics, and Statistics, Case Western Reserve University, Cleveland, OH, USA
e-mail: wxg49@case.edu

Y. Lou
Department of Mathematics & School of Data Science and Society, University of North Carolina at Chapel Hill, Chapel Hill, NC, USA
e-mail: yflou@unc.edu

J. Qin (✉)
Department of Mathematics, University of Kentucky, Lexington, KY, USA
e-mail: jing.qin@uky.edu

M. Yan
School of Data Science, The Chinese University of Hong Kong, Shenzhen (CUHK-Shenzhen), Shenzhen, China
e-mail: yanming@cuhk.edu.cn

C. Garcia-Cardona, H. Lee (eds.), *Advances in Data Science*, Association for Women in Mathematics Series 37, https://doi.org/10.1007/978-3-031-87804-6_4

due to its capability of dealing with big data with irregular and complex graph structures from various applications, such as natural language processing [21], traffic prediction [31], climate change monitoring [28], and epidemic prediction [10]. Graph signal recovery aims to recover a collection of signals with certain smoothness assumptions defined on a graph from partial and/or noisy observations. Unlike signals defined in traditional Euclidean spaces, the intricate geometry of the underlying graph domain must be considered when processing and recovering graph signals. Graph signals typically exhibit smoothness either locally or globally over the graph.

There are some challenges in graph signal recovery when exploiting the underlying graph structure to improve signal reconstruction accuracy. First, the topology of a graph desires a comprehensive representation involving many graph components, such as structural properties, connectivity patterns, vertex/edge density, and distribution. Second, it may be insufficient to describe the smoothness of graph signals by simply restricting the similarity of signal values locally. Moreover, the growth of graph size leads to a significant computational burden. To address them, various techniques have been developed, including graph-based regularization methods [4, 5, 17, 18], spectral graph theory [6, 24, 32, 35], and optimization algorithms [1, 15].

1.1 Time-Varying Graph Signal Recovery

A time-varying or spatial-temporal graph signal can be considered as a sequence of signals arranged chronologically, where each signal at a specific time instance is defined on a static or dynamically changing spatial graph.

Consider an undirected unweighted graph $G = (V, E)$, where V is a set of n vertices and E is a set of edges. We assume a collection of time-varying graph signals $\{\mathbf{x}_t\}_{t=1,\dots,m}$ with $\mathbf{x}_t \in \mathbb{R}^n$ are defined on V with a time index t. Let $X = [\mathbf{x}_1, \dots, \mathbf{x}_m] \in \mathbb{R}^{n\times m}$ be the dataset represented in matrix. The pairwise connections on the graph G can be modeled by an adjacency matrix A, where the (i, j)-th entry of A is one if there is an edge between vertices i and j and zero otherwise. This binary adjacency matrix can be extended to the non-binary case for a weighted graph, where each entry indicates the similarity between two vertices. Throughout the paper, we use a standard K nearest neighbor (KNN) approach with an integer K, based on the Euclidean distance of data points to construct the adjacency matrix.

Given an adjacency matrix A, we further define the graph Laplacian matrix, $L = M - A \in \mathbb{R}^{n\times n}$, where M is a diagonal matrix with its diagonal element $M_{ii} = \sum_j A_{ij}$. The graph Laplacian serves as a matrix representation of the graph structure and can be used to describe some important characteristics of a graph, such as node connectivity and similarity. For example, geographic locations in the form of coordinates, i.e., longitude and latitude, are typically used to calculate the pairwise distance and, thereby, the graph Laplacian for geospatial data. For some datasets without obvious graph domains, a preprocessing step of graph learning can be implemented; see [33] for a comprehensive review of graph learning techniques.

Time-varying graph signal recovery aims to recover an underlying matrix from its partially observed entries that are possibly polluted by additive noise. Mathematically, a forward model is $Y = J \circ X + \mathcal{N}$, where Y is the observed data, $J \in \{0, 1\}^{n \times m}$ is a sampling matrix, and $\mathcal{N}$ is a random noise. In this work, we focus on recovering time-varying signals, represented by the matrix X, from incomplete noisy data Y defined on static spatial graphs in the sense that the vertex set and the edges do not change over time. In addition, we adopt a symmetrically normalized graph Laplacian that is pre-computed based on geographic locations.

1.2 Related Works

The recovery of graph signals from partial observations is an ill-posed problem due to missing information. Graph regularization plays a crucial role in developing a recovery model for time-varying signals by enforcing temporal correlation and/or describing the underlying graph topology. An intuitive approach for recovering time-varying graph signals is to apply interpolation methods to fill in the missing entries, such as natural neighborhood interpolation (NNI) [30]. Numerous recovery models with diverse smoothness terms have been proposed to further preserve the underlying geometry. For example, graph smoothing (GS) [22] characterizes the smoothness of the signal using the graph Laplacian of X. Alternatively, temporal smoothness is incorporated in time-varying graph signal recovery (TGSR) [27] by formulating the graph Laplacian of DX, where D is a first-order temporal difference operator. The combination of the graph Laplacian of X and the Tikhonov regularity of DX was considered in [25]. In contrast, the graph Laplacian of DX with an additional low-rank regularity of X was formulated as low-rank differential smoothness (LRDS) [20]. In the Tikhonov regularization, $\|XD\|_F^2 = \mathrm{tr}(XDD^TX^T)$ implies that DD^T is treated as the temporal graph Laplacian. In [11], the graph Laplacian matrix L is replaced by $(L + \epsilon I)^r$, where I is the identity matrix and $r \geq 1$ for a high-order Sobolev spatial-temporal smoothness. Its main advantage lies in faster convergence, as this approach does not necessitate extensive eigenvalue decomposition or matrix inversion. Recently, another low-rank and graph-time smoothness (LRGTS) method has been proposed in [19], where the sum of the nuclear norm and the Tikhonov regularizer on the second-order temporal smoothness are adopted to promote the low-rankness and the temporal smoothness, respectively.

All the aforementioned models can be unified into one minimization framework:

$$\min_X \frac{1}{2}\|Y - J \circ X\|_F^2 + \frac{\alpha}{2}\mathrm{tr}(D_\theta^T X^T L_s X D_\theta) + \beta R(X) + \frac{\gamma}{2}\mathrm{tr}(X L_t X^T), \quad (1)$$

where D_θ is a θ-th order temporal difference operator; L_s and L_t are the spatial and temporal graph Laplacian matrices, respectively; $R(X)$ is the regularization term

Table 1 Comparison of related works and proposed methods

Method	Optimization model
GS [22]	$\min_X \frac{1}{2}\,\|Y - J \circ X\|_F^2 + \frac{\alpha}{2}\,\mathrm{tr}(X^T L X)\ (\theta = \beta = \gamma = 0\,)$
Tikhonov [25]	$\min_X \frac{1}{2}\,\|Y - J \circ X\|_F^2 + \frac{\alpha}{2}\,\mathrm{tr}(X^T L X) + \frac{\gamma}{2}\,\|X D_1\|_F^2$ $(\theta = \beta = 0,\ L_t = D_1 D_1^T\,)$
TGSR [27]	$\min_X \frac{1}{2}\,\|Y - J \circ X\|_F^2 + \frac{\alpha}{2}\,\mathrm{tr}(D_1^T X^T L X D_1)\ (\,\theta = 1, \beta = \gamma = 0)$
LRDS [20]	$\min_X \frac{1}{2}\,\|Y - J \circ X\|_F^2 + \frac{\alpha}{2}\,\mathrm{tr}(D_1^T X^T L X D_1) + \beta\|X\|_* \quad (\theta = 1, \gamma = 0)$
Sobolev [11]	$\min_X \frac{1}{2}\,\|Y - J \circ X\|_F^2 + \frac{\alpha}{2}\,\mathrm{tr}(D_1^T X^T (L + \epsilon I)^r X D_1)$ $(\theta = 1,\ L_s = \tilde{L},\ \beta = \gamma = 0)$
LRGTS [19]	$\min_X \frac{1}{2}\,\|Y - J \circ X\|_F^2 + \frac{\alpha}{2}\,\mathrm{tr}(X^T L X) + \beta\,\|X\|_* + \frac{\gamma}{2}\,\|X D_2\|_F^2$ $(\theta = 0,\ L_t = D_2 D_2^T)$
Proposed L2	$\min_X \frac{1}{2}\|Y - J \circ X\|_F^2 + \frac{\alpha}{2}\,\mathrm{tr}(D_\theta^T X^T (L + \epsilon I)^r X D_\theta) + \beta\,\|X\|_{\mathrm{erf}}\ (\gamma = 0)$ where $\|X\|_{\mathrm{erf}}$ is an ERF-weighted nuclear norm
Proposed L1	$\min_X \|Y - J \circ X\|_1 + \frac{\alpha}{2}\,\mathrm{tr}(D_\theta^T X^T (L + \epsilon I)^r X D_\theta) + \beta\,\|X\|_{\mathrm{erf}}\ (\gamma = 0)$ where $\|X\|_{\mathrm{erf}}$ is an ERF-weighted nuclear norm

applied to X describing its characteristics; and $\alpha \geq 0,\ \beta \geq 0,\ \gamma \geq 0$ are three parameters. Two common choices of θ are (1) $\theta = 0$ that corresponds to $D_\theta = I$ and (2) $\theta = 1$ used in TGSR. Additionally, L_s can be a transformed version of the classical graph Laplacian L, e.g., $L_s = (L + \epsilon I)^r$ as used in the Sobolev method [11], where ϵ is positive and $r \geq 1$ can be non-integer. The temporal graph Laplacian can be constructed by using the τ-th order temporal difference operator, i.e., $L_t = D_\tau D_\tau^T$, for which case the temporal Laplacian can be expressed via the Frobenius norm $\mathrm{tr}(X D_\tau D_\tau^T X^T) = \|X D_\tau\|_F^2$, e.g., Tikhonov with $\tau = 1$ and LRGTS with $\tau = 2$. The regularization $R(X)$ can be chosen as the nuclear norm of X if the underlying time-varying graph signal X is of low rank. Various models utilize different choices of D_θ, L_s, L_t and the regularization R.

Following the general framework (Eq. 1), we propose a novel low-rank regularization $R(X)$ based on the error function (ERF) [14] for sparse signal recovery (see Sect. 2.3). In addition, to handle the non-Gaussian type of noise such as Laplace noise, we propose a variant model in which the Frobenius norm-based data fidelity term is replaced with the ℓ_1-norm data fidelity (see Sect. 2.4). In Table 1, we provide a summary of the proposed models and relevant works pertaining to the general framework outlined in Eq. (1).

Leveraging the recent growth in deep learning, some time-varying graph signal recovery methods include unrolling technique [16], graph neural network (GNN) [3], and joint sampling and reconstruction of time-varying graph signals [34]. In this work, we are dedicated to developing unsupervised time-varying graph signal recovery algorithms that do not involve or rely on data training.

1.3 Contributions

The major contributions of this work are described as follows.

1. We develop a generalized time-varying graph signal recovery framework encompassing several state-of-the-art works as special cases. We also develop two new models with an ERF-based regularization.
2. The proposed models combine high-order temporal smoothness and graph structures with the temporal correlation exploited by iteratively reweighted nuclear norm regularization.
3. We propose efficient algorithms for solving the proposed models. Convergence analysis is provided to show that the proposed Algorithm 1 (ℓ_2 case) generates a sequence that converges to a stationary point of the problem.
4. We conduct various numerical experiments, utilizing both synthetic and real-world datasets (specifically PM2.5 and sea surface temperature data), to validate the effectiveness of the proposed algorithm.

1.4 Organization

The subsequent sections of this paper are structured as follows. In Sect. 2, we introduce a pioneering framework for recovering time-varying graph signals, leveraging Sobolev smoothness and ERF regularization. Additionally, we put forth an efficient algorithm based on the alternating direction method of multipliers (ADMM) and iterative reweighting scheme. A comprehensive convergence analysis of the proposed Algorithm 1 is provided. In Sect. 3, we present numerical experiments conducted on synthetic and real-world datasets sourced from environmental and epidemic contexts. Finally, Sect. 4 encapsulates our conclusions and outlines potential avenues for future research.

2 Proposed Method

2.1 Error-Function-Weighted Nuclear Norm Regularization

To enhance the low-rankness of a matrix, weighted nuclear norm minimization (WNNM) has been developed with promising performance in image denoising [13]. Specifically, the weighted nuclear norm (WNN) is defined as

$$\|L\|_{\mathbf{w},*} := \sum_i w_i \sigma_i(L), \tag{2}$$

where $\sigma_i(L)$ is the i-th singular value of L in the decreasing order and the weight vector $\mathbf{w} = (w_i)$ is in the nondecreasing order with $w_i \geq 0$ being the i-th entry. Choosing the weights is challenging in sparse and low-rank signal recovery problems. Iteratively reweighted L1 (IRL1) [2] was proposed for the sparse recovery problem, where the weight $\mathbf{w}$ is updated based on the previous estimate. IRL1 can solve many problems with complicated sparse regularizations, exhibiting improved sparsity and convergence speed.

In this work, we introduce a novel ERF-weighted nuclear norm based on the ERF regularizer [14] and use linearization to obtain WNN. For any real matrix X with n singular values $\sigma_1(X) \geq \ldots \geq \sigma_n(X)$, the ERF-weighted nuclear norm is

$$\|X\|_{\text{erf}} = \sum_{i=1}^{n} \int_0^{\sigma_i(X)} e^{-t^2/\sigma^2} dt, \tag{3}$$

where σ serves as a filtering parameter. In particular, when $\sigma \to 0^+$, $\frac{\|X\|_{\text{erf}}}{\sigma} \to \frac{\sqrt{\pi}}{2}\text{rank}(X)$ and hence it can enforce the low-rankness. To solve the ERF-nuclear norm regularized minimization problem, we use iterative reweighting (linearization) to get WNN with adaptive weights.

2.2 Fractional-Order Derivative

Inspired by the Grünwald-Letnikov fractional derivative [26], we introduce the total θ-th order temporal forward difference matrix with a zero boundary condition, as shown below:

$$D_\theta = \begin{bmatrix} C(0) & & & & \\ \vdots & \ddots & & & \\ C(k) & \cdots & C(0) & & \\ & \ddots & & \ddots & \\ & & C(k) & \cdots & C(0) \end{bmatrix} \in \mathbb{R}^{m\times m}. \tag{4}$$

Here, the coefficients $\{C(i)\}_{i=0}^{k}$ are defined as

$$C(i) = \frac{\Gamma(\theta+1)}{\Gamma(i+1)\Gamma(\theta+1-i)}, \quad 0 \leq i \leq k,$$

where $\Gamma(x)$ is the Gamma function. Notice that if θ is a positive integer, k can be deterministic. For example, if $\theta = 1$, then $k = 1$, and we have $C(0) = 1$ and $C(1) = -1$, which is reduced to the first-order finite difference case. If $\theta = 2$, then it reduces to the temporal Laplacian operator. Generally, if $\theta = n$, then only the first

$n + 1$ coefficients $\{C(i)\}_{i=0}^{n}$ are nonzero and thereby $k = n + 1$. For any fractional value θ, we have to choose the parameter k. In our experiments, for instance, we choose $k = 3$ for $\theta = 1.8$. Compared to integer-order derivatives that only consider local properties, fractional-order derivatives can more accurately describe complex systems such as those with long-range temporal or spatial dependence. The difference matrix (Eq. 4) is built upon the zero boundary condition, while other types of boundary conditions, e.g., Newmann and periodic boundary conditions, can also be used. Alternatively, we can use low-order difference schemes for boundary conditions, e.g., the first-order forward difference based on the first $m-1$ time points and the zeroth order for the last time point.

2.3 Proposed Algorithm 1

We propose the following ERF regularized time-varying graph signal recovery model:

$$\min_{X} \frac{1}{2} \|Y - J \circ X\|_F^2 + \frac{\alpha}{2} \operatorname{tr}(D_\theta^T X^T (L + \epsilon I)^r X D_\theta) + \beta \|X\|_{\text{erf}}. \tag{5}$$

Here, we use the Frobenius norm to define the data fidelity term for Gaussian noise, the Sobolev smoothness of time-varying graph signals [11] as the graph regularization, and the ERF-based regularization defined in Eq. (3) for temporal low-rank correlation.

We apply ADMM with linearization to solve the problem (Eq. 5). First, we introduce an auxiliary variable Z to rewrite the problem (Eq. 5) into an equivalent constrained problem:

$$\min_{X,Z} \frac{1}{2} \|Y - J \circ X\|_F^2 + \frac{\alpha}{2} \operatorname{tr}(D_\theta^T X^T (L + \epsilon I)^r X D_\theta) + \beta \|Z\|_{\text{erf}}, \text{ s.t. } X = Z.$$

Since the proximal operator of $\|\cdot\|_{\text{erf}}$ is difficult to compute, we apply linearization on the ERF term to obtain a WNN when solving the subproblem for Z. The ADMM iterates as follows:

$$\begin{aligned}
w_i &\leftarrow \exp\left(-\sigma_i^2(X)/\sigma^2\right), \qquad \text{for } i = 1, \ldots, m \\
Z &\leftarrow \operatorname*{argmin}_{Z} \beta \|Z\|_{\mathbf{w},*} + \frac{\rho}{2} \left\|X - Z + \widehat{Z}\right\|_F^2 \\
X &\leftarrow \operatorname*{argmin}_{X} \frac{1}{2} \|J \circ X - Y\|_F^2 + \frac{\alpha}{2} \operatorname{tr}(D_\theta^T X^T (L + \epsilon I)^r X D_\theta) + \frac{\rho}{2} \left\|X - Z + \widehat{Z}\right\|_F^2 \\
\widehat{Z} &\leftarrow \widehat{Z} + (X - Z),
\end{aligned} \tag{6}$$

where $\rho > 0$ is a stepsize that affects the convergence. Refer to Theorem 1 for more details. We derive closed-form solutions for both Z- and X-subproblems in Eq. (6). Specifically for the Z-subproblem, it can be updated via the singular value thresholding operator, i.e.,

$$Z = SVT(X + \widehat{Z}) = U\,\text{shrink}(\Sigma, \text{diag}(\beta\mathbf{w}/\rho))V^T, \tag{7}$$

where $U\Sigma V^T$ is the singular value decomposition of $X + \widehat{Z}$ and $\text{diag}(\cdot)$ is a diagonalization operator turning a vector into a diagonal matrix with entries of the vector sitting on the diagonal. Here, the shrink operator $\text{shrink}(x, \xi) = \text{sign}(x) * \max(|x| - \xi, 0)$ is implemented entrywise, where $\text{sign}(x)$ returns the sign of x if $x \neq 0$ and zero otherwise.

In the X-subproblem, we can rewrite the second term of the objective function as

$$\begin{aligned}\text{tr}(D_\theta^T X^T (L + \varepsilon I)^r X D_\theta) &= \left\| (L + \varepsilon I)^{r/2} X D_\theta \right\|_F^2 \\ &= \left\| (D_\theta^T \otimes (L + \varepsilon I)^{r/2}) \,\text{vec}(X) \right\|_2^2 := \|A\,\text{vec}(X)\|_2^2,\end{aligned}$$

where $\otimes$ is the Kronecker product. Thus, the X-subproblem has the closed-form solution as

$$X = \text{mat}[(\widehat{J} + \alpha A^T A + \rho I)^{-1}(\widehat{J}^T Y + \rho\,\text{vec}(Z - \widehat{Z}))], \tag{8}$$

where $\widehat{J} = \text{diag}(\text{vec}(J))$. Note that $\widehat{J}^T Y = Y$ since $\widehat{J}$ is a diagonal matrix with binary entries in the diagonal, whose nonzero entries correspond to the sampled spatial points. Furthermore, considering that the matrix $\widehat{J} + \alpha A^T A + \rho I$ is symmetric and positive definite, we perform its Cholesky factorization as $\widehat{J} + \alpha A^T A + \rho I = \tilde{L}\tilde{L}^T$. Subsequently, we leverage forward/backward substitution as a substitute for matrix inversion, thereby reducing computational time. The pseudo-code of the proposed approach for minimizing the model (Eq. 5) is given in Algorithm 1.

Algorithm 1 Robust time-varying graph signal recovery with high-order smoothness and adaptive low-rankness

---

Input: graph Laplacian L, parameters α, β, ρ, spatial Laplacian parameters ϵ and r, ERF parameter σ, Fractional-order derivative parameters $\theta > 0$ and integer $k \geq 1$.
Output: X
Initialize: X, $\widehat{Z}$
while The stopping criteria is satisfied **do**
  compute the weights $w_i's$
  update Z via Eq. (7)
  update X via Eq. (8)
  $\widehat{Z} \leftarrow \widehat{Z} + (X - Z)$
end while

2.4 Proposed Algorithm 2

In real-world applications, the type of noise could be unknown, and it is possible to encounter a mixture of different types of noise. To enhance the robustness against noise, we propose the second model:

$$\min_{X} \; \|Y - J \circ X\|_1 + \frac{\alpha}{2}\,\mathrm{tr}(D_\theta^T X^T (L + \epsilon I)^r X D_\theta) + \beta \, \|X\|_{\mathrm{erf}} \tag{9}$$

Compared with Eq. (5), this new model utilizes the ℓ_1-norm data fidelity to accommodate various types of noise. Because of the ℓ_1 term, we need to introduce an additional variable V to make the subproblems easy to solve. The constrained formulation equivalent to Eq. (9) is

$$\min_{\substack{J \circ X - Y = V \\ X = Z}} \; \|V\|_1 + \frac{\alpha}{2}\,\mathrm{tr}(D_\theta^T X^T (L + \epsilon I)^r X D_\theta) + \beta \, \|Z\|_{\mathrm{erf}}\,.$$

Therefore, the ADMM with linearization [7, 9, 12, 23] on the ERF term has the following subproblems:

$$\begin{aligned}
V &\leftarrow \underset{V}{\mathrm{argmin}}\, \|V\|_1 + \frac{\rho_1}{2}\left\|J \circ X - Y - V + \widehat{V}\right\|_F^2 \\
Z &\leftarrow \underset{Z}{\mathrm{argmin}}\, \|Z\|_{\mathbf{w},*} + \frac{\rho_2}{2}\left\|X - Z + \widehat{Z}\right\|_F^2 \\
X &\leftarrow \underset{X}{\mathrm{argmin}}\, \frac{\alpha}{2}\,\mathrm{tr}(D_\theta^T X^T (L + \epsilon I)^r X D_\theta) + \frac{\rho_1}{2}\left\|J \circ X - Y - V + \widehat{V}\right\|_F^2 \\
&\quad + \frac{\rho_2}{2}\left\|X - Z + \widehat{Z}\right\|_F^2
\end{aligned} \tag{10}$$

For the V-subproblem, we get the closed-form solution expressed via the shrinkage operator:

$$V = \mathrm{shrink}(\widehat{J}^T (Y + V - \widehat{V}), 1/\rho_1). \tag{11}$$

Similar to Algorithm 1, the solution of the Z-subproblem is given by Eq. (7) with ρ replaced by ρ_2. For the X-subproblem, we get the closed-form solution:

$$X = \mathrm{mat}[(\rho_1 \widehat{J} + \alpha A^T A + \rho_2 I)^{-1} (\rho_1 \widehat{J}^T (Y + V - \widehat{V}) + \rho_2 \,\mathrm{vec}(Z - \widehat{Z}))]. \tag{12}$$

The entire algorithm is described in Algorithm 2.

Algorithm 2 Robust time-varying graph signal recovery with high-order smoothness and adaptive low-rankness

Input: graph Laplacian L, parameters α, β, ρ_1, ρ_2, spatial Laplacian parameters ϵ and r, ERF parameter σ, Fractional-order derivative parameters $\theta > 0$ and integer $k \geq 1$.
Output: X
Initialize: X, $\widehat{V}$, $\widehat{Z}$
while The stopping criteria is satisfied **do**
  compute the weights $w_i' s$
  update V via Eq. (11)
  update Z via Eq. (7)
  update X via Eq. (12)
  $\widehat{V} \leftarrow \widehat{V} + (J \circ X - Y - V)$
  $\widehat{Z} \leftarrow \widehat{Z} + (X - Z)$
end while

2.5 Convergence Analysis of Algorithm 1

For simplicity, we define

$$f(X) := \frac{1}{2}\|Y - J \circ X\|_F^2 + \frac{\alpha}{2}\operatorname{tr}(D_\theta^T X^T (L + \epsilon I)^\gamma X D_\theta)$$

and hence the augmented Lagrangian function is given by

$$\mathcal{L}(X, Z, \widehat{Z}) = f(X) + \beta\|Z\|_{\text{erf}} + \rho\langle \widehat{Z}, X - Z\rangle + \frac{\rho}{2}\|X - Z\|_F^2.$$

The function f is convex and continuously differentiable. In addition, ∇f is Lipschitz continuous with a constant L_f.

Theorem 1 *Let $\rho > L_f$ and $\{(X^k, Z^k, \widehat{Z}^k)\}$ be a sequence generated from Algorithm 1; then, the sequence is bounded, and any limit point of the sequence is a stationary point of the problem (Eq. 5).*

Proof Consider one iteration of Algorithm 1; the update of Z^{k+1} gives

$$\begin{aligned}
&\mathcal{L}(X^k, Z^{k+1}, \widehat{Z}^k) - \mathcal{L}(X^k, Z^k, \widehat{Z}^k) \\
=& \beta\|Z^{k+1}\|_{\text{erf}} + \frac{\rho}{2}\|X^k - Z^{k+1} + \widehat{Z}^k\|_F^2 - \beta\|Z^k\|_{\text{erf}} - \frac{\rho}{2}\|X^k - Z^k + \widehat{Z}^k\|_F^2 \\
\leq& \beta\|Z^{k+1}\|_{w^k,*} - \beta\|Z^k\|_{w^k,*} + \frac{\rho}{2}\|X^k + \widehat{Z}^k - Z^{k+1}\|_F^2 - \frac{\rho}{2}\|X^k + \widehat{Z}^k - Z^k\|_F^2 \\
\leq& -\frac{\rho}{2}\|Z^{k+1} - Z^k\|_F^2.
\end{aligned} \tag{13}$$

The first inequality holds because the error function is concave for positive values. The second inequality is valid because Z^{k+1} is the optimal solution of the Z-subproblem.

Then, we consider the updates of X^{k+1} and $\widehat{Z}^{k+1}$, which together give

$$
\begin{aligned}
&\mathcal{L}(X^{k+1}, Z^{k+1}, \widehat{Z}^{k+1}) - \mathcal{L}(X^k, Z^{k+1}, \widehat{Z}^k) \\
=& f(X^{k+1}) + \rho\langle \widehat{Z}^{k+1}, X^{k+1} - Z^{k+1}\rangle + \frac{\rho}{2}\|X^{k+1} - Z^{k+1}\|_F^2 \\
&- f(X^k) - \rho\langle \widehat{Z}^k, X^k - Z^{k+1}\rangle - \frac{\rho}{2}\|X^k - Z^{k+1}\|_F^2 \\
=& f(X^{k+1}) - f(X^k) + \rho\langle \widehat{Z}^{k+1}, X^{k+1} - X^k\rangle \\
&+ \rho\|\widehat{Z}^{k+1} - \widehat{Z}^k\|_F^2 - \frac{\rho}{2}\|X^{k+1} - X^k\|_F^2,
\end{aligned}
$$

where the last equality uses the update $\widehat{Z}^{k+1} = \widehat{Z}^k + X^{k+1} - Z^{k+1}$. Since f is smooth, the updates of X^{k+1} and $\widehat{Z}^{k+1}$ show that $\rho\widehat{Z}^{k+1} + \nabla f(X^{k+1}) = 0$. The convexity and smoothness of f give $f(X^{k+1}) + \langle \nabla f(X^{k+1}), X^k - X^{k+1}\rangle + \frac{1}{2L_f}\|\nabla f(X^{k+1}) - \nabla f(X^k)\|^2 \le f(X^k)$. Therefore, we have

$$
\begin{aligned}
&\mathcal{L}(X^{k+1}, Z^{k+1}, \widehat{Z}^{k+1}) - \mathcal{L}(X^k, Z^{k+1}, \widehat{Z}^k) \\
\le& \left(\max\left(\frac{1}{\rho} - \frac{1}{2L_f}, 0\right) L_f^2 - \frac{\rho}{2}\right)\|X^{k+1} - X^k\|_F^2. \qquad (14)
\end{aligned}
$$

If $\rho > L_f$, then $\max\left(\frac{1}{\rho} - \frac{1}{2L_f}, 0\right) {L_f}^2 - \frac{\rho}{2} < 0$.

Combing Eqs. (13) and (14), we see that $\mathcal{L}(X^k, Z^k, \widehat{Z}^k)$ is decreasing. Furthermore, if $\rho > L_f$, we have

$$
\begin{aligned}
& f(X^k) + \beta\|Z^k\|_{\text{erf}} + \rho\langle \widehat{Z}^k, X^k - Z^k\rangle + \frac{\rho}{2}\|X^k - Z^k\|_F^2 \\
=& f(X^k) + \beta\|Z^k\|_{\text{erf}} - \langle \nabla f(X^k), X^k - Z^k\rangle + \frac{\rho}{2}\|X^k - Z^k\|_F^2 \\
\ge& f(Z^k) + \beta\|Z^k\|_{\text{erf}} + \frac{\rho - L_f}{2}\|X^k - Z^k\|_F^2 \ge 0, \qquad (15)
\end{aligned}
$$

where the last inequality comes from the Lipschitz continuity of ∇f. So, $\mathcal{L}(X^k, Z^k, \widehat{Z}^k)$ is bounded from below. Therefore, $\mathcal{L}(X^k, Z^k, \widehat{Z}^k)$ converges and

$$
\lim_{k\to\infty}(X^{k+1} - X^k) = 0, \quad \lim_{k\to\infty}(Z^{k+1} - Z^k) = 0. \qquad (16)
$$

Since ∇f is Lipschitz continuous, we can get

$$\lim_{k\to\infty} \widehat{Z}^{k+1} - \widehat{Z}^k = X^k - Z^k = 0. \tag{17}$$

Next, we show that $(X^k, Z^k, \widehat{Z}^k)$ is bounded. We have shown in Eq. (15) that

$$\mathcal{L}(X^k, Z^k, \widehat{Z}^k) \geq f(Z^k) + \beta \|Z^k\|_{\text{erf}} + \frac{\rho - L_f}{2} \|X^k - Z^k\|_F^2.$$

Therefore, when $\rho > L_f$, the boundedness of $\mathcal{L}(X^k, Z^k, \widehat{Z}^k)$ gives the boundedness of $f(Z^k) + \beta\|Z^k\|_{\text{erf}}$ and $\|X^k - Z^k\|_F^2$. Thus, sequences $\{X^k\}$ and $\{Z^k\}$ are also bounded. Because $\rho\widehat{Z}^k = -\nabla f(X^k)$, the sequence $\{\widehat{Z}^k\}$ is also bounded.

Since the sequence $\{(X^k, Z^k, \widehat{Z}^k)\}$ is bounded, there exists a convergent subsequence, that is, $(X^{k_i}, Z^{k_i}, \widehat{Z}^{k_i}) \to (X^\star, Z^\star, \widehat{Z}^\star)$. The limits (Eqs. 16 and 17) show that $(X^{k_i+1}, Z^{k_i+1}, \widehat{Z}^{k_i+1}) \to (X^\star, Z^\star, \widehat{Z}^\star)$. Then, we have that $X^\star = Z^\star$ and $\beta\partial\|Z^\star\|_{\text{erf}} - \rho\widehat{Z}^\star = 0$. Thus, $X^\star$ is a stationary point of the original problem (Eq. 5). Since it holds for any convergent subsequence, any limit point of the sequence is a stationary point of Eq. (5). □

3 Numerical Experiments

In this section, we conduct various numerical experiments on synthetic and real data to demonstrate the performance of our proposed methods. In particular, we compare our methods—Algorithm 1 and Algorithm 2—with other related states of the art, including natural NNI [30], GS [22], Tikhonov [25], TGSR [27], LRDS [20], and Sobolev [11]. To evaluate the reconstruction quality, we adopt the root mean square error (RMSE) as a comparison metric, defined as follows:

$$\text{RMSE} = \frac{\|X - \widehat{X}\|_F}{\sqrt{nm}}, \tag{18}$$

where $\widehat{X}$ is the approximation of the ground truth graph signal $X \in \mathbb{R}^{n\times m}$ defined on a spatial-temporal graph with n nodes and m time instances. All the numerical experiments are implemented on MATLAB R2021a in a desktop computer with Intel CPU i9-9960X RAM 64GB and GPU Dual Nvidia Quadro RTX5000 with Windows 10 Pro.

3.1 Synthetic Data

Following the work of [27], we generate $N = 100$ nodes randomly from the uniform distribution in a 100×100 square area. The graph weight is determined using K-nearest neighbors. Specifically, the weight between any two nodes is inversely

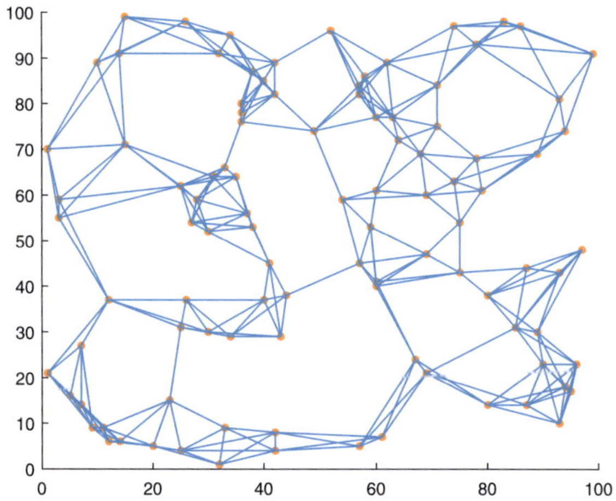

Fig. 1 The graph is constructed by KNN with $K = 5$. The weight between any two nodes is inversely proportional to the square of their Euclidean distance

proportional to the square of their Euclidean distance. We consider $K = 5$ and visualize the corresponding graph in Fig. 1.

Denote the weight matrix by W, its degree matrix M, and the graph Laplacian L has eigen-decomposition $L = U\Lambda U^T$, where $\Lambda = \mathrm{diag}(0, \lambda_2, \cdots, \lambda_N)$. We further define $L^{-1/2} = U\Lambda^{-1/2}U^T$ where $\Lambda^{-1/2} = \mathrm{diag}(0, \lambda_2^{-1/2}, \cdots, \lambda_N^{-1/2})$. Starting from x_1, we generate the time-varying graph signal

$$x_t = x_{t-1} + L^{-1/2} f_t, \quad \text{for } t = 2, \cdots, T, \tag{19}$$

where f_t is an i.i.d. Gaussian signal rescaled to $\|f_t\|_2 = \kappa$ and κ corresponds to a temporal smoothness of the signal. Stacking $\{x_t\}$ as a column vector, we obtain a data matrix $X = [x_1, x_2, \cdots, x_T]$. We generate a *low-rank* data matrix obtained by starting with an empty matrix X and repeating $X \leftarrow [X, x_1, \cdots, x_{10}, x_{10}, x_9, \cdots, x_1]$ 10 times, thus also getting a 100×200 data matrix. The measurement noise at each node is i.i.d. Gaussian noise $\mathcal{N}(0, \eta^2)$, where η is the standard deviation.

Parameter Tuning For the proposed Algorithm 1, we fix the following parameters: $k = 3$ and $\theta = 1.8$ in the definition of fractional-order derivative (Eq. 4); $\sigma = 10^3$ in the definition of the ERF regularization (Eq. 3); $\epsilon = 0.1$ and $r = 3$ in the Sobolev graph Laplacian; and the step size $\rho = 10^{-6}$ in the ADMM iterations (Eq. 6). In each set of experiments, we carefully tune two parameters (α, β) that determine the weights for the spatial-temporal smoothness and the low-rankness, respectively, in the proposed model (Eq. 5). We choose the best combination of (α, β) among $\alpha \in \{0, 10^{-5}, 10^{-4}, 10^{-3}, 10^{-2}, 10^{-1}, 1, 10\}$ and $\beta \in \{0, 10^{-8}, 10^{-7}, 10^{-6}, 10^{-5}, 10^{-4}, 10^{-3}, 10^{-2}, 10^{-1}, 1, 10\}$. As demonstrated in Table 1, some competing methods are special cases of the proposed models, and hence, we only tune the parameters α, β for these methods while keeping other parameters fixed.

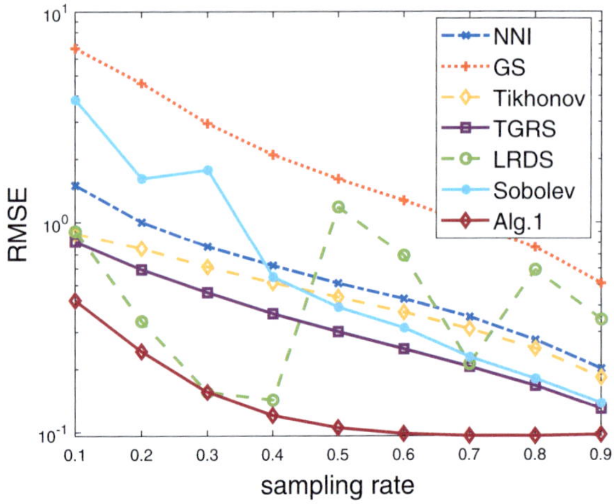

Fig. 2 RMSE versus sampling rates. Averaged over 50 trials

Reconstruction Errors with Respect to Sampling Rates We begin by evaluating the performance of competing methods under different sampling rates. The smoothness level is set as $\kappa = 1$, while the standard deviation of the Gaussian noise is $\eta = 0.1$. The reconstruction performance is evaluated via RMSE, defined in Eq. (18), showing that the recovery errors of all the methods decrease with the increase of the sampling rates. The comparison results are visualized in Fig. 2. The proposed method achieves significant improvements over the competing methods. Surprisingly, LRDS, equipped with the nuclear norm, does not yield stable reconstruction performance in the low-rank case.

Reconstruction Errors with Respect to Noise Levels We then investigate the recovery performance under different noise levels by setting the noise variance $\eta^2 = \{0.01, 0.1, 0.2, 0.4, 0.6, 0.8, 1\}$. In this set of experiments, we fix the sampling rate as 40% and smoothing level $\kappa = 1$. The noise level affects the magnitude of the least-squares fit, and as a result, we adjust the search window of $\alpha \in \{0, 10^{-3}, 10^{-2}, 10^{-1}, 1, 10, 10^2, 10^3, 10^4$. The parameter β remains the same: $\beta \in \{0, 10^{-8}, 10^{-7}, 10^{-6}, 10^{-5}, 10^{-4}, 10^{-3}, 10^{-2}, 10^{-1}, 1, 10\}$. The results are presented in Fig. 3, demonstrating the superior performance of the proposed Algorithm 1 under various noise levels.

3.2 Real Data

In the real data experiments, we first test the daily mean particulate matter (PM) 2.5 concentration dataset from California provided by the US Environmental Protection Agency https://www.epa.gov/outdoor-air-quality-data. We used the data captured daily from 93 sensors in California for the first 200 days in 2015. The constructed graph is depicted in Fig. 4. In Fig. 5, we compare the average

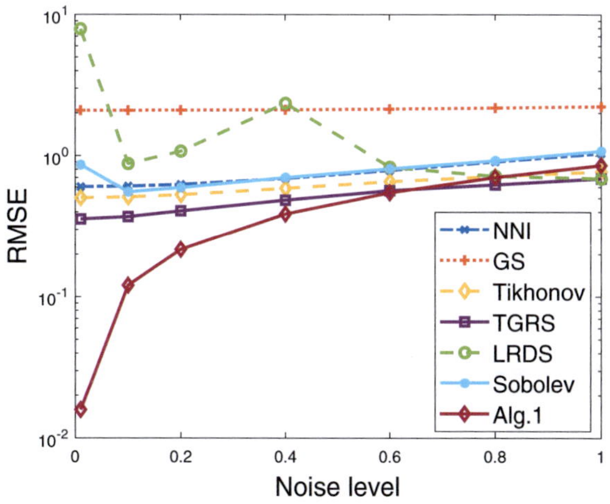

Fig. 3 RMSE versus noise level: $\eta^2 = \{0.01, 0.1, 0.2, 0.4, 0.6, 0.8, 1\}$. Averaged over 50 trials

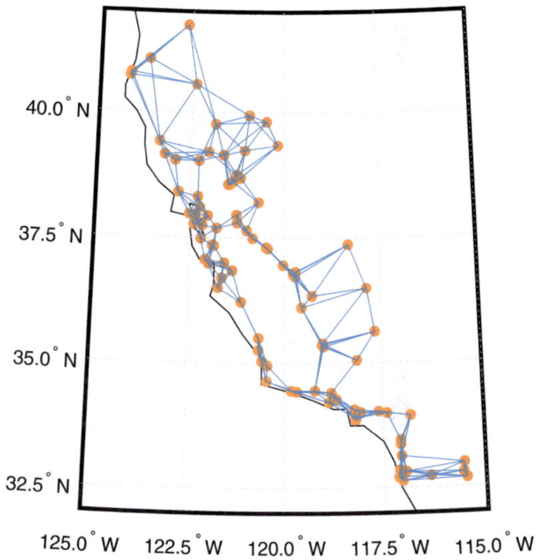

Fig. 4 Graph with the places in California for the PM 2.5 concentration data. The graph was constructed with KNN for $K = 5$

recovery accuracy of all the comparing methods over 50 trials when the sampling rates are 0.1, 0.15, 0.2, 0.25, 0.3, 0.35, 0.4, 0.45. In Table 2, we also compare the performance of Algorithm 1 and Algorithm 2, which shows Algorithm 2 can improve the accuracy of Algorithm 1 under some sampling rates with longer time in general.

Next, we test the sea surface temperature dataset, which was captured monthly by the NOAA Physical Sciences Laboratory (PSL). The dataset can be downloaded from the PSL website https://psl.noaa.gov/. We use a subset of 200 time points on the Pacific Ocean within 400 months. The constructed graph is illustrated in Fig. 6. We see from Fig. 7 that the proposed algorithm outperforms other methods significantly and consistently across all sampling rates. In Table 3, we also compare

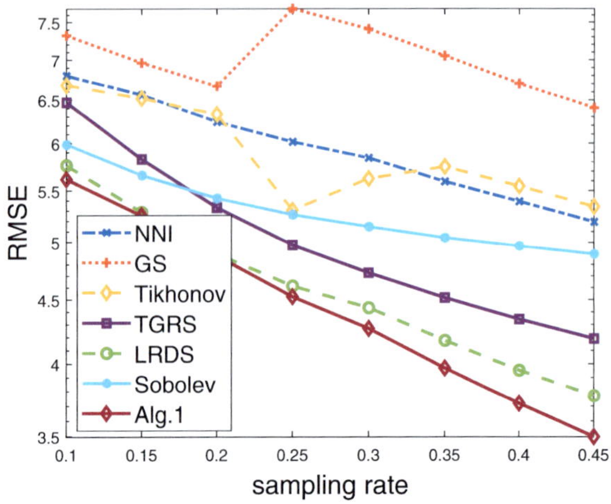

Fig. 5 Average recovery accuracy comparison on the PM2.5 data

Table 2 Performance comparison of Algorithm 1 and Algorithm 2 for the PM2.5 data. The running time for Algorithm 1 is about 22 ~23 seconds, while Algorithm 2 uses about 46 ~ 48 seconds

	Alg.1		Alg.2	
Sampling rate	RMSE	Time (s)	RMSE	Time (s)
0.10	5.7321	22.95	5.6915	46.49
0.15	5.4770	22.63	6.4992	45.39
0.20	5.0427	23.83	5.9730	48.50
0.25	6.0358	23.37	5.6976	47.44
0.30	5.6065	23.70	5.3809	47.86
0.35	5.1920	23.55	5.1535	47.72
0.40	5.2398	23.56	4.7758	47.59
0.45	5.2283	23.80	5.0913	48.17

the performance of Algorithm 1 and Algorithm 2, which indicates Algorithm 2 can improve the accuracy of Algorithm 1 under certain sampling rates but with more computational time in general.

3.3 *Discussions*

Using the sea surface temperature data, we conduct an ablation study of the proposed model (Eq. 5) without the smoothing regularization by setting $\alpha = 0$ or without the low-rank ERF term by setting $\beta = 0$. We plot the RMSE curves with respect to the sampling rates and the noise levels in Fig. 8, showing that the ERF regularization has a larger influence on the performance compared to the Sobolev-base graph Laplacian regularization.

Using the same sea surface temperature data, we investigate whether the proposed model (Eq. 5) is sensitive to the parameters (r, ϵ) in defining the Sobolev-graph Laplacian and σ^2 in defining the ERF regularization. Figure 9 shows that the

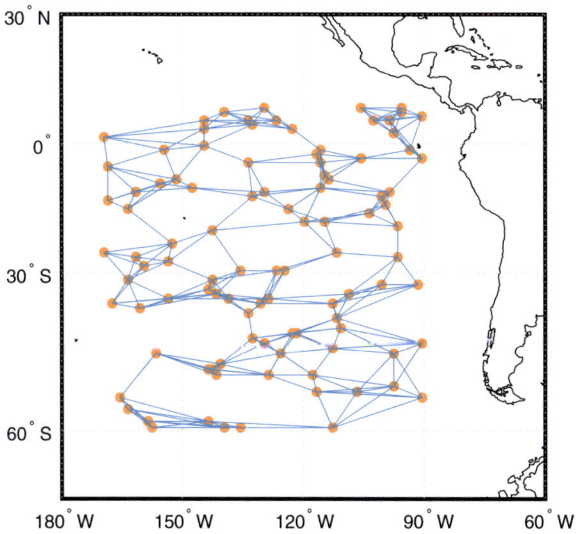

Fig. 6 Graph with the places in the sea for the sea surface temperature data. The graph was constructed with KNN for $K = 10$

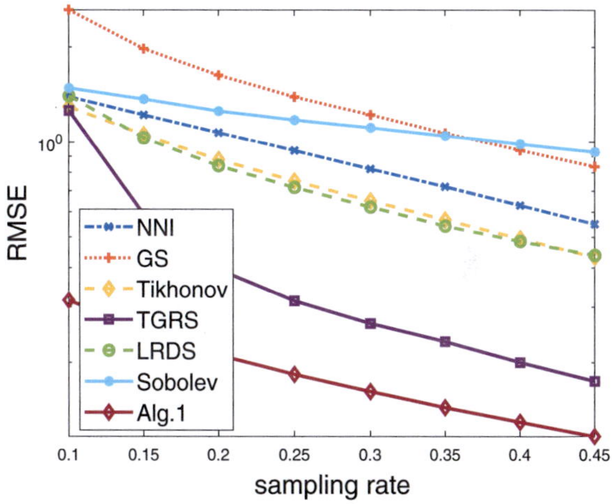

Fig. 7 Average recovery accuracy comparison on the sea surface temperature data

proposed approach is not sensitive to various degrees of smoothness controlled by r and ϵ. Although the ERF regularization plays an important role in the recovery performance, as illustrated in the ablation study, the proposed model is not sensitive to the choice σ^2 as long as it is larger than 10,000.

In addition, we compare the proposed Algorithm 1 and Algorithm 2 using the sea surface temperature data and show the results in Tables 2 and 3. One can see that the two algorithms lead to similar RMSE, but Algorithm 2 is slower overall.

Table 3 Performance comparison of Algorithm 1 and Algorithm 2 for the sea surface temperature data. The running time for Algorithm 1 is about 2 ∼4 seconds, while Algorithm 2 uses about 6 ∼ 23 seconds

	Algorithm 1		Algorithm 2	
Sampling rate	RMSE	Time (s)	RMSE	Time (s)
0.10	0.3148	3.97	0.3163	22.99
0.15	0.2497	3.37	0.2483	17.13
0.20	0.2110	3.08	0.2109	13.52
0.25	0.1832	2.87	0.1857	11.27
0.30	0.1617	2.74	0.1666	9.62
0.35	0.1438	2.63	0.1450	5.77
0.40	0.1294	2.54	0.1291	5.66
0.45	0.1166	2.46	0.1153	5.57

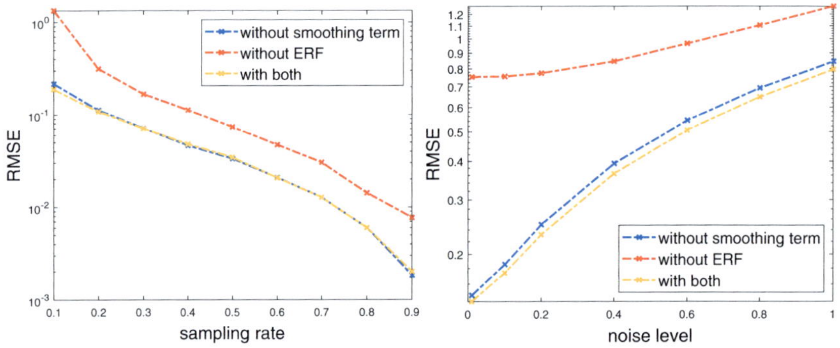

Fig. 8 Ablation study sampling rates (left) and noise levels (right) on the sea surface temperature data

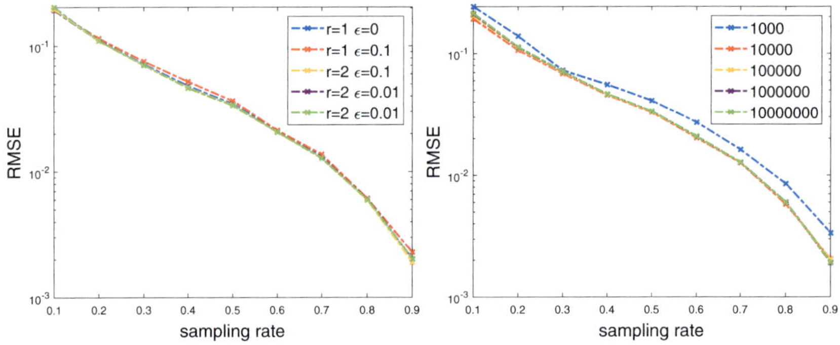

Fig. 9 Sensitivity analysis with respect to varying the graph Laplacian (left) and σ^2 in ERF (right) on the sea surface temperature data

We therefore prefer to use Algorithm 1 unless the data is heavily polluted by the non-Gaussian type of noise, such as Laplace noise.

4 Conclusions and Future Work

In this paper, we exploit high-order smoothness across the temporal domain and adaptive low-rankness for time-varying graph signal recovery. In particular, we propose a novel graph signal recovery model based on a hybrid graph regularization involving a general order temporal difference, together with an error-function-weighted nuclear norm. We also derive an effective optimization algorithm with guaranteed convergence by adopting a reweighting scheme and the ADMM framework. Numerical experiments have demonstrated their efficiency and performance in terms of accuracy. However, the graph Laplacian is a computational bottleneck in our workflow, especially when the graph contains a large number of nodes. The acceleration of the weight calculation via sparse or low-rank approximations [8] will be left in future work. In addition, we will explore using high-order difference schemes to create a temporal Laplacian and low-rankness for recovering graph signals with dynamic graph topology.

Acknowledgments The authors would like to thank the support from the American Institute of Mathematics during 2019–2022 for making this collaboration happen. WG, YL, and JQ would also like to thank the Women in Data Science and Mathematics Research Workshop (WiSDM) hosted by UCLA in 2023 for the support of continuing this collaboration. YL is partially supported by NSF CAREER 2414705. JQ is partially supported by the NSF grant DMS-1941197. MY was partially supported by the Guangdong Key Lab of Mathematical Foundations for Artificial Intelligence, the Shenzhen Science and Technology Program ZDSYS20211021111415025, and the Shenzhen Stability Science Program.

References

1. Berger, P., Hannak, G., Matz, G.: Graph signal recovery via primal-dual algorithms for total variation minimization. IEEE J. Sel. Topics Signal Process. **11**(6), 842–855 (2017)
2. Candes, E.J., Wakin, M.B., Boyd, S.P.: Enhancing sparsity by reweighted ℓ_1 minimization. J. Fourier Anal. Appl. **14**(5), 877–905 (2008)
3. Castro-Correa, J.A., Giraldo, J.H., Mondal, A., Badiey, M., Bouwmans, T., Malliaros, F.D.: Time-varying signals recovery via graph neural networks. In: ICASSP 2023–2023 IEEE International Conference on Acoustics, Speech and Signal Processing (ICASSP), pp. 1–5. IEEE (2023)
4. Chen, F., Cheung, G., Zhang, X.: Manifold graph signal restoration using gradient graph laplacian regularizer. IEEE Trans. Signal Process. **72**, 744–761 (2024)
5. Chen, S., Sandryhaila, A., Moura, J.M., Kovačević, J.: Signal recovery on graphs: Variation minimization. IEEE Trans. Signal Process. **63**(17), 4609–4624 (2015)
6. Domingos, J., Moura, J.M.: Graph fourier transform: A stable approximation. IEEE Trans. Signal Process. **68**, 4422–4437 (2020)
7. Fang, E.X., He, B., Liu, H., Yuan, X.: Generalized alternating direction method of multipliers: new theoretical insights and applications. Math. Programm. Comput. **7**(2), 149–187 (2015)
8. Fowlkes, C., Belongie, S., Chung, F., Malik, J.: Spectral grouping using the Nyström method. IEEE Trans. Pattern Anal. Mach. Intell. **26**(2), 214–225 (2004)
9. Gabay, D., Mercier, B.: A dual algorithm for the solution of nonlinear variational problems via finite element approximation. Comput. Math. Appl. **2**(1), 17–40 (1976)

10. Giraldo, J.H., Bouwmans, T.: On the minimization of sobolev norms of time-varying graph signals: estimation of new coronavirus disease 2019 cases. In: 2020 IEEE 30th International Workshop on Machine Learning for Signal Processing (MLSP), pp. 1–6. IEEE (2020)
11. Giraldo, J.H., Mahmood, A., Garcia-Garcia, B., Thanou, D., Bouwmans, T.: Reconstruction of time-varying graph signals via sobolev smoothness. IEEE Trans. Signal Inf. Process. Networks **8**, 201–214 (2022)
12. Glowinski, R., Marroco, A.: On the approximation by finite elements of order one, and resolution, penalisation-duality for a class of nonlinear dirichlet problems. ESAIM: Math. Model. Numer. Anal. Math. Model. Numer. Anal. **9**(R2), 41–76 (1975)
13. Gu, S., Zhang, L., Zuo, W., Feng, X.: Weighted nuclear norm minimization with application to image denoising. In: Proceedings of the IEEE Conference on Computer Vision and Pattern Recognition, pp. 2862–2869 (2014)
14. Guo, W., Lou, Y., Qin, J., Yan, M.: A novel regularization based on the error function for sparse recovery. J. Sci. Comput. **87**(1), 1–22 (2021)
15. Jiang, J., Tay, D.B., Sun, Q., Ouyang, S.: Recovery of time-varying graph signals via distributed algorithms on regularized problems. IEEE Trans. Signal Inf. Process. Networks **6**, 540–555 (2020)
16. Kojima, H., Noguchi, H., Yamada, K., Tanaka, Y.: Restoration of time-varying graph signals using deep algorithm unrolling. In: ICASSP 2023–2023 IEEE International Conference on Acoustics, Speech and Signal Processing (ICASSP), pp. 1–5. IEEE (2023)
17. Kroizer, A., Eldar, Y.C., Routtenberg, T.: Modeling and recovery of graph signals and difference-based signals. In: 2019 IEEE Global Conference on Signal and Information Processing (GlobalSIP), pp. 1–5. IEEE (2019)
18. Li, X.P., Yan, Y., Kuruoglu, E.E., So, H.C., Chen, Y.: Robust recovery for graph signal via ℓ_0-norm regularization. IEEE Signal Process. Lett. **30**, 1322–1326 (2023)
19. Liu, J., Lin, J., Qiu, H., Wang, J., Nong, L.: Time-varying signal recovery based on low rank and graph-time smoothness. Digital Signal Process. **133**, 103821 (2023)
20. Mao, X., Qiu, K., Li, T., Gu, Y.: Spatio-temporal signal recovery based on low rank and differential smoothness. IEEE Trans. Signal Process. **66**(23), 6281–6296 (2018)
21. Mills, M.T., Bourbakis, N.G.: Graph-based methods for natural language processing and understanding - a survey and analysis. IEEE Trans. Syst. Man Cybern. Syst. **44**(1), 59–71 (2013)
22. Narang, S.K., Gadde, A., Sanou, E., Ortega, A.: Localized iterative methods for interpolation in graph structured data. In: 2013 IEEE Global Conference on Signal and Information Processing, pp. 491–494. IEEE (2013)
23. Ouyang, Y., Chen, Y., Lan, G., Pasiliao Jr, E.: An accelerated linearized alternating direction method of multipliers. SIAM J. Imag. Sci. **8**(1), 644–681 (2015)
24. Ozturk, C., Ozaktas, H.M., Gezici, S., Koç, A.: Optimal fractional fourier filtering for graph signals. IEEE Trans. Signal Process. **69**, 2902–2912 (2021)
25. Perraudin, N., Loukas, A., Grassi, F., Vandergheynst, P.: Towards stationary time-vertex signal processing. In: 2017 IEEE International Conference on Acoustics, Speech and Signal Processing (ICASSP), pp. 3914–3918. IEEE (2017)
26. Podlubny, I.: Fractional differential equations. Math. Sci. Eng. **198**, 41–119 (1999)
27. Qiu, K., Mao, X., Shen, X., Wang, X., Li, T., Gu, Y.: Time-varying graph signal reconstruction. IEEE J. Sel. Topics Signal Process. **11**(6), 870–883 (2017)
28. Sandryhaila, A., Moura, J.M.: Discrete signal processing on graphs. IEEE Trans. Signal Process. **61**(7), 1644–1656 (2013)
29. Shuman, D.I., Narang, S.K., Frossard, P., Ortega, A., Vandergheynst, P.: The emerging field of signal processing on graphs: Extending high-dimensional data analysis to networks and other irregular domains. IEEE Signal Process. Mag. **30**(3), 83–98 (2013)
30. Sibson, R.: A brief description of natural neighbour interpolation. Interpret. Multivariate Data, 21–36 (1981)
31. Tremblay, N., Borgnat, P.: Graph wavelets for multiscale community mining. IEEE Trans. Signal Process. **62**(20), 5227–5239 (2014)

32. Varma, R., Chen, S., Kovačević, J.: Spectrum-blind signal recovery on graphs. In: 2015 IEEE 6th International Workshop on Computational Advances in Multi-Sensor Adaptive Processing (CAMSAP), pp. 81–84. IEEE (2015)
33. Xia, F., Sun, K., Yu, S., Aziz, A., Wan, L., Pan, S., Liu, H.: Graph learning: A survey. IEEE Trans. Artif. Intell. **2**(2), 109–127 (2021)
34. Xiao, Z., Fang, H., Tomasin, S., Mateos, G., Wang, X.: Joint sampling and reconstruction of time-varying signals over directed graphs. IEEE Trans. Signal Process. **71**, 2204–2219 (2023)
35. Ying, W., Rui, X., Xiaoyang, M., Qiang, F., Jinshuai, Z., Runze, Z.: Spectral graph theory-based recovery method for missing harmonic data. IEEE Trans. Power Delivery **37**(5), 3688–3697 (2021)

Graph-Directed Topic Models of Text Documents

Arjuna Flenner and Cristina Garcia-Cardona

1 Introduction

One goal of analyzing large collections of text documents is to build representations that facilitate the interpretation and recovery of information. This in turn requires the extraction of relevant features, which should encode interesting aspects of the observed data. This feature extraction characterization and quantification is difficult to accomplish in practice. The seminal work by Blei et al. in [5] introduced topic modeling using the Dirichlet process. This model used the Dirichlet process as a sparsity inducing prior while treating the words as simple Poisson count data [43]. Fundamental to topic modeling is the assumption that the data, or its expected value, can be represented as a linear combination of basis vectors if no other prior information is available. In the multivariate statistical arena, this is a subset of factor analysis. In this work, we integrate the idea of feature extraction using topic modeling with undirected graphs.

As shown in [43], a basic structure of topic modeling is a Poisson factor analysis model that effectively models count data. Consider a collection of N documents where each document is composed of a collection of words. Let $\mathbf{d}_n \in \mathbb{Z}^L$ represent a vector of word counts with a dictionary of size L. The Poisson factor analysis model assumes

A. Flenner (✉)
Advanced Technology Office, GE Aerospace, Grand Rapids, Grand Rapids, MI, USA
e-mail: Arjuna.Flenner@ge.com

C. Garcia-Cardona
Los Alamos National Laboratory, Los Alamos, NM, USA
e-mail: cgarciac@lanl.gov

C. Garcia-Cardona, H. Lee (eds.), *Advances in Data Science*, Association for Women in Mathematics Series 37, https://doi.org/10.1007/978-3-031-87804-6_5

$$\mathbb{E}\left[\mathbf{d}_n\right] = \sum_{k=1}^{K} \boldsymbol{\psi}_k b_{nk}, \tag{1}$$

where the observed data $\mathbf{d}_n$ is represented in terms of ℓ_1 normalized *basis vectors*, or topics, $\boldsymbol{\psi}_k$ (factors) and the coefficients b_{nk} describe the *weights* of the combination (factor loading). In other words, the basis vectors capture patterns present in the data set, i.e., a good set of features, and the learning representation task is to compute them, together with the associated weights. In general, the number of components in the combination, denoted by K, characterizes the complexity of the model, and a key insight in [5] was the use of the Dirichlet process to automatically determine the value of K.

Many different extensions to the basic topic modeling approach have been proposed [4, 12, 16–20, 23, 28, 31, 38, 39]. We add to this body of work by introducing a general model to include graphs in topic models and deriving an efficient Gibbs sampling method for posterior inference. Our approach is similar in nature to the graph-based clustering approaches in computational linguistics [1, 10, 35]; however, instead of segmenting the graph, we explore the utilization of a graph structure as a regularizer, with the hope of driving the learning procedure such that entities that are connected in the graph structure end up having similar representations. The advantages of using a graph are twofold. First, the graph enables the integration of information from different sources into our learning algorithms to influence the model priors. Second, the graph allows to take into account information encoded via relationships between the data that do not come from a metric or distance function. Thus, information such as interactions or common group memberships, explicitly encoded by networks of connections, or social networks, can also be incorporated as part of the learning procedure.

Note that in contrast with [24], we are not trying to integrate topic modeling and dictionary learning. Instead, we are trying to integrate graph information into representation learning models. In the topic modeling case, we show how the graph-directed model yields topics that are more descriptive and whose distribution is more balanced through the corpus of documents. This type of directed learning proves to be effective also for the case of dictionary learning, where we show that using the graph structure to encode a priori relations between observations allows for more distinctive basis vectors and, at the same time, lower average reconstruction errors.

To make the representation learning tractable, we build stochastic models and embed them in a Bayesian framework such that the model parameters are learned by maximizing the posterior distribution given the data observations and the assumed priors. The computations are carried out using a Hamiltonian Monte Carlo (HMC) sampling method, whose energy-based formulation facilitates the information integration, in particular the graph encoded priors.

The document is structured as follows. Section 2 summarizes the previous work. The representation learning problem and the graph model are introduced in Sect. 3, while Sect. 4 describes the computation procedure. Section 5 illustrates the applications to topic modeling and dictionary learning and compares performance with other methods. Finally, Sect. 6 includes the conclusion and perspectives for future work.

2 Previous Work

A simple model as the linear combination of basis vectors described by expression (1) is expressive and flexible enough, but the problem of finding the unknown basis, weights, and number of components is not well-defined. Most of the time, the learning task tries to build more structure into the problem by exploiting the inherent range of the data at hand, as in nonnegative matrix factorization [11], or assuming some conditions over the basis vectors. These different assumptions are the essential characteristics of the different methods. In the factor analysis field, different assumptions are used to decompose the data into a few factors and estimate associated weights. The basic procedure is to try to establish a stochastic generative model to describe the data set and provide a framework for learning the parameters of the model.

However, many applications of factor analysis neglect the discrete nature of count data [40]. A case in point is the description of a corpus of text documents. In general, the corpus is given in terms of the times a word from the (corpus) vocabulary appears per document, and the learning task is expressed as the construction of topic models [4, 5, 17, 23, 38]. In the language of topic modeling literature, the set of basis vectors correspond to *topics*, and each topic is assimilated to a probability distribution of the words in the corpus vocabulary. Thus, more probable distributions are the ones that are compatible with the observed count of words. In [5], each document is a mixture of topics, and the topics a distribution of words. The priors in both cases are symmetric Dirichlet distributions, leading to the well-known latent Dirichlet allocation or LDA model. Several posterior works have studied variants of the LDA model. These include correlated topic models [23], where the basis elements are assumed correlated, while the words per topic are still assumed independent and dynamic topic models [4] where the topics are allowed to slowly change over time. The work of Wallace et al. [38] studies the influence that handling of stop words, number of topics selected, and Dirichlet priors have in the resulting LDA topic model and shows how the performance is improved when an asymmetric Dirichlet prior is used for the document-topic distributions.

Computationally, there are two main approaches: variational and Markov chain Monte Carlo (MCMC) methods. The variational approach finds the parameters of a family of functions that approximate the posterior distribution given the observed topics [5]. Variational Bayesian methods are efficient if the resulting optimization is easy to compute as is the case of the conjugate priors used in the original work of Blei. Markov Chain Monte Carlo methods obtain samples from the posterior distribution [6], but complex models often mix slowly. Again, conjugate priors are often exploited in a Gibbs sampling scheme to simplify computations. Stick-breaking techniques have been employed [30] to generate efficient MCMC strategies. Stick breaking has also been used to extend the models as in the spatial modeling with stick breaking in [29, 30, 36] where a kind of spatial dependence of the basis vector component is assumed by using a similarity kernel and a stick-breaking construction.

Two non-Bayesian approaches that are similar in spirit to topic modeling are the nonnegative matrix and tensor factorization [11, 22] and nonnegative sparse representations [9, 41, 42]. The matrix and tensor factorization methods are most effective when combined with a word weighting [33] and thus do not represent the documents as pure count data. To obtain a more compact representation, the nonnegative sparse matrix or tensor factorization includes a ℓ_0 or ℓ_1 sparsity promoting regularization [33]. We do not consider these approaches in this work.

We note that the Gaussian Markov random fields (GMRF) approach by Rue [32] has been used by Mimno et al. [26] to regularize the topic weights. Their model resembles the model in this work, but instead of creating a graph between the topic weighs they define a mean for the Gaussian process.

3 Background

We are learning representations of the data by integrating Poisson factor analysis and graphical methods. In this section, we give a brief overview of Poisson factor analysis and graphical models appropriate for this work.

3.1 Poisson Factor Analysis

In the case that the information available corresponds to count data, as, for example, in text documents where each document is represented as a count of the number of times a specific word appears in the document, Poisson factor analysis (PFA) is a natural model. For clarity, define the following notation. Given a corpus of D text document,

- $d \in \{1, \ldots, D\}$ indexes each of the documents in the corpus.
- N_d stands for the number of words in document d.
- $w \in \{1, \ldots, W\}$ indexes each of the distinct words in the corpus vocabulary.
- $h_{dw} \in \mathbb{Z}^+$ is the **observed** number of times that word w is present in document d.
- $\mathbf{h}_d$ represents the **observed** histogram of words for document d.

Poisson factor analysis assumes that the set of integer observations $h_{dw} \in \mathbb{Z}^+$ comes from a Poisson distribution:

$$h_{dw} \sim \text{Poisson}\left(\sum_{k}^{K} \psi_{kw}\, b_{dk}\right), \tag{2}$$

$$\mathbb{E}[h_{dw}] = \sum_{k}^{K} \psi_{kw}\, b_{dk}, \tag{3}$$

$$\psi_{kw} \geq 0, \qquad \sum_{w} \psi_{kw} = 1, \qquad b_{dk} \geq 0.$$

Due to the additive property of the Poisson distribution [21], this model implicitly assumes the existence of a decomposition of the form $h_{dw} = \sum_{k=1}^{K} h_{dwk}$, where $h_{dwk} \sim \text{Poisson}(\psi_{kw} b_{dk})$. This suggests the following stochastic generative model for each of the documents $\boldsymbol{x}_d$ in the corpus:

$$\begin{aligned}
\boldsymbol{x}_d &\sim \prod_{l=1}^{N_d} P\left(z_l \middle| \boldsymbol{\pi}_d\right)\, P\left(w_l \middle| z_l, \boldsymbol{\psi}_1, \ldots, \boldsymbol{\psi}_K\right) \\
z_l &\sim \text{Multinomial}(\boldsymbol{\pi}_d), \\
w_l \middle| z_l &\sim \text{Multinomial}(\boldsymbol{\psi}_l), \\
h_{dwk} &\overset{i.i.d}{\sim} \text{Poisson}(\psi_{kw}\, b_{dk}), \\
T_{dk} &= \sum_{w} h_{dwk}, \quad \pi_{dk} = \frac{b_{dk}}{\sum_l b_{dl}}
\end{aligned}$$

where $\boldsymbol{\pi}_d$ represents the topic distribution in document d (consequently, the component π_{dk} is the probability of topic k in document d); $\boldsymbol{\psi}_k$, the word distribution in topic k; z_l is an indicator variable for the topic assignment of the l-th word; w_l, the l-th word drawn; h_{dwk}, the count of words of type w appearing in topic k in document d; and T_{dk}, the count of words of topic k appearing in document d. Note that the last equalities are necessary to keep the consistency between the generative model and the PFA formulation. Hence, the learning task corresponds to estimating the model parameters ψ_{kw} and b_{dk}. Note that to complete the specification of the model, it is necessary to define priors for these parameters, as well as setting the maximum number of topics K.

3.2 Graphical Models

Graphical models are often used to describe joint probability distributions of multiple variables [8]. A generic graph, denoted by $G(V, E)$, can be regarded as a node (vertex) set V and a collection of edges E that connect the nodes. The nodes in the graph are in one-to-one correspondence to random variables in the model, while edges in the graph encode dependency relationships between the nodes (random variables) they connect. The graph can be undirected, in which case the edges denote dependence between the corresponding nodes, or directed, in which case the conditional dependence is restricted to incoming edges.

This kind of models allows for inference and estimation of local marginal distributions, likelihood of a particular random variable, or the most probable configuration of the model, among other summary statistics, in the case of directed acyclic graphs and random variables living in a discrete probability space [8]. However, there is no guarantee of convergence for cases of arbitrary graph configurations or for random variables drawn from continuous probability space. Hence, approximate inference methods, such as loopy belief propagation, Monte Carlo Markov chain (MCMC) sampling, or variational Bayes, are commonly used [8].

However, an alternative take on the variable dependence representation with graphs can be constructed. Instead of correspondence between nodes and random variables, a correspondence between nodes and observations can be established. Specifically, each element in the node set $V = \{v_n\}_{n=1}^N$ is associated with a data sample $\mathbf{x}_n$, and an edge E_{ij} between the i-th and j-th nodes exists if sample i is related to sample j and does not exist otherwise. Note that this allows to encode known interactions between data samples. In this work, we only consider simple connections as interactions. It is easy to extend this work to the case where a quantitative dependency such as metric information given in terms of a similarity measurement is used. For example, one can consider the following:

$$W_{ij} = \begin{cases} 1 & \text{if } x_i \text{ is related to } x_j \\ 0 & \text{otherwise} \end{cases} . \tag{4}$$

3.2.1 Graph Laplacian and Graph Energy

Let's define the degree of node i as

$$d_i = \sum_j W_{ij}. \tag{5}$$

Thus, by definition of W_{ij}, d_i measures how strong is the relation between sample $\mathbf{x}_i$ and the rest of the samples in the data set.

If $\mathbf{W}$ is the matrix of edge weights W_{ij}, and $\mathbf{D}$, a $N \times N$ diagonal matrix with diagonal elements $D_{ii} = d_i$, the graph Laplacian can be written as the matrix

$$\mathbf{L} = \mathbf{D} - \mathbf{W}. \tag{6}$$

A state vector $\boldsymbol{\phi}_j = (\phi_{j1}, \ldots, \phi_{jK})^T$ can be associated with each of the $j \in \{1, \ldots, N\}$ nodes in the graph. The graph Laplacian allows to define the energy of the graph using the quadratic form:

$$\langle \mathbf{\Phi}, \mathbf{L}\, \mathbf{\Phi} \rangle = \frac{1}{2} \sum_{i=1}^{N} \sum_{j=1}^{N} W_{ij} \|\boldsymbol{\phi}_i - \boldsymbol{\phi}_j\|^2 \tag{7}$$

$$= \frac{1}{2} \sum_{k=1}^{K} \sum_{i=1}^{N} \sum_{j=1}^{N} W_{ij} (\phi_{ik} - \phi_{jk})^2,$$

with matrix $\mathbf{\Phi} = (\boldsymbol{\phi}_1, \ldots, \boldsymbol{\phi}_N)$, where each column corresponds to the state of a node in the graph. Note that this form of energy penalizes the differences in state for nodes that are closely related (edge with a large weight W_{ij}). Then, a state of minimal energy is characterized by a homogeneous state of strongly connected nodes. This does not exclude the trivial case where all the nodes have the same state. Other energy functions, based on p-Laplacian, can be used [7]. They are similar to the quadratic form but use an exponent p, with $1 \leq p < 2$.

As will be shown in the next section, previous information about the relationships of data points, encoded in terms of a weighted or unweighted graph, can be included in the computations of the model parameters by incorporating a graph energy term, expressed as a function of the graph Laplacian. A more detailed discussion of graph energies and Laplacians can be found in [2, 3, 13–15, 25].

4 Model Computations: Hamiltonian Monte Carlo

Gibbs sampling and variational methods are the dominating computational techniques for probabilistic topic models. We adapt a Gibbs or block Gibbs sampling method where each of the variables is updated in blocks. The addition of graphs imposes an additional computational difficulty since it is not part of a conjugate family in our model. Furthermore, we found a Metropolis-Hastings algorithm to mix slowly. For these reasons, we adapted a Hamiltonian Monte Carlo sampling technique. The clearest strategy to deriving the Hamiltonian Monte Carlo method is to define potential energy functions for our distributions. This section briefly discusses Hamiltonian Monte Carlo computations and derives the potential energy of our topic models given the documents.

Our goal is to compute the topics $\boldsymbol{\psi}_k$ and topic weights b_{dk} for the given corpus. Recall Bayes' rule:

$$P(\boldsymbol{\theta} \big| \mathcal{D}) = \frac{P(\mathcal{D} \big| \boldsymbol{\theta})\ P(\boldsymbol{\theta})}{P(\mathcal{D})} \propto P(\mathcal{D} \big| \boldsymbol{\theta})\ P(\boldsymbol{\theta}), \tag{8}$$

where $\boldsymbol{\theta}$ represents the parameters, $\mathcal{D}$ the observations, $P(\mathcal{D} \big| \boldsymbol{\theta})$ the likelihood, and $P(\boldsymbol{\theta})$ the parameters' prior distribution. The main computational difficulty in finding the posterior probability distribution is the normalization $P(\mathcal{D})$. The Hamiltonian Monte Carlo computational technique allows us to sample from a posterior distribution by establishing a correspondence between such a distribution and the energy function of the probability distribution. Thus, sampling from the distribution becomes sampling from the *canonical* distribution of the system [6, 27]. The probability density for the state $\boldsymbol{q}$ under the canonical distribution is defined by

$$P(\boldsymbol{q}) \propto \exp\left(-U(\boldsymbol{q})\right), \tag{9}$$

where $U(\boldsymbol{q})$ is the *potential* energy function.

The probability $P(\boldsymbol{q})$ for the state $\boldsymbol{q}$ corresponds, in turn, to the posterior probability function we want to sample from ($P(\boldsymbol{\theta}|\mathcal{D})$). Consequently, the state $\boldsymbol{q}$ of the physical system is equivalent to the set of parameters $\boldsymbol{\theta}$ we are sampling, and computing any dynamical evolution over $\boldsymbol{q}$ immediately translates into computing a dynamical evolution for $\boldsymbol{\theta}$. At the same time, this dynamical evolution corresponds to a sampling over the parameter space. The longer the dynamical evolution simulated, the less correlated the states and, therefore, the better sampling over the parameter space.

To allow the use of dynamical methods, a *momentum* variable $\boldsymbol{p}$ is introduced [27]. This momentum variable has as many components as components in the state $\boldsymbol{q}$. The canonical distribution over the joint space of $\boldsymbol{q}$ and $\boldsymbol{p}$ is defined as

$$P(\boldsymbol{q}, \boldsymbol{p}) \propto \exp\left(-H(\boldsymbol{q}, \boldsymbol{p})\right), \tag{10}$$

where $H(\boldsymbol{q}, \boldsymbol{p}) = U(\boldsymbol{q}) + K(\boldsymbol{p})$ is the *Hamiltonian* function giving the total energy and $K(\boldsymbol{p})$ the *kinetic* energy. Typically,

$$K(\boldsymbol{p}) = \sum \frac{p_i^2}{2}. \tag{11}$$

(Here, the masses are not considered explicitly; instead, they become part of the step size in the dynamical updates).

4.1 Dynamical Updates

The system evolution is simulated by means of the *Hamiltonian* dynamics:

$$\begin{aligned}
\frac{dq_i}{dt} &= +\frac{\partial H}{\partial p_i} = p_i, \\
\frac{dp_i}{dt} &= -\frac{\partial H}{\partial q_i} = -\frac{\partial U}{\partial q_i}.
\end{aligned}$$

This dynamics is approximated by a *leapfrog* discretization using finite time steps. A leapfrog step can be expressed as

$$p_i\left(t+\frac{\epsilon}{2}\right) = p_i(t) - \frac{\epsilon}{2}\frac{\partial U}{\partial q_i}(q(t))$$
$$q_i(t+\epsilon) = q_i(t) + \epsilon\ p_i\left(t+\frac{\epsilon}{2}\right)$$
$$p_i(t+\epsilon) = p_i\left(t+\frac{\epsilon}{2}\right) - \frac{\epsilon}{2}\frac{\partial U}{\partial q_i}(q(t+\epsilon)),$$

with ϵ representing the stepsize. This leapfrog update is applied for a specified number of steps L to simulate the evolution of the system for a time $\Delta t = \epsilon L$. Since the leapfrog procedure only approximates the evolution of the Hamiltonian dynamics, a systematic error is introduced in the system update. This error is eliminated by adding a step corresponding to the Metropolis algorithm [27]. Thus, the states obtained after the Metropolis algorithm constitute the samplings over the parameter space.

Note that in order to compute the dynamical updates, it is necessary to compute the partial derivatives of U with respect to q_i.

4.2 Defining Potential Energy Functions

When the models are written in terms of probability densities, it is often easy to describe the computations in terms of energy. According to Eq. (9), the potential energy can be defined in terms of the probability density $P(\boldsymbol{\theta})$, by

$$U(\boldsymbol{\theta}) = -\log(P(\boldsymbol{\theta})). \tag{12}$$

Thus, all the previously defined generative models can be written in terms of an energy functional as described next.

4.3 Topic Modeling

The generative model for the topic modeling problem can be expressed in terms of a potential energy given by

$$\begin{aligned} U\left(\psi_{kw}, b_{dk}\middle|h_{dw}\right) &= -\log P\left(\psi_{kw}, b_{dk}\middle|h_{dw}\right) \\ &= -\log P\left(h_{dw}\middle|\psi_{kw}, b_{dk}\right) \\ &\quad -\log P\left(\psi_{kw}\middle|b_{dk}\right) \\ &= -\sum_d\sum_w\sum_k h_{dwk}\log(\psi_{kw}) \\ &\quad -\sum_d\sum_k T_{dk}\log(\pi_{dk}) \end{aligned}$$

$$
\begin{aligned}
T_{dk} &= \sum_w h_{dwk} \\
\pi_{dk} &= \frac{b_{dk}}{\sum_l b_{dl}} \\
\text{such that} \quad & \psi_{kw} \geq 0, \ \sum_w \psi_{kw} = 1, \\
& b_{dk} \geq 0, \quad \pi_{dk} \geq 0, \\
& \sum_k \pi_{dk} = 1.
\end{aligned}
$$

The model can be completed by assuming a specific form for the prior distributions. Here, we assume two different model priors leading to two different topic models.

4.3.1 Latent Dirichlet Allocation (LDA)

When the priors for the topic distribution over the documents and the word distribution over topics are assumed as Dirichlet distributions, the model corresponds to the latent Dirichlet allocation (LDA) model [5]. The Dirichlet distribution is a probability distribution over the simplex and has energy

$$
-\log P(\boldsymbol{x}) = -\sum_k (\alpha_k - 1)\log(x_k) + f(\alpha_k), \tag{13}
$$

with α_k the hyperparameters of the Dirichlet distribution. For a symmetric distribution, $\alpha_k = \alpha > 0$.

If the energies of the Dirichlet priors for the word distribution over topic k, $\boldsymbol{\psi}_k = (\psi_{k1}, \ldots, \psi_{kW})^T$ for $k \in \{1, \ldots, K\}$, and for the topic distribution over document d, $\boldsymbol{\pi}_d = (\pi_{d1}, \ldots, \pi_{dK})^T$ for $d \in \{1, \ldots, D\}$, are included in the PFA formulation, the energy for the complete model can be expressed by

$$
\begin{aligned}
U(\psi_{kw}, b_{dk}) = &- \sum_d \sum_w \sum_k h_{dwk} \log(\psi_{kw}) \\
&- \sum_d \sum_k T_{dk} \log(\pi_{dk}) \\
&- \sum_w \sum_k (\alpha - 1) \log(\psi_{kw}) \\
&- \sum_d \sum_k (\beta - 1) \log(\pi_{dk}),
\end{aligned}
$$

with restrictions $\psi_{kw} \geq 0$, $\pi_{dk} \geq 0$, $\sum_k \pi_{dk} = 1$, and $\sum_w \psi_{kw} = 1$.

The advantage of using Dirichlet priors is that due to the Dirichlet-multinomial conjugacy, $\boldsymbol{\psi}$ and $\boldsymbol{\pi}$ can be marginalized, which simplifies the computations.

4.3.2 Graph-Directed Topic Modeling

Analogously to the LDA model, we assume a Dirichlet distribution as a prior for the word distribution over topics, $\boldsymbol{\psi}$. This allows to exploit the Dirichlet-multinomial conjugacy for $\boldsymbol{\psi}$.

In contrast, for the case of the prior of topic distribution over documents $\boldsymbol{\pi}$, we replace the Dirichlet prior by a graphical model encoding prior information about the documents. In particular, we want that documents with strong connections end up having similar topic distributions. This bias over $\boldsymbol{\pi}$ can be enforced by introducing a term that measures how close are the topic distributions of strongly related documents. As described before, such term can be written via the graph Laplacian. We use the state $\boldsymbol{\phi}_d = (\phi_{d1}, \ldots, \phi_{dK})^T$ for the state of node d in the topic modeling problem and associated matrix $\boldsymbol{\Phi} = (\boldsymbol{\phi}_1, \ldots, \boldsymbol{\phi}_N)$. In order for these states to represent valid topic distributions, we use an approach similar to Mimno et al. [26]. With the help of the logistic function, we map the state $\boldsymbol{\phi}_d$ of node d to the multinomial parameter π_{dk} that describes the probability of topic k being included in document d:

$$\pi_{dk} = \frac{b_{dk}}{\sum_l b_{dl}} = \frac{\exp(\phi_{dk})}{\sum_l \exp(\phi_{dl})}. \tag{14}$$

Therefore, the potential energy with Dirichlet prior for word distribution over topics and graph-energy term for topic distribution over documents can be written as

$$\begin{aligned} U\left(\psi_{kw}, b_{dk}\right) = &- \sum_d \sum_w \sum_k h_{dwk} \log(\psi_{kw}) \\ &- \sum_d \sum_k T_{dk} \log(\pi_{dk}) \\ &- \sum_w \sum_k (\alpha - 1) \log(\psi_{kw}) \\ &+ \langle \boldsymbol{\Phi}, \mathbf{L}\, \boldsymbol{\Phi} \rangle \,. \end{aligned} \tag{15}$$

For the graph-based energy term, we exploit the quadratic form (Eq. 8) to define a probability density over $\boldsymbol{\Phi}$ by $P(\boldsymbol{\Phi}|\mathbf{L}) = \exp(-\langle \boldsymbol{\Phi}, \mathbf{L}\boldsymbol{\Phi} \rangle - \log Z(\mathbf{W}))$. Note that since we know the weight matrix $\mathbf{W}$, we are not interested in learning it and we do not need to know the partition function $Z(\mathbf{W})$ explicitly.

The final computations use block Gibbs sampling for the variables z_l and $\boldsymbol{\psi}_k$ as in [5] and Hamiltonian Monte Carlo sampling for the variables $\boldsymbol{\pi}_k$ using the leapfrog technique in Sect. 4 with the energy function in Eq. (15).

5 Results

To demonstrate the utility of the graph-directed topic modeling, we apply the formulation to two data sets as described next. In both cases, we compare the results obtained with the LDA model [5].

5.1 Toy Example

A simple data set is constructed following Griffiths and Steyvers work [17]. A set of ten topics $\boldsymbol{\psi}_k$, $k = 1, \ldots, 10$ corresponding to horizontal and vertical bars in a 5×5 grid is defined (Fig. 1). Random combinations of these topics are constructed generating different documents. Each pixel in the image corresponds to a unique word. A sample of documents can be found in Fig. 2.

Recovered topics for LDA model and the graph-directed topic model can be seen in Figs. 3 and 4, respectively. A comparison of the topic mixture per document can be found in Fig. 5. In this case, an unweighted graph is constructed arbitrary such that documents 1–50 and documents 51–100 are connected.

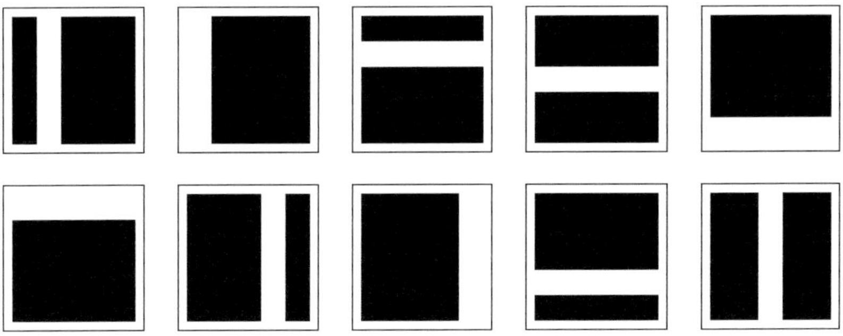

Fig. 1 Graphical representation of ten topics, each containing 25 pixels in a 5 x 5 grid. Each pixel in the image corresponds to a unique "word"

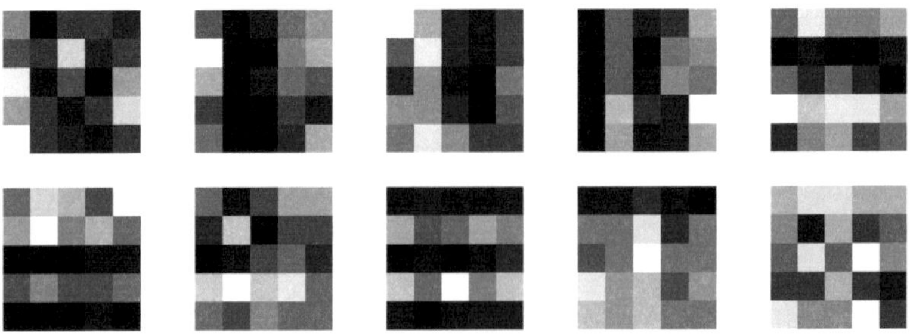

Fig. 2 Documents: 25 words represented as 5 x 5 pixel images. These correspond to random weighted combinations of the topics in the previous figure. White, high count of word; black, 0 count of word

Fig. 3 Topics calculated with the LDA model. The original topics are essentially recovered by the LDA model

Fig. 4 Topics calculated with the graph-directed topic model. The original topics, or linear combinations, are recovered by the graph-directed topic model

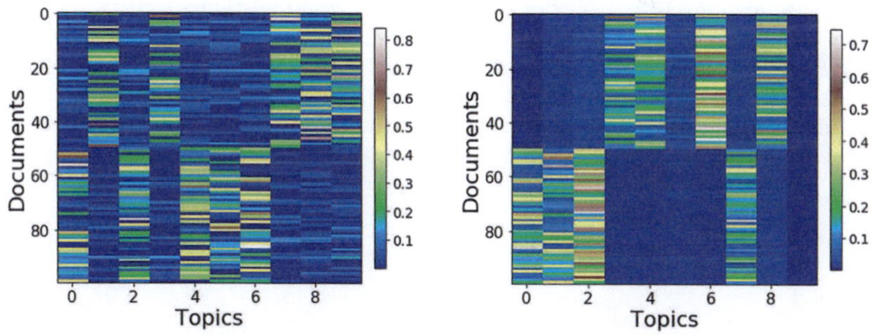

(a) LDA (b) LDA + Graph

Fig. 5 Toy example: topic mixture per document. Colors indicate weight of topic in the document. (**a**) LDA model. (**b**) Graph-directed topic model. The graph-directed topic model recovers the same topics for documents 1–50 and documents 51–100, as expected. The LDA model recovers more noisy distributions for the 1–50 and 51–100 subgroups. The graph-directed topic model does not make use of topic 10, the least expressive of the topics found (see Fig. 4)

5.2 Enron Data

A subset of Enron data set is used for the topic model task. We note that stemming can influence the discovered topic models [34] and stemming was not used in this work. In this case, an unweighted graph is constructed such that documents that share the same folder are connected. A number of topics $K = 30$, roughly the double of available folders, are specified (Fig. 6).

As noted in [38], one of the problems of topic models is that most frequent words tend to dominate all topics. However, the graph-directed representation is able to construct representations that are more descriptive and more robust to frequent words. Likewise, the topic distribution is more balanced through the corpus of documents (Fig. 7). Lists of words for the two most important topics in LDA and graph-directed models are displayed in Fig. 8. The lists include the probability of the word in the topic.

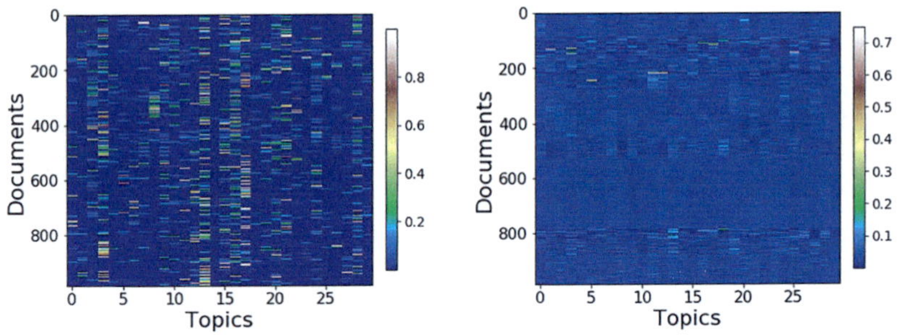

(a) LDA (b) LDA + Graph

Fig. 6 Enron data: topic mixture per document. Colors indicate weight of topic in the document. (**a**) LDA model. (**b**) Graph-directed topic model. The graph-directed topic model tends to recover more unique mixtures of topics per document, i.e., mixtures that include less topics, with relatively more weight per topic. The LDA model recovers more noisy topic distributions per document

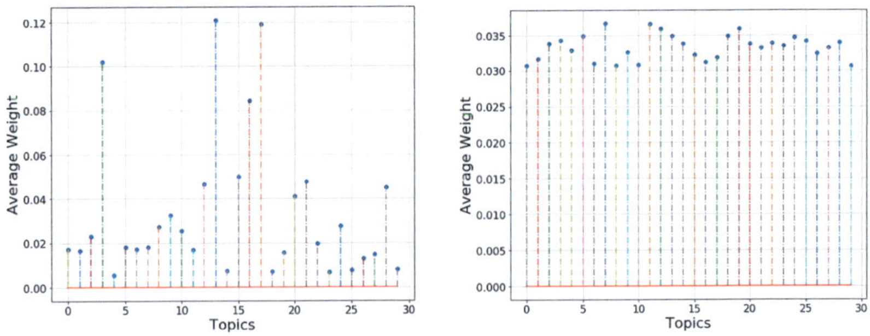

(a) LDA (b) LDA + Graph

Fig. 7 Enron data: average weight of topic in the corpus. (**a**) LDA model. (**b**) Graph-directed topic model. All the topics have similar influence in the corpus when using the graph-directed topic model. In contrast, few topics dominate the corpus when using the LDA model

LDA		LDA + Graph	
Word	Word Probability	Word	Word Probability
please	0.0332594	enron	0.0828999
louise	0.0263467	dynegy	0.0347854
will	0.0260858	said	0.0331599
thanks	0.0208687	company	0.0315345
pm	0.0203469	stock	0.0156047

(a) Most probable topic

LDA		LDA + Graph	
Word	Word Probability	Word	Word Probability
will	0.0287065	ceo	0.0320795
get	0.0172239	president	0.0303183
louise	0.016661	chairman	0.0179897
message	0.0137341	new	0.0174865
week	0.0136215	year	0.0142156

(b) Second most probable topic

LDA		LDA + Graph	
Word	Word Probability	Word	Word Probability
will	0.0187032	don	0.0376162
also	0.0092416	tax	0.0365836
may	0.00902156	use	0.030388
risk	0.0074813	may	0.0259625
meeting	0.0073346	reserve	0.0215371

(c) Third most probable topic

LDA		LDA + Graph	
Word	Word Probability	Word	Word Probability
message	0.0275497	energy	0.0343342
original	0.0262532	will	0.029487
please	0.0178263	power	0.0281406
will	0.0178263	gas	0.0195234
louise	0.01707	california	0.0164266

(d) Fourth most probable topic

Fig. 8 Enron data: word probabilities for the most important topics. LDA topics use more generic words. Graph-directed topics give more probability to specialized words, hence yielding more insightful representations

6 Conclusion

Graph-directed representations for the unsupervised learning methods of topic modeling and dictionary learning problems have been implemented. A common approach to topic modeling and dictionary learning is to include a sparsity inducing prior into the model, such as the Dirichlet prior in a Bayesian setting or a ℓ_0 or ℓ_1 regularization term in an optimization approach. Without any other prior information, the sparsity prior can yield insightful representations.

Our results illustrate how a graph enforces known binary relationships in the data set, such that strongly connected data samples yield more informative representations. For example, by including the information that emails are sorted into folders by subjects, a more informative topic representation can be obtained. As Fig. 7 illustrates, even though we are using a Bayesian sparsity inducing prior, the graph learns a less sparse model. However, as Fig. 8 shows, this model is more representative of the information content.

In this work, we used a quadratic energy function in Eq. (8) in order to integrate the graphical model with our topic models. A possible extension of this work is to use other graph ℓ_p energy functions by replacing the square of the difference with the pth power of the difference. The computational complexity remains the same for $1 <= p < \infty$; however, it is well known that graph segmentation is often improved when $p = 1$ [3].

A further extension of this work is the integration of graphical models with attention [37]. Note that Eq. (14) is the commonly used softmax function, which is often used to define attention. Using Eq. (15), it is straightforward to include a graphical model with attention.

Competing Interests The authors have no conflicts of interest to declare that are relevant to the content of this chapter.

References

1. Ambwani, G., Davis, A.R.: Contextually-mediated semantic similarity graphs for topic segmentation. In: Proceedings of the 2010 Workshop on Graph-based Methods for Natural Language Processing, pp. 60–68. Association for Computational Linguistics (2010)
2. Bertozzi, A.L., Flenner, A.: Diffuse interface models on graphs for classification of high dimensional data. Multiscale Model. Simul. **10**(3), 1090–1118 (2012)
3. Bertozzi, A.L., Flenner, A.: Diffuse interface models on graphs for classification of high dimensional data. SIAM Rev. **58**(2), 293–328 (2016)
4. Blei, D.M., Lafferty, J.D.: Dynamic topic models. In: Proceedings of the 23rd International Conference on Machine Learning, pp. 113–120. ACM (2006)
5. Blei, D.M., Ng, A.Y., Jordan, M.I.: Latent Dirichlet allocation. J. Mach. Learn. Res. **3**, 993–1022 (2003)
6. Brooks, S., Gelman, A., Jones, G., Meng, X.L.: Handbook of Markov Chain Monte Carlo. CRC Press, Boca Raton, FL (2011)
7. Bühler, T., Hein, M.: Spectral clustering based on the graph p-Laplacian. In: Bottou, L., Littman, M. (eds.) Proceedings of the 26th International Conference on Machine Learning, pp. 81–88. Omnipress, Montreal, Canada (2009)
8. Cevher, V., Indyk, P., Carin, L., Baraniuk, R.G.: Sparse signal recovery and acquisition with graphical models. IEEE Signal Process. Mag. **27**(6), 92–103 (2010)
9. Chen, X., Qi, Y., Bai, B., Lin, Q., Carbonell, J.G.: Sparse latent semantic analysis. In: Proceedings of the 2011 SIAM International Conference on Data Mining, pp. 474–485. SIAM (2011)
10. Chen, Z., Ji, H.: Graph-based clustering for computational linguistics: A survey. In: Proceedings of the 2010 Workshop on Graph-Based Methods for Natural Language Processing, pp. 1–9. Association for Computational Linguistics (2010)
11. Cichocki, A., Zdunek, R., Phan, A.H., Amari, S.i.: Nonnegative Matrix and Tensor Factorizations: Applications to Exploratory Multi-way Data Analysis and Blind Source Separation. Wiley.com (2009)
12. De Nicola, G., Sischka, B., Kauermann, G.: Mixture models and networks: The stochastic blockmodel. Statist. Model. **22**(1-2), 67–94 (2022)
13. Garcia-Cardona, C., Flenner, A., Percus, A.G.: Multiclass diffuse interface models for semi-supervised learning on graphs. In: Proceedings of the 2nd International Conference on Pattern Recognition Applications and Methods. SciTePress (2013)
14. Garcia-Cardona, C., Flenner, A., Percus, A.G.: Multiclass semi-supervised learning on graphs using ginzburg-landau functional minimization. In: Pattern Recognition Applications and Methods, pp. 119–135. Springer (2015)
15. Garcia-Cardona, C., Merkurjev, E., Bertozzi, A.L., Flenner, A., Percus, A.G.: Multiclass data segmentation using diffuse interface methods on graphs. IEEE Trans. Pattern Anal. Mach. Intell. **36**(8), 1600–1613 (2014)
16. Gerlach, M., Pexioto, T., Altmann, E.: A network approach to topic models. Sci. Adv. **4**(7), (2018). https://doi.org/10.1126/sciadv.aaq1360
17. Griffiths, T.L., Steyvers, M.: Finding scientific topics. PNAS **101**, 5228–5235 (2004)
18. Gutiérrez, E.D., Shutova, E., Lichtenstein, P., de Melo, G., Gilardi, L.: Detecting cross-cultural differences using a multilingual topic model. Trans. Assoc. Comput. Linguist. **4**, 47–60 (2016). https://www.transacl.org/ojs/index.php/tacl/article/view/755
19. Kim, J., Kim, D., Oh, A.: Joint modeling of topics, citations, and topical authority in academic corpora. Trans. Assoc. Comput. Linguist. **5**, 191–204 (2017). https://www.transacl.org/ojs/index.php/tacl/article/view/1061
20. King, B., Jha, R., Radev, D.R.: Heterogeneous networks and their applications: Scientometrics, name disambiguation, and topic modeling. Trans. Assoc. Comput. Linguist. **2**, 1–14 (2014). https://www.transacl.org/ojs/index.php/tacl/article/view/110
21. Kingman, J.F.C.: Poisson Processes, vol. 3. Oxford University Press, Oxford (1992)

22. Kolda, T.G., Bader, B.W.: Tensor decompositions and applications. SIAM Rev. **51**(3), 455–500 (2009)
23. Lafferty, J.D., Blei, D.M.: Correlated topic models. In: Advances in Neural Information Processing Systems, pp. 147–154 (2005)
24. Li, L., Zhou, M., Sapiro, G., Carin, L.: On the integration of topic modeling and dictionary learning. In: Proceedings of the 28th International Conference on Machine Learning (ICML-11), pp. 625–632 (2011)
25. Merkurjev, E., Garcia-Cardona, C., Bertozzi, A.L., Flenner, A., Percus, A.G.: Diffuse interface methods for multiclass segmentation of high-dimensional data. Appl. Math. Lett. **33**, 29–34 (2014)
26. Mimno, D., Wallach, H.M., McCallum, A.: Gibbs sampling for logistic normal topic models with graph-based priors. In: NIPS Workshop on Analyzing Graphs. Whistler, BC (2008)
27. Neal, R.M.: Bayesian Learning for Neural Networks, Lecture Notes in Statistics, vol. 118. Springer, New York (1996)
28. Nguyen, D.Q., Billingsley, R., Du, L., Johnson, M.: Improving topic models with latent feature word representations. Trans. Assoc. Comput. Linguist. **3**, 299–313 (2015). https://www.transacl.org/ojs/index.php/tacl/article/view/582
29. Paisley, J., Carin, L.: Hidden markov models with stick-breaking priors. IEEE Trans. Signal Process. **57**(10), 3905–3917 (2009)
30. Paisley, J.W., Blei, D.M., Jordan, M.I.: Stick-breaking beta processes and the poisson process. In: International Conference on Artificial Intelligence and Statistics, pp. 850–858 (2012)
31. Paul, M.J., Dredze, M.: Sprite: Generalizing topic models with structured priors. Trans. Assoc. Comput. Linguist. **3**, 43–57 (2015). https://www.transacl.org/ojs/index.php/tacl/article/view/403
32. Rue, H., Held, L.: Gaussian Markov Random Fields: Theory and Applications. CRC Press, Boca Raton, FL (2005)
33. Salton, G., Buckley, C.: Term-weighting approaches in automatic text retrieval. Inf. Process. Manag. **24**(5), 513–523 (1988)
34. Schofield, A., Mimno, D.: Comparing apples to apple: The effects of stemmers on topic models. Trans. Assoc. Comput. Linguist. **4**, 287–300 (2016). https://www.transacl.org/ojs/index.php/tacl/article/view/868
35. Seeker, W., Çetinoğlu, Ö.: A graph-based lattice dependency parser for joint morphological segmentation and syntactic analysis. Trans. Assoc. Comput. Linguist. **3**, 359–373 (2015). https://www.transacl.org/ojs/index.php/tacl/article/view/631
36. Teh, Y.W., Görür, D., Ghahramani, Z.: Stick-breaking construction for the indian buffet process. In: Proceedings of the International Conference on Artificial Intelligence and Statistics (2007)
37. Vaswani, A., Shazeer, N., Parmar, N., Uszkoreit, J., Jones, L., Gomez, A.N., Kaiser, L.u., Polosukhin, I.: Attention is all you need. In: Guyon, I., Luxburg, U.V., Bengio, S., Wallach, H., Fergus, R., Vishwanathan, S., Garnett, R. (eds.) Advances in Neural Information Processing Systems, vol. 30. Curran Associates, Inc. (2017)
38. Wallach, H.M., Mimno, D.M., McCallum, A.: Rethinking LDA: Why priors matter. In: Bengio, Y., Schuurmans, D., Lafferty, J., Williams, C., Culotta, A. (eds.) Advances in Neural Information Processing Systems 22, pp. 1973–1981. Curran Associates, Inc. (2009). http://papers.nips.cc/paper/3854-rethinking-lda-why-priors-matter.pdf
39. Wang, J., Bansal, M., Gimpel, K., Ziebart, B.D., Yu, C.T.: A sense-topic model for word sense induction with unsupervised data enrichment. Trans. Assoc. Comput. Linguist. **3**, 59–71 (2015). https://www.transacl.org/ojs/index.php/tacl/article/view/485
40. Wedel, M., Böckenholt, U., Kamakura, W.A.: Factor models for multivariate count data. J. Multivariate Anal. **87**, 356–369 (2003)
41. Wright, J., Ma, Y., Mairal, J., Sapiro, G., Huang, T.S., Yan, S.: Sparse representation for computer vision and pattern recognition. Proc. IEEE **98**(6), 1031–1044 (2010). https://doi.org/10.1109/JPROC.2010.2044470

42. Wright, J., Yang, A.Y., Ganesh, A., Sastry, S.S., Ma, Y.: Robust face recognition via sparse representation. IEEE Trans. Pattern Anal. Mach. Intell. **31**(2), 210–227 (2009). https://doi.org/10.1109/TPAMI.2008.79
43. Zhou, M., Hannah, L.A., Dunson, D.B., Carin, L.: Beta-negative binomial process and poisson factor analysis. J. Mach. Learn. Res. **22**, 1462–1471 (2012)

Linear Independent Component Analysis in Wasserstein Space

Shiying Li, Caroline Moosmüller, and Chuxiangbo Wang

1 Introduction

Independent component analysis (ICA) is a computational and statistical technique used to uncover independent components from multivariate data, also known as blind source separation [8, 18, 19]. It was first introduced in [1] and has gained much interest since then; see, e.g., [2, 13, 17]. Specifically, [7] highlighted the potential of ICA in mathematics and statistics. ICA has since become an essential tool in various fields, including signal separation of biological data [10, 28], MRI data [26], and audio and image noise reduction [16, 24].

The classical linear ICA problem assumes n independent random variables, which have been "mixed" by the application of an orthogonal matrix, and one only has access to N observations of this mixing process. From these observations, the aim is to identify the independent components and the matrix. This problem is usually formulated in Euclidean space, i.e., the independent components and the observations are elements of some $\mathbb{R}^k$. In this paper, we study a version of linear ICA in the Wasserstein space, which is the space of probability measures; see [33]. In particular, we assume that the observed data consists of probability measures or point-clouds, which have been obtained by a linear mixing through Euclidean independent components. This setup is motivated by applications in which an instance of data is not naturally interpreted as a vector in some $\mathbb{R}^k$, but rather as a probability measure or point-cloud. Examples include imaging data [29], text documents [36], gene expression data [5, 22], and flow cytometry [3, 37].

S. Li
Department of Mathematics, University of Nebraska at Lincoln, Lincoln, NE, USA
e-mail: sli82@unl.edu

C. Moosmüller (✉) · C. Wang
Department of Mathematics, University of North Carolina at Chapel Hill, Chapel Hill, NC, USA
e-mail: cmoosm@unc.edu; chuxianw@unc.edu

C. Garcia-Cardona, H. Lee (eds.), *Advances in Data Science*, Association for Women in Mathematics Series 37, https://doi.org/10.1007/978-3-031-87804-6_6

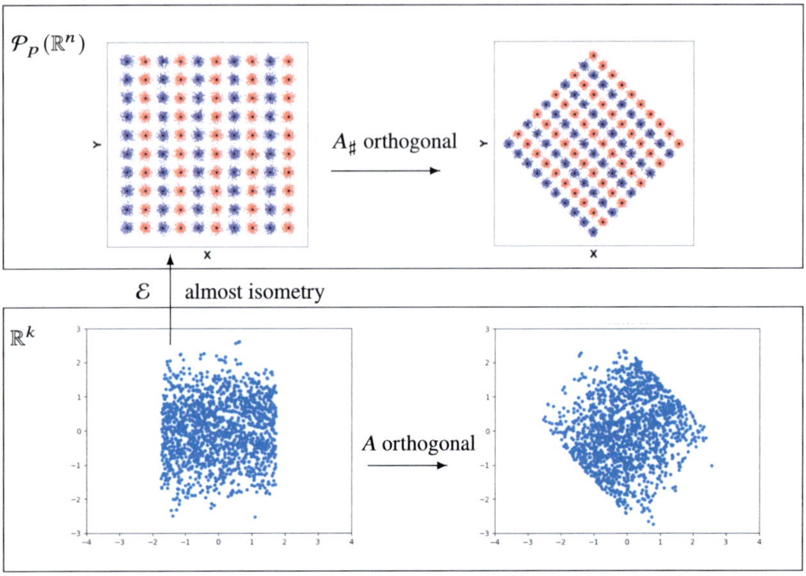

Fig. 1 Setup for linear ICA in Wasserstein space. *Bottom panel ("classical" linear setting in* $\mathbb{R}^k$*)*: Independent components are drawn from $\mathcal{U}(-\sqrt{3}, \sqrt{3}) \times \mathcal{N}(0, 1)$ (left plot). An unknown orthogonal transformation A is applied (right plot). The eigenvectors of the Laplacian built from the observed data (right plot) are used to uncover the independent components [31]. *Top panel (proposed setting in* $\mathcal{P}_p(\mathbb{R}^n)$*)*: An almost isometric map $\mathcal{E} : \mathbb{R}^k \to \mathcal{P}_p(\mathbb{R}^n)$ is assumed; in this example, the map is $\mathcal{E}(\omega) = \mathcal{N}(\omega, cI)$ for $c > 0$ fixed. An unknown push-forward operator $A_\sharp$ based on orthogonal transformation A is applied to obtain the observable data in Wasserstein space (right plot). The eigenvectors of the Laplacian built from the observed data (right plot) with Wasserstein distances are used to uncover the independent components (bottom left); see Sect. 3. *Note:* The plots in the top panel are sketches for visualization purposes, i.e., we sampled a uniform grid with only a small number of points so that the Gaussians are visible. The blue-red coloring scheme is for visualization purposes only. For the actual numerical experiments, we sampled the means of the Gaussian from the bottom left plot; see Sect. 4 for details

While there exists a large body of literature on linear ICA (in the Euclidean setting), we follow the ideas of [31], which uses the eigenvectors of a graph Laplacian built from the observed data to identify the independent components and the mixing matrix. This method naturally adapts to our setting, as we only need to reinterpret the graph Laplacian for point-cloud data. Essentially, we replace the Euclidean distance by the Wasserstein distance when building the graph Laplacian, which has shown success in other methods as well [5, 21–23, 35].

The contributions of this paper are twofold. We first describe a natural setting for linear ICA in Wasserstein space, where the observed data consists of probability measures. This idea mimics the classical linear ICA in Euclidean space and is outlined in Fig. 1. We then show that our method is successful in identifying the independent components as long as the observed point-cloud data is "close to" (almost isometric to) Euclidean data by using results on eigenvector perturbations.

Here, we use a version of the classical Davis-Kahan theorem to derive eigenvector perturbation results [9, 34]. Improved bounds are possible for data with more structures; see, e.g., [11]. We present toy examples where the observed data are rotated Gaussians, and the independent components are their means.

The paper is organized as follows. Section 2 presents the preliminaries on spectral linear ICA as introduced by [31] and gives a basic introduction to optimal transport and the Wasserstein distance. In Section 3 we show a natural setting for linear ICA in the Wasserstein space and provide the main result on recovery of the independent components in the almost isometric setting. Section 4 contains numerical toy examples to showcase our proposed method.

2 Preliminaries

2.1 Linear ICA via the Graph Laplacian

For the linear ICA problem, we follow the setup and results from [31]. Here, we briefly summarize the main results needed.

The linear ICA problem is formulated as follows. Let $S = (S_1, S_2, \ldots, S_n)$ be n unknown independent components (random variables) with zero mean and unit variance. Let $A \in \mathbb{R}^{n\times n}$ be an unknown orthogonal mixing matrix. Consider observations of these random variables, denoted as $\bar{S}_i \in \mathbb{R}^N$. The observed data under the mixing matrix A is given by

$$X = A\bar{S}^T. \tag{1}$$

To recover the independent components S from X, [31] interprets the observed data points $x_1, \ldots, x_N$ (column vectors of X) as the nodes of a graph. The weights of this graph are defined by

$$W_{i,j} = e^{\frac{-\|x_i - x_j\|^2}{2h}}, \tag{2}$$

where $\|\cdot\|$ denotes the Euclidean distance between x_i and x_j and h is the width parameter of the kernel. From this weight matrix, the normalized graph Laplacian,

$$L = I - D^{-1}W, \tag{3}$$

is constructed, where D is the diagonal degree matrix defined by $D = \mathrm{diag}\left(\sum_{j=1}^N W_{i,j}\right)$.

It is proved in [31] that the eigenvectors of the graph Laplacian approximate the independent components S_i. The main argument concerns the convergence of the graph Laplacian L to the backward Fokker-Planck operator as the number of samples $N \to \infty$ and the fact that the Fokker-Planck operator separates into n one-dimensional operators when S_i are independent; see [31].

2.2 Optimal Transport and Wasserstein Space

In this paper, we focus on recovering the underlying independent components when probability measures or point-clouds undergo a linear mixing transformation. The optimal transport (OT) theory [20, 25] provides a natural framework for comparing probability measures. We introduce the necessary background here and refer the readers to [30, 33] for a thorough treatment of the subject. For an overview of the computational aspects of OT, see [27].

Let $\mathcal{P}(\mathbb{R}^n)$ be a set of Borel probability measures on $\mathbb{R}^n$. Consider the space of probability measures with bounded p ($p \geq 1$) moments, denoted by $\mathcal{P}_p(\mathbb{R}^n)$ where

$$\mathcal{P}_p(\mathbb{R}^n) := \left\{ \mu \in \mathcal{P}(\mathbb{R}^n) : \int_{\mathbb{R}^n} \|x\|^p \, d\mu(x) < +\infty \right\}. \tag{4}$$

For two probability measures $\alpha, \beta \in \mathcal{P}_p(\mathbb{R}^n)$, the 2-Wasserstein distance is defined as

$$W_p(\alpha, \beta) = \min_{\pi \in \Pi(\alpha,\beta)} \left(\int_{\mathbb{R}^n \times \mathbb{R}^n} \|x - y\|^p \, d\pi(x, y) \right)^{\frac{1}{p}}, \tag{5}$$

where $\Pi(\alpha, \beta)$ denotes the set of transport plans (couplings) between α and β, i.e., $\pi \in \Pi(\alpha, \beta)$ is a probability measure on $\mathbb{R}^n \times \mathbb{R}^n$ with first marginal α and second marginal β. Here, we refer to the metric space $(\mathcal{P}_p(\mathbb{R}^n), W_p)$ as the Wasserstein space.

In the case when α is absolutely continuous, the minimizer π^* of Eq. (5) is unique and of the form $(\mathrm{id}, T^*)_\sharp \alpha$, where T^* is called the optimal transport map between α and β (see, e.g., [32, Theorem 2.12]). Here, $\sharp$ denotes push-forward operation between probability measures. Specifically, given $g : \mathbb{R}^n \to \mathbb{R}^m$, $g_\sharp \alpha$, often referred to as the push-forward measure of α by g, is a measure in $\mathcal{P}(\mathbb{R}^m)$ defined via

$$g_\sharp \alpha(B) := \alpha(g^{-1}(B)), \ \forall \text{ Borel sets } B \subseteq \mathbb{R}^m. \tag{6}$$

3 Linear ICA in Wasserstein Space

We now describe a natural setup for linear ICA in the Wasserstein space. This is similar to ideas related to manifold learning in the Wasserstein space; see, e.g., [6, 14, 15]. As introduced in Sect. 2.1, the linear ICA problem for Euclidean data can be solved by analyzing the spectral properties of the normalized graph Laplacian L (Eq. (3)). In the context of Wasserstein space, this process involves analyzing a Wasserstein-based graph Laplacian by leveraging the optimal transport (OT) framework and, more specifically, using the 2-Wasserstein distance to compare probability measures.

Consider $\mathcal{E} : \Omega \to \mathcal{P}_p(\mathbb{R}^n)$ where $\Omega \subseteq \mathbb{R}^n$ represents a set of governing parameters and the map $\mathcal{E}$ describes a nonlinear process of generating probability measures from the parameters. Assume $\{\omega_j\}_{j=1}^N$ are the underlying parameters sampled from n independent components $S = (S_1, \ldots, S_n)$ from Ω and consider observed data of the form $\beta_j := A_\sharp \mathcal{E}(\omega_j), j = 1, \ldots, N$. The probability measures β_j are obtained via the push-forward of $\mathcal{E}(\omega_j)$ by a mixing orthogonal transformation A. Throughout this paper, we abuse notation by letting A represent both the orthogonal matrix and the linear transformation it induces. As the Euclidean distance is invariant under orthogonal transformations, the W_p distance is invariant under the push-forward via orthogonal transformations. See the Appendix for the proof of the following:

Lemma 1 *Let $p \geq 1$. Let $\alpha, \beta \in \mathcal{P}_p(\mathbb{R}^n)$ and A be a $n \times n$ orthogonal matrix. Then, $W_p(\alpha, \sigma) = W_p(A_\sharp\alpha, A_\sharp\sigma)$ for any $\alpha, \sigma \in \mathcal{P}_p(\mathbb{R}^n)$.*

As in the linear setting (Sect. 2.1), the task is to uncover the independent components $S_1, \ldots, S_n$ from the observed data $\{\beta_j\}_{j=1}^N$. See Fig. 1 for an overview of this construction. We will focus on the following settings:

(i) (Almost isometric $\mathcal{E}$): there exists some $\eta > 0$ such that

$$\left| W_p^2(\mathcal{E}(\omega), \mathcal{E}(\kappa)) - \|\omega - \kappa\|^2 \right| \leq \eta, \; \forall\, \omega, \kappa \in \Omega$$

(ii) (Special case: isometric $\mathcal{E}$): $W_p(\mathcal{E}(\omega), \mathcal{E}(\kappa)) = \|\omega - \kappa\|, \forall\, \omega, \kappa \in \Omega$.

Example 1 To illustrate the Wasserstein ICA setup, we give a basic example, which is discussed in more detail in Sect. 4 and is visualized in Fig. 1. Denote by $\mathcal{N}(m, \Sigma)$ the Gaussian in $\mathbb{R}^n$ with mean m and covariance Σ. A possible map $\mathcal{E}$ is $\omega \mapsto \mathcal{N}(\omega, cI)$, where $c > 0$ is fixed. This is a simple way of generating probability measures from parameters. The observed data would then be $A_\sharp\mathcal{N}(\omega, cI) = \mathcal{N}(A\omega, cI)$, i.e., pushing these Gaussians by an orthogonal transformation is equivalent to applying A to their means. Moreover, in this case, $(\Omega, \|\cdot\|)$ is isometric to $(\mathcal{E}(\Omega), W_2)$, where $\Omega \subseteq \mathbb{R}^n$.

Remark 1 We assume that $\Omega \subseteq \mathbb{R}^n$ and that support space for the probability measures is $\mathbb{R}^n$. It is not necessary for those spaces to have the same dimension n; this is mostly for convenience of presentation. The results that follow still hold if the dimensions differ.

To recover the independent components $S_1, \ldots, S_n$, the idea is to utilize the graph Laplacian with Wasserstein distances between the observed probability measures $\{\beta_j\}_{j=1}^N$ rather than the Euclidean distance used in Eq. (3). This is natural since we are dealing with objects in the metric space $(\mathcal{P}_p(\mathbb{R}^n), W_p)$. In particular, we construct the normalized graph Laplacian:

$$L_{W_p} = I - \widetilde{D}^{-1}\widetilde{W}, \tag{7}$$

where

$$\widetilde{W}_{ij} = e^{\frac{-W_p(\beta_i,\beta_j)^2}{2h}}, \tag{8}$$

and $\widetilde{D}$ is the degree matrix associated with $\widetilde{W}$.

Our goal is to understand to which extent the eigenvectors of L_{W_p} approximate the independent components under the assumption that the parameter space $(\Omega, \|\cdot\|)$ is "almost" isometric to $(\mathcal{E}(\Omega), W_p)$. The recovery result follows from combining eigenvector perturbation results with results from [31]. The main theorem of this paper concerns the eigenvector perturbation under an almost isometry assumption.

Theorem 1 (Almost isometric $\mathcal{E}$) *Let $p \geq 1$. Let $S_1, \ldots, S_n$ be (real-valued) independent random variables and let A denote an $n \times n$ orthogonal mixing matrix or the orthogonal transformation it induces. Assume that $\Omega \subseteq \mathbb{R}^n$ such that $S \in \Omega$,*[1] *where $S = (S_1, \ldots, S_n)$. Let $\mathcal{E} : \Omega \to \mathcal{P}_p(\mathbb{R}^n)$. Assume that there exists $\eta \geq 0$ such that*

$$\left| W_p^2(\mathcal{E}(\omega), \mathcal{E}(\kappa)) - \|\omega - \kappa\|^2 \right| \leq \eta, \ \forall\, \omega, \kappa \in \Omega. \tag{9}$$

Let $\{\omega_j\}_{j=1}^N$ be N instances of S and $\beta_j := A_\sharp \mathcal{E}(\omega_j)$. Let L and L_{W_p} be the normalized graph Laplacian associated with $\{A\omega_j\}_{j=1}^N$ (see Eq. (3)) and associated with $\{\beta_j\}_{j=1}^N$ (see Eq. (7)), respectively. Let $\lambda_1 \leq \ldots \leq \lambda_N$ and $\widetilde{\lambda}_1 \leq \ldots \leq \widetilde{\lambda}_N$ be the eigenvalues of L and L_{W_p}, respectively. Fix $1 \leq j \leq N$, and assume that $\delta_j := \min\{\lambda_j - \lambda_{j-1}, \lambda_{j+1} - \lambda_j\} > 0$. Then, for the eigenvectors ϕ_j and $\widetilde{\phi}_j$ satisfying $L\phi_j = \lambda_j \phi_j$ and $L_{W_p}\widetilde{\phi}_j = \widetilde{\lambda}_j \widetilde{\phi}_j$, the following holds:

$$\cos\angle(\phi_j, \widetilde{\phi}_j) \geq 1 - \frac{\varepsilon_{max}^{1/2}}{2} r_D \left(b + 2^{3/2}\delta_j^{-1}\left(a\varepsilon_{min}^{-1} r_D + b r_D \varepsilon_{min}^{-1/2} + b r_D^{1/2}\right)\right)^2. \tag{10}$$

Here, $\varepsilon_{min} = e^{-\frac{\eta}{2h}}$ and $\varepsilon_{max} = e^{\frac{\eta}{2h}}$, $a = \max\{|\varepsilon_{max} - 1|, |\varepsilon_{min} - 1|\}$, $b = \max\{|\varepsilon_{max}^{-1/2} - 1|, |\varepsilon_{min}^{-1/2} - 1|\}$, and $r_D = \frac{D_{max}}{D_{min}}$ with $D_{max} = \max_i D_{ii}$, $D_{min} = \min_i D_{ii}$. Here, D is the degree matrix associated with $\{\omega_j\}_{j=1}^N$.

Proof We start by comparing the distances used in the kernel in the Wasserstein and Euclidean settings (see Eqs. (8) and (2)). The former uses the Wasserstein distance of the observed measures, i.e., $W_p^2(\beta_i, \beta_j)$, while the latter uses the Euclidean distances between the mixed parameters, i.e., $\|A\omega_i - A\omega_j\|^2$. Since A is an orthogonal matrix, by Lemma 1, we have

$$W_p(\beta_i, \beta_j) = W_p(A_\sharp \mathcal{E}(\omega_i), A_\sharp \mathcal{E}(\omega_j)) = W_p(\mathcal{E}(\omega_i), \mathcal{E}(\omega_j)).$$

[1] Here, we abuse notation and do not differentiate the measurable function S from its function value. $S \in \Omega$ means that the function values of S are in the set Ω.

Since $\|A\omega_i - A\omega_j\| = \|\omega_i - \omega_j\|$, it follows from Eq. (9) that

$$\left| W_p^2(\beta_i, \beta_j) - \left\|A\omega_i - A\omega_j\right\|^2 \right| \leq \eta, \ \forall\, i, j = 1, \ldots, N. \tag{11}$$

Denoting by $\widetilde{W}$, W the weight matrices associated with L_{W_p} and L, respectively, we have that

$$e^{-\eta/2h} \leq \frac{\widetilde{W}_{ij}}{W_{ij}} \leq e^{\eta/2h}. \tag{12}$$

The relationship between the corresponding eigenvectors of L_{W_p} and L then follows from an eigenvector perturbation result, which we summarize in the Appendix; see Proposition 1. □

Remark 2 Theorem 1 also holds when W_p in (9) is replaced by any nonnegative function $D : \mathcal{P}_p(\mathbb{R}^n) \times \mathcal{P}_p(\mathbb{R}^n) \to \mathbb{R}^+$ satisfying $D(A_\sharp\alpha, A_\sharp\sigma) = D(\alpha, \sigma)$ for any orthogonal transformation $A : \mathbb{R}^n \to \mathbb{R}^n$. One example for another distance satisfying this property is the total variation distance $D(\alpha, \sigma) = \sup_{B\in\mathcal{B}(\mathbb{R}^n)} |\alpha(B) - \sigma(B)|$, where $\mathcal{B}(\mathbb{R}^n)$ denotes the Borel sets of $\mathbb{R}^n$. Similarly, the Kullback-Leibler (KL) divergence also satisfies this property, as it is nonnegative and invariant under orthogonal transformations.

Remark 3 When $\mathcal{E}$ defines an isometry, i.e., $\eta = 0$ in Eq. (9), we have that $\varepsilon_{\min} = \varepsilon_{\max} = 1$, which implies that $a = b = 0$ and hence $\angle(\phi_j, \widetilde{\phi}_j) = 0$. Therefore, as expected in the isometric case, there is no difference between the Wasserstein and the Euclidean settings; hence, the angle between the eigenvectors is 0 (see Corollary 1).

Remark 4 In general, the lower bound of $\cos\angle(\phi_j, \widetilde{\phi}_j)$ given by the RHS of Eq. (10) depends on an interplay between the constants $\varepsilon_{\min}, \varepsilon_{\max}, \delta_j$, and r_D. In particular, when $\mathcal{E}$ is an almost isometry, i.e., the perturbation η between the distances is "small" such that $\varepsilon_{\min}, \varepsilon_{\max} \approx 1$, one can expect that $\cos\angle(\phi_j, \widetilde{\phi}_j) \approx 1$ (or equivalently, $\angle(\phi_j, \widetilde{\phi}_j) \approx 0$), as long as δ_j is reasonably large and r_D is reasonably small. However, the numerical experiments seem to be more robust than what this lower bound can predict. Remark 7 shows that the independent components can be recovered, even when the RHS in Eq. (10) is below -1, in which case this bound is not useful in predicting the recovery performance. We leave the improvement of this lower bound for future work.

Remark 5 When $0 \leq W_p^2(\mathcal{E}(\omega), \mathcal{E}(\kappa)) - \|\omega - \kappa\|^2 \leq \eta$, we have that the constant $\varepsilon_{\max} \leq 1$, and hence $a \leq 1$.

Corollary 1 (Special Case: Isometric $\mathcal{E}$) *Let $S, A, \Omega, \mathcal{E}$, and $\{\omega_j\}_{j=1}^N$ be as defined in Theorem 1. Assume that*

$$W_p(\mathcal{E}(\omega), \mathcal{E}(\kappa)) = \|\omega - \kappa\|, \ \forall\, \omega, \kappa \in \Omega. \tag{13}$$

Then, $L_{W_p} = L$ in Theorem 1.

Proof In this case, $\widetilde{W}_{ij} = W_{ij}$, which implies $L_{W_p} = L$. □

Remark 6 (Recovery of independent components) From [31, Section 4.1], we know that in the Euclidean setting, the eigenvectors of the graph Laplacian L approximate the independent components $S_1, \ldots, S_n$ up to errors coming from the sampling process (in the limit $N \to \infty$). Our Theorem 1 now states that in the almost-isometric setting, the eigenvectors of the Wasserstein-based Laplacian L_{W_p} approximate the eigenvectors of L up to the error (Eq. (10)). Putting this together, in the almost-isometric setting, the Wasserstein ICA recovers the independent components $S_1, \ldots, S_n$ up to these two approximation errors combined.

4 Examples and Numerical Experiments

In the following two numerical experiments, we use point-clouds drawn from Gaussian distributions that use independent sources as means, with fixed (isometric case; see Example 3) and varying (almost isometric case; see Example 4) covariance matrices (Figs. 2a, 3a). These Gaussians then undergo an unknown orthogonal transformation (Figs. 2b, 3b). The Wasserstein-based ICA method is then applied to the observed "linearly mixed" point-clouds, and the recovery of the independent components (the means) is presented.

Figures 2 and 3 are illustrations of the two settings we consider (isometric and almost-isometric). We note that as with Fig. 1, these are sketches for visualization purposes and do not represent the actual data used to carry out the numerical experiments. The reason for using sketches only is to make sure individual

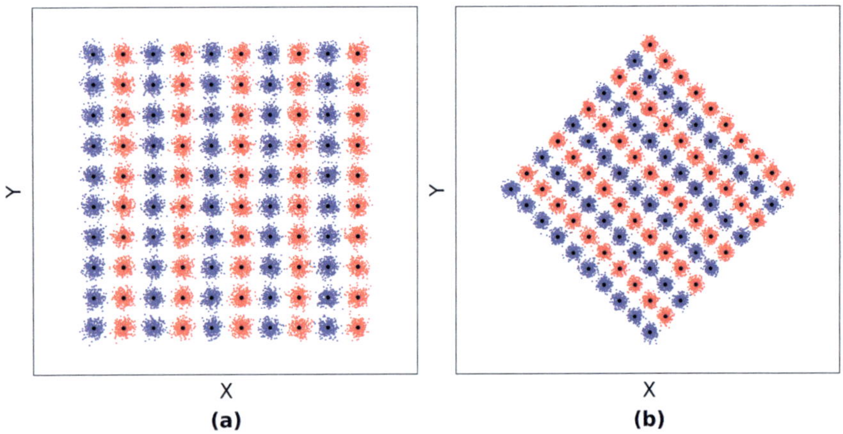

Fig. 2 An illustration of the isometric case (Sect. 4.1). This is a sketch for illustration purposes; the actual numerical setup is described in Sect. 4.1. (**a**) Independent components (the Gaussian means) are sampled on a square, and Gaussians with these means and the same covariance (are multiple of I) are considered. (**b**) Gaussians from (**a**) are transformed with an orthogonal matrix A

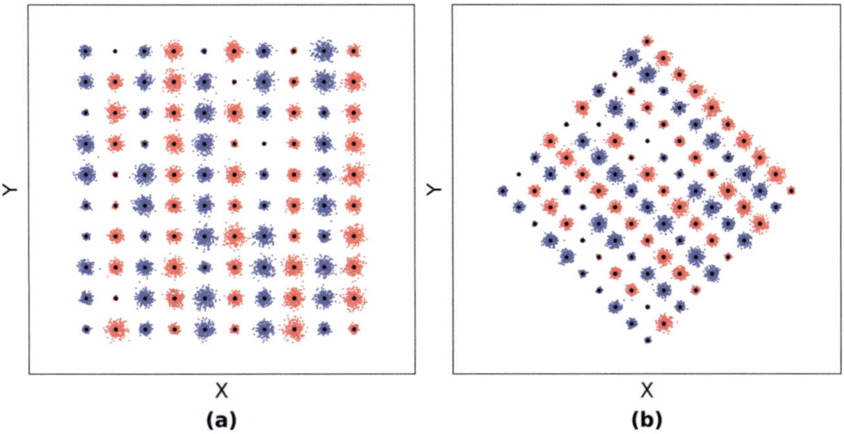

Fig. 3 An illustration of the almost-isometric case (Sect. 4.2). This is a sketch for illustration purposes; the actual numerical setup is described in Sect. 4.2. (**a**) Independent components (the Gaussian means) are sampled on a square, and Gaussians with these means and the covariances of varying sizes are considered. (**b**) Gaussians from (**a**) are transformed with an orthogonal matrix A

Gaussians are visible (when means are densely and nonuniformly sampled, different Gaussians easily overlap making it hard to identify individual instances of data).

4.1 Isometric Case

We first look at the case when $\mathcal{E}$ defines an isometry from the parameter space $(\Omega, \|\cdot\|)$ to the space of probability measures $(\mathcal{P}_p(\mathbb{R}^n), W_p)$ for $p = 2$, i.e., when $\eta = 0$ in Theorem 1. In this case, the Wasserstein ICA problem reduces to the Euclidean linear ICA problem, as the graph Laplacian using the Wasserstein distances of observed measures coincides with the graph Laplacian in the parameter space. The same approximation and recovery results for the independent components hence follow from [31]; see Corollary 1 and Remark 6.

One way of generating measures for which isometry (e.g., Eq. (13)) holds can be obtained by the translation of a base measure.

Example 2 (Isometric $\mathcal{E}$) Let $\alpha_0 \in \mathcal{P}_2(\mathbb{R}^n)$ and $\Omega \subseteq \mathbb{R}^n$. Define $\mathcal{E} : \Omega \to \mathcal{P}_2(\mathbb{R}^n)$ by $\mathcal{E}(\omega) = T_{\omega\sharp}\alpha_0$, where $T_\omega(x) = x - \omega$ is a translation. It follows that $\mathcal{E}$ is an isometry since $W_2(\mathcal{E}(\omega), \mathcal{E}(\kappa)) = W_2(T_{\omega\sharp}\alpha_0, T_{\kappa\sharp}\alpha_0) = \|\omega - \kappa\|$.

We now consider a related example that is built from an example in [31].

Example 3 Consider $n = 2$, and generate parameters in Ω by $S = (S_1, S_2)$ with the independent components S_1, S_2 given by

$$S_1 \sim \mathcal{U}(-\sqrt{3}, \sqrt{3}), \quad S_2 \sim \mathcal{N}(0, 1), \tag{14}$$

where $\mathcal{U}(-\sqrt{3}, \sqrt{3})$ denotes uniform distribution on $[-\sqrt{3}, \sqrt{3}]$ and $\mathcal{N}(0, 1)$ is the standard normal distribution. In all the numerical experiments performed, we have applied a filter on samples from S to remove the isolated outliers, similar to [31].

Let $\{\omega_j\}_{j=1}^N$ be N instances of S and generate point-clouds $\widehat{\beta}_j$ sampled from $A_\sharp \mathcal{N}(\omega_j, cI)$, $j = 1, \ldots, N$. Here, we choose $c = 0.003$ and

$$A = \begin{bmatrix} \cos(\pi/4) & -\sin(\pi/4) \\ \sin(\pi/4) & \cos(\pi/4) \end{bmatrix}, \tag{15}$$

which describes the orthogonal mixing matrix. The number of point-clouds is $N = 600$, each of which contains 30 points.

The eigenvalues $0 = \lambda_1 \leq \lambda_2 \leq \ldots \leq \lambda_N$ of the normalized graph Laplacian L_{W_2} are computed from $\{\widehat{\beta}_j\}_{j=1}^N$ using the kernel (Eq. (2)) based on the W_2-distance, i.e., $\widetilde{W}_{ij} = e^{\frac{-W_2(\widehat{\beta}_i, \widehat{\beta}_j)^2}{2h}}$, where the Wasserstein distances $W_2(\widehat{\beta}_i, \widehat{\beta}_j)$ are computed using the Python Optimal Transport (POT) package [12] and h is set to 0.2. Here, $\widetilde{D}$ is the degree matrix associated with the weight matrix $\widetilde{W}$. We use the eigenvectors $\widetilde{\phi}_2$ and $\widetilde{\phi}_3$, which correspond to the first two nontrivial eigenvalues to recover the independent components. In Fig. 4c and d, we plot $\widetilde{\phi}_2$ and $\widetilde{\phi}_3$ and color them by the original independent components S_1 (Fig. 4c) and S_2 (Fig. 4d), respectively. We observe that the independent components are recovered since the eigenvectors are in one-to-one correspondence with the independent components (Fig. 4a, b), as is expected from Corollary 1.

The term "one-to-one correspondence" means that $\widetilde{\phi}_2$ is an increasing function of S_1 and that $\widetilde{\phi}_3$ is an increasing function of S_2. In particular, $\widetilde{\phi}_2$ is independent of S_2 and similarly, $\widetilde{\phi}_3$ is independent of S_1. Visually, this is demonstrated by the coloring in Fig. 4.

4.2 *Almost-Isometric Case*

A more interesting case is when η in Theorem 1 is small, i.e., the almost-isometric case. One way of generating measures such that Eq. (9) holds is by varying an isotropic Gaussian by its mean and variance.

Example 4 Let $\Omega \subseteq \mathbb{R}^n$. Let $c : \Omega \to [c_1, c_2]$ where $0 < c_1 < c_2$. Define $\mathcal{E} : \Omega \to \mathcal{P}_p(\mathbb{R}^n)$ by $\mathcal{E}(\omega) = \mathcal{N}(\omega, c(\omega)I)$. Using the Wasserstein distance formula for Gaussians (see Lemma 3), we have

$$W_2^2(\mathcal{E}(\omega), \mathcal{E}(\kappa)) - \|\omega - \kappa\|^2 = n\Big(\sqrt{c(\omega)} - \sqrt{c(\kappa)}\Big)^2 \tag{16}$$

$$\leq n(\sqrt{c_2} - \sqrt{c_1})^2 \tag{17}$$

$$\leq \min\{n(c_2 - c_1), \frac{n(c_2 - c_1)^2}{4c_1}\}, \tag{18}$$

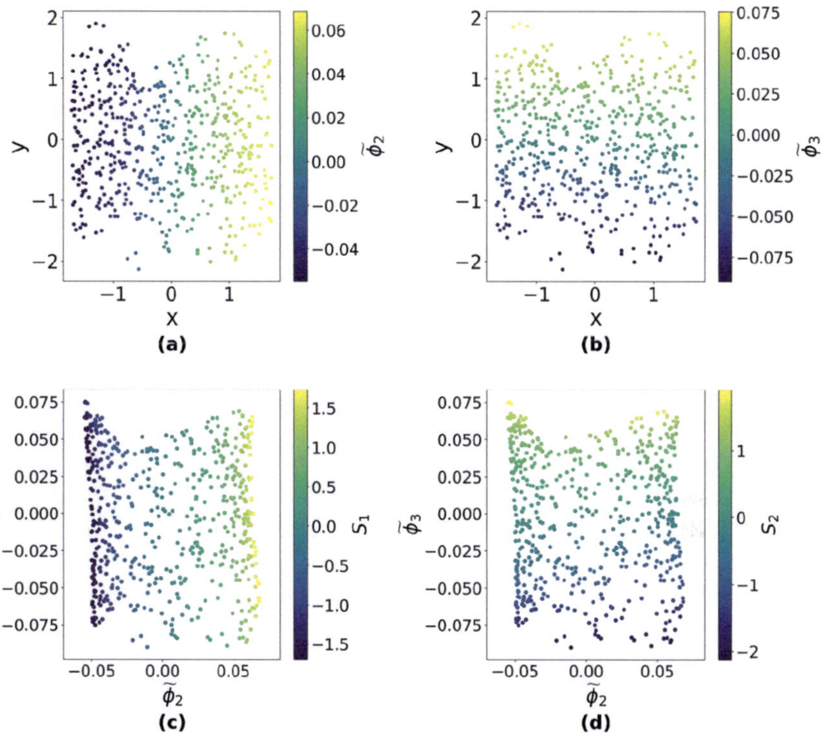

Fig. 4 Illustration of the one-to-one correspondence between the eigenvectors of the graph Laplacian and the independent components described in Example 3 (isometric example). (**a**) The independent components $S = (S_1, S_2)$ are colored by the first nontrivial eigenvector $\widetilde{\phi}_2$. (**b**) The independent components $S = (S_1, S_2)$ are colored by the second nontrivial eigenvector $\widetilde{\phi}_3$. (**c**) The eigenvectors $(\widetilde{\phi}_2, \widetilde{\phi}_3)$ are colored by the first independent component S_1. (**d**) The eigenvectors $(\widetilde{\phi}_2, \widetilde{\phi}_3)$ are colored by the second independent component S_2

where the last inequality follows from the simple facts that $\sqrt{c_2} - \sqrt{c_1} \leq \sqrt{c_2 - c_1}$ and $\sqrt{c_2} - \sqrt{c_1} \leq \frac{c_2 - c_1}{2\sqrt{c_1}}$. Similar to Eq. (12), we get

$$e^{-\eta/2h} \leq \frac{\widetilde{W}_{ij}}{W_{ij}} \leq 1, \tag{19}$$

where $\eta = \min\{n(c_2 - c_1), \frac{n(c_2-c_1)^2}{4c_1}\}$ can be made small by choosing $c_2 - c_1$ small. By Proposition 1, the relationship between the eigenvectors of L_{W_2} and L is given by Eq. (10) with constants $a = |1 - e^{-\eta/2h}| \leq 1$ and $b = |1 - e^{\eta/4h}|$.

In our numerical experiments, we again choose empirical measures corresponding to point-clouds $\widehat{\beta}_j$ sampled from the observed Gaussians $A_\sharp \mathcal{N}\Big(\omega_j, c(\omega_j)I\Big)$, where

ω_j are instances of parameters sampled from the independent component vector $S = (S_1, S_2, \ldots, S_n)$.[2]

We follow a similar numerical setup as in Example 3. The dimension is $n = 2$, the parameters $\{\omega_j\}_{j=1}^N$ are generated from $S = (S_1, S_2)$ specified in Eq. (14), $N = 600$, A is in Eq. (15), and $h = 0.2$. Each point-cloud $\widehat{\beta}_j$ contains 30 points sampled from $A_\sharp \mathcal{N}\Big(\omega_j, c(\omega_j)I\Big)$, where $c(\omega_j)$ is chosen uniformly in [0.998, 1.002].

As in Example 3, the eigenvectors $\widetilde{\phi}_2$ and $\widetilde{\phi}_3$ of L_{W_2} corresponding to the first two nontrivial eigenvalues are used for the independent component recovery.

In Fig. 5, we plot $\widetilde{\phi}_2$ and $\widetilde{\phi}_3$ colored by the original independent components S_1 (Fig. 5c) and S_2 (Fig. 5d), respectively. In Fig. 5a, b, the original parameters $S = (S_1, S_2)$ are colored by $\widetilde{\phi}_2$ (Fig. 5a) and $\widetilde{\phi}_3$ (Fig. 5b), respectively. We observe that the eigenvectors are in one-to-one correspondence with the independent components, as expected from Remark 6.

To obtain the theoretical bound in Eq. (10) from Theorem 1, we estimate $\varepsilon_{\min}$, $\varepsilon_{\max}$ by Eq. (16) and hence use the min and max of $\left(\sqrt{c(\omega_i)} - \sqrt{c(\omega_j)}\right)^2$. We observe that $\varepsilon_{\min}, \varepsilon_{\max} \approx 1$ for the chosen parameter interval. The remaining constants $r_D, \delta_j, j = 2, 3$ are computed directly using W and L (see Eq. (3)) associated with $\{A\omega_j\}_{j=1}^N$. Taking the average of multiple numerical outputs, we obtain

$$\cos \angle(\phi_2, \widetilde{\phi}_2) \geq 0.993, \quad \cos \angle(\phi_3, \widetilde{\phi}_3) \geq 0.989, \tag{20}$$

with standard deviation 0.006 and 0.012. The two angles are around 6.8° and 8.5°, respectively. Based on the chosen example, small angles were expected; see Remark 4.

The preceding example shows that when the covariance of the Gaussians varies by small constants, i.e., when $\frac{\max c(\omega)}{\min c(\omega)} \approx 1$, then the independent components S_1 and S_2 are well approximated by the eigenvectors of the graph Laplacian and the error established in Theorem 1 (Eq. (10)) can be explicitly computed; compare Eq. (20). Even when the error bound (Eq. (10)) is not meaningful (e.g., when the lower bound is negative), the Wasserstein ICA method may still be successful in recovering the independent components. We now discuss one such case.

Remark 7 Following the exact same setup as Example 4, we choose $c(\omega_j)$ uniformly from [0.00003, 0.3] such that $\frac{\max c(\omega)}{\min c(\omega)} \approx 10^4$, which indicates a significant size difference in $\{\widehat{\beta}_j\}_{j=1}^N$. The error established in Eq. (10) is computed but exceeds the range of cosine function due to small $\varepsilon_{\min}$ defined in Theorem 1 and is thus not insightful. However, the first two nontrivial eigenvectors $\widetilde{\phi}_2$ and $\widetilde{\phi}_3$ computed from L_{W_2} (Eq. (7)) are nevertheless in one-to-one correspondence with the independent components S_1 and S_2, as illustrated by Fig. 6 in the Appendix.

[2] Note here $\mathcal{E}(\omega_j)$ is the empirical measure of a point-cloud sampled from $\mathcal{N}(\omega_j, c(\omega_j)I)$.

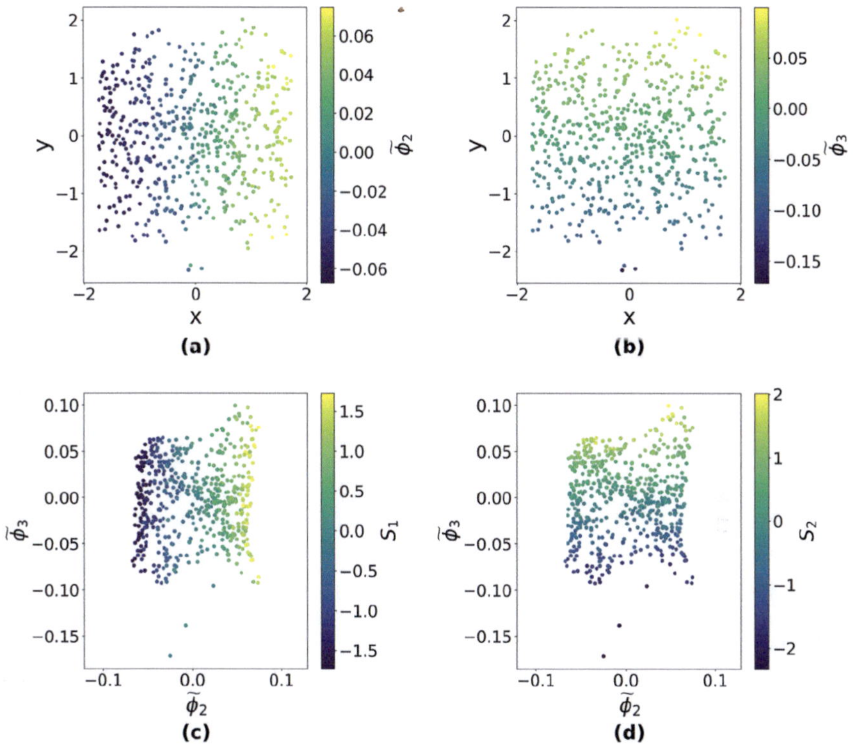

Fig. 5 Illustration of the one-to-one correspondence between the eigenvectors of the graph Laplacian and the independent components described in Example 4 (almost-isometric example). (**a**) Independent components $S = (S_1, S_2)$ colored by the first nontrivial eigenvector $\widetilde{\phi}_2$, (**b**) independent components $S = (S_1, S_2)$ colored by the second nontrivial eigenvector $\widetilde{\phi}_3$, (**c**) eigenvectors $(\widetilde{\phi}_2, \widetilde{\phi}_3)$ colored by the first independent component S_1, and (**d**) eigenvectors $(\widetilde{\phi}_2, \widetilde{\phi}_3)$ colored by the second independent component S_2

5 Discussion

We have presented a framework for applying linear independent component analysis when the observed data consists of probability measures or point-clouds. Our method mimics the classical Euclidean setting and shows that when the observed point-cloud data is almost isometric to Euclidean data, comparable recovery results can be achieved. We consider this paper a first step toward the development of a complete theory for ICA in the Wasserstein space. Topics of future interest concern going beyond the almost-isometry assumption and studying nonlinear ICA problems.

Acknowledgments The authors thank the anonymous reviewers for their constructive comments.

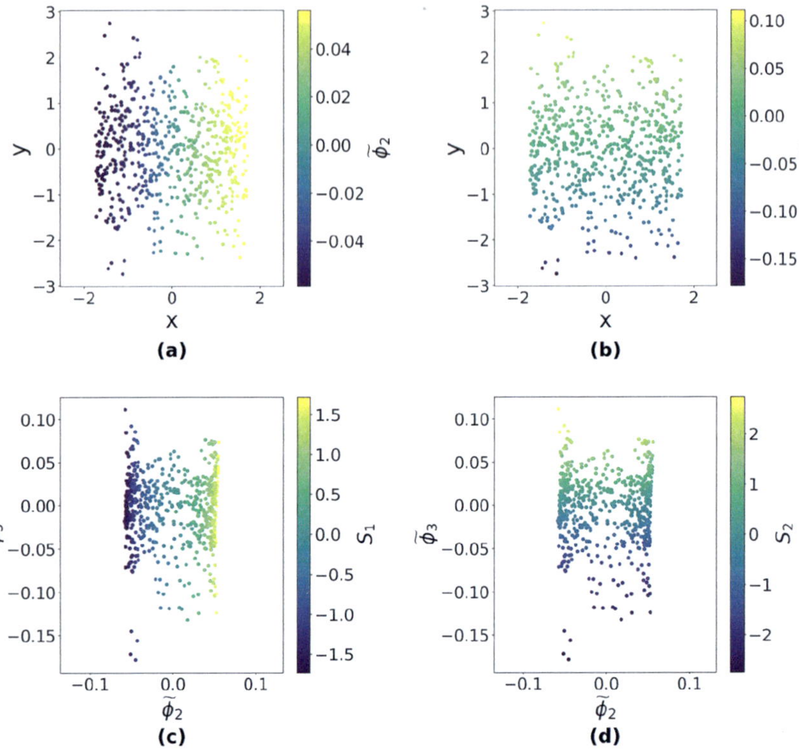

Fig. 6 Illustration of the one-to-one correspondence between the eigenvectors of the graph Laplacian and the independent components described in Remark 7. (**a**) Independent components $S = (S_1, S_2)$ colored by the first nontrivial eigenvector $\widetilde{\phi}_2$, (**b**) independent components $S = (S_1, S_2)$ colored by the second nontrivial eigenvector $\widetilde{\phi}_3$, (**c**) eigenvectors $(\widetilde{\phi}_2, \widetilde{\phi}_3)$ colored by the first independent component S_1, and (**d**) eigenvectors $(\widetilde{\phi}_2, \widetilde{\phi}_3)$ colored by the second independent component S_2

Competing Interests Shiying Li and Caroline Moosmüller have received a research grant from NSF DMS (award 2410140). Caroline Moosmüller has furthermore received a research grant from NSF DMS (award 2306064) and a seed grant from the School of Data Science and Society at UNC.

Appendix

Proposition 1 *Let $W, \widetilde{W}$ be $N \times N$ weight matrices built from Eqs. (2) and (8). Let $L = I - D^{-1}W$ and $\widetilde{L} = I - \widetilde{D}^{-1}\widetilde{W}$ be the corresponding normalized graph Laplacians, with D and $\widetilde{D}$ being the associated degree matrices, respectively. Assume that $\widetilde{W}_{ij} = \varepsilon_{ij} W_{ij}$ such that*

$$0 < \varepsilon_{min} \leq \varepsilon_{ij} \leq \varepsilon_{max}, \quad i, j = 1, \ldots, N. \tag{21}$$

Suppose $\lambda_1 \leq \ldots \leq \lambda_N$ *and* $\widetilde{\lambda}_1 \leq \ldots \leq \widetilde{\lambda}_N$ *are the eigenvalues of* L *and* $\widetilde{L}$, *respectively. Fix* $j \in \{1, \ldots, N\}$, *and assume that* $\delta_j := \min\{\lambda_j - \lambda_{\widetilde{j-1}}, \lambda_{\widetilde{j+1}} - \lambda_j\} > 0$. *Then, for eigenvectors* ϕ_j *and* $\widetilde{\phi}_j$ *satisfying* $L\phi_j = \lambda_j \phi_j$ *and* $\widetilde{L}\widetilde{\phi}_j = \widetilde{\lambda}\widetilde{\phi}_j$,

$$\cos \angle(\phi_j, \widetilde{\phi}_j) \geq 1 - \frac{\varepsilon_{max}^{1/2}}{2} r_D \left(b + 2^{3/2}\delta_j^{-1}\left(a\varepsilon_{min}^{-1} r_D + br_D\varepsilon_{min}^{-1/2} + br_D^{1/2}\right)\right)^2. \tag{22}$$

where $a = \max\{|\varepsilon_{max} - 1|, |\varepsilon_{min} - 1|\}$, $b = \max\{|\varepsilon_{max}^{-1/2} - 1|, |\varepsilon_{min}^{-1/2} - 1|\}$, *and* $r_D = \frac{D_{max}}{D_{min}}$ *with* $D_{max} = \max_i D_{ii}$, $D_{min} = \min_i D_{ii}$. *Here, D is the degree matrix associated with W*.

Proof Let $M = D^{-1}W$ and $\widetilde{M} = \widetilde{D}^{-1}\widetilde{W}$. Since L and M (similarly, $\widetilde{L}$ and $\widetilde{M}$) have the same eigenvectors, it is equivalent to analyze the eigenvectors for M and $\widetilde{M}$. We first look at eigenvectors for the symmetric matrices $S = D^{1/2}MD^{-1/2}$ and $\widetilde{S} = \widetilde{D}^{1/2}\widetilde{M}\widetilde{D}^{-1/2}$. It is not hard to verify that if V is an orthogonal matrix whose columns are eigenvectors of S, then the columns of $D^{-1/2}V$ are eigenvectors of M corresponding to the same eigenvalue. Without loss of generality, assume that $\phi_j = D^{-1/2}v_j$ and $\widetilde{\phi}_j = \widetilde{D}^{-1/2}\widetilde{v}_j$, where v_j and $\widetilde{v}_j$ are unit eigenvectors of S and $\widetilde{S}$, corresponding to eigenvalues λ_j and $\widetilde{\lambda}_j$, respectively. We will first bound $\|\widetilde{v}_j - v_j\|$ using Corollary 2. Observe that $S = D^{-1/2}WD^{-1/2}$ and $\widetilde{S} = \widetilde{D}^{-1/2}\widetilde{W}\widetilde{D}^{-1/2}$. By a direct computation, we have the following bounds:

$$\|D^{-1/2}\| \leq D_{\min}^{-1/2}, \tag{23}$$

$$\|\widetilde{D}^{-1/2}\| \leq \varepsilon_{\min}^{-1/2} D_{\min}^{-1/2}, \tag{24}$$

$$\|W\| \leq \|D\|\|D^{-1}W\| \leq D_{\max}, \tag{25}$$

$$\|\widetilde{W} - W\| \leq \sqrt{\|\widetilde{W} - W\|_1 \|\widetilde{W} - W\|_\infty} = \|\widetilde{W} - W\|_1 \leq aD_{\max}, \tag{26}$$

where $a = \max\{|\varepsilon_{\max} - 1|, |\varepsilon_{\min} - 1|\}$. Here, $\|\cdot\|$ denotes the matrix 2-norm, and we have used the fact that $\|D^{-1}W\| \leq 1$ in Eq. (24) (since DW^{-1} is nonnegative and row stochastic) and the fact that $\widetilde{W} - W$ is symmetric in Eq. (26). Similarly, since

$$(\varepsilon_{\max}^{-1/2} - 1)(D_{ii})^{-1/2} \leq (\widetilde{D}^{-1/2})_{ii} - (D^{-1/2})_{ii} \leq (\varepsilon_{\min}^{-1/2} - 1)(D_{ii})^{-1/2},$$

we obtain

$$\|\widetilde{D}^{-1/2} - D^{-1/2}\| \leq bD_{\min}^{-1/2}, \tag{27}$$

where $b = \max\{|\varepsilon_{\max}^{-1/2} - 1|, |\varepsilon_{\min}^{-1/2} - 1|\}$. By the triangle inequality,

$$\begin{aligned} \|\widetilde{S} - S\| \leq & \|\widetilde{D}^{-1/2}\widetilde{W}\widetilde{D}^{-1/2} - D^{-1/2}W\widetilde{D}^{-1/2}\| \\ & + \|D^{-1/2}W\widetilde{D}^{-1/2} - D^{-1/2}WD^{-1/2}\|. \end{aligned} \tag{28}$$

For the first term in Eq. (28), we have

$$\begin{aligned} \|\widetilde{D}^{-1/2}\widetilde{W}\widetilde{D}^{-1/2} - D^{-1/2}W\widetilde{D}^{-1/2}\| &\leq \|\widetilde{D}^{-1/2}\widetilde{W} - D^{-1/2}W\|\|\widetilde{D}^{-1/2}\| \\ &\leq \left(\|\widetilde{D}^{-1/2}\widetilde{W} - \widetilde{D}^{-1/2}W\| + \|\widetilde{D}^{-1/2}W - D^{-1/2}W\|\right)\|\widetilde{D}^{-1/2}\| \\ &\leq \|\widetilde{D}^{-1/2}\|^2\|\widetilde{W} - W\| + \|\widetilde{D}^{-1/2} - D^{-1/2}\|\|W\|\|\widetilde{D}^{-1/2}\| \\ &\leq a\varepsilon_{\min}^{-1}D_{\min}^{-1}D_{\max} + br_D\varepsilon_{\min}^{-1/2}, \end{aligned}$$

where the last inequality follows from Eqs. (24)–(27). For the second term in Eq. (28), we have

$$\begin{aligned} \|D^{-1/2}W\widetilde{D}^{-1/2} - D^{-1/2}WD^{-1/2}\| &\leq \|D^{-1/2}W\|\|\widetilde{D}^{-1/2} - D^{-1/2}\| \\ &\leq \|D^{1/2}\|\|D^{-1}W\|(bD_{\min}^{-1/2}) \\ &\leq bD_{\max}^{1/2}D_{\min}^{-1/2}. \end{aligned}$$

Hence,

$$\|\widetilde{S} - S\| \leq a\varepsilon_{\min}^{-1}r_D + br_D\varepsilon_{\min}^{-1/2} + br_D^{1/2}, \tag{29}$$

where $r_D = \frac{D_{\max}}{D_{\min}}$.

Without loss of generality, assume that $\widetilde{v}_j^T v_j \geq 0$ (otherwise reverse the direction of one of the vectors). Then, by Corollary 2, we have

$$\|\widetilde{v}_j - v_j\| \leq \frac{2^{3/2}\|\widetilde{S} - S\|}{\delta_j} \leq 2^{3/2}\delta_j^{-1}\left(a\varepsilon_{\min}^{-1}r_D + br_D\varepsilon_{\min}^{-1/2} + br_D^{1/2}\right). \tag{30}$$

Here, we have used the fact S has the same "eigenvalue gaps" (δ_j's) as L (in the reversed order) since $\{1 - \lambda_j\}_{j=1}^N$ are eigenvalues of S. It follows that

$$\begin{aligned} \|\widetilde{\phi}_j - \phi_j\| &= \|\widetilde{D}^{-1/2}\widetilde{v}_j - D^{-1/2}v_j\| \\ &\leq \|\widetilde{D}^{-1/2}\widetilde{v}_j - D^{-1/2}\widetilde{v}_j\| + \|D^{-1/2}\widetilde{v}_j - D^{-1/2}v_j\| \\ &\leq \|\widetilde{D}^{-1/2} - D^{-1/2}\| + \|D^{-1/2}\|\|\widetilde{v}_j - v_j\| \\ &\leq D_{\min}^{-1/2}\left(b + 2^{3/2}\delta_j^{-1}\left(a\varepsilon_{\min}^{-1}r_D + br_D\varepsilon_{\min}^{-1/2} + br_D^{1/2}\right)\right). \end{aligned} \tag{31}$$

Moreover, it is not hard to show that $\|D^{-1/2}v\| \geq D_{\max}^{-1/2}\|v\|$ for any $v \in \mathbb{R}^N$, which implies that $\|\phi_j\| = \|D^{-1/2}v_j\| \geq D_{\max}^{-1/2}$ as $\|v_j\| = 1$. Similarly, $\|\widetilde{\phi}_j\| \geq (\varepsilon_{\max} D_{\max})^{-1/2}$. Let θ_j be the angle between $\widetilde{\phi}_j$ and ϕ_j. Then,

$$\begin{aligned}\cos\theta_j &\geq 1 - \frac{\|\widetilde{\phi}_j - \phi_j\|^2}{2\|\widetilde{\phi}_j\|\|\phi_j\|} \\ &\geq 1 - \frac{\varepsilon_{\max}^{1/2}}{2} r_D \left(b + 2^{3/2}\delta_j^{-1}\left(a\varepsilon_{\min}^{-1} r_D + b r_D \varepsilon_{\min}^{-1/2} + b r_D^{1/2}\right)\right)^2.\end{aligned}$$

□

Proof of Lemma 1 Let $\alpha, \sigma \in \mathcal{P}_p(\mathbb{R}^n)$. It suffices to show that $W_p(\alpha, \sigma) \geq W_p(A_\sharp\alpha, A_\sharp\sigma)$ for any orthogonal matrix A, which implies the reversed inequality by starting with $W_p(A_\sharp\alpha, A_\sharp\sigma)$ and applying A^{-1}. Let γ be an optimal transport plan between α and σ. It is not hard to see that $(A, A)_\sharp\gamma$ is a transport plan between $A_\sharp\alpha$ and $A_\sharp\sigma$, where $(A, A) : \mathbb{R}^n \times \mathbb{R}^n \to \mathbb{R}^n \times \mathbb{R}^n$ is defined by $(A, A)(x, y) = (Ax, Ay)$. Indeed, let $\pi_1, \pi_2 : \mathbb{R}^n \times \mathbb{R}^n \to \mathbb{R}^n$ be the projection functions mapping (x, y) to its first and second coordinates, respectively. Since $\pi_1 \circ (A, A) = A \circ \pi_1$, it follows that $\pi_{1\sharp}\left((A, A)_\sharp \gamma\right) = A_\sharp(\pi_{1\sharp}\gamma) = A_\sharp\alpha$. Similarly, $\pi_{2\sharp}\left((A, A)_\sharp \gamma\right) = A_\sharp\sigma$. By the change of variable formula, we have

$$\begin{aligned}W_p^p(A_\sharp\alpha, A_\sharp\sigma) &\leq \int_{\mathbb{R}^n \times \mathbb{R}^n} \|\widetilde{x} - \widetilde{y}\|^p d\big((A, A)_\sharp\gamma\big)(\widetilde{x}, \widetilde{y}) \\ &= \int_{\mathbb{R}^n \times \mathbb{R}^n} \|Ax - Ay\|^p d\gamma(x, y) \\ &= \int_{\mathbb{R}^n \times \mathbb{R}^n} \|x - y\|^p d\gamma(x, y) \\ &= W_p^p(\alpha, \sigma).\end{aligned}$$

□

Lemma 2 ([34, Corollary 3]) *Let $\Sigma, \hat{\Sigma} \in \mathbb{R}^{n\times n}$ be symmetric, with eigenvalues $\lambda_1 \geq \cdots \geq \lambda_p$ and $\hat{\lambda}_1 \geq \cdots \geq \hat{\lambda}_p$, respectively. Fix $j \in \{1, \ldots, n\}$, and assume that $\min(\lambda_{j-1}-\lambda_j, \lambda_j-\lambda_{j+1}) > 0$, where $\lambda_0 := \infty$ and $\lambda_{n+1} := -\infty$. If $v, \hat{v} \in \mathbb{R}^n$ satisfy $\Sigma v = \lambda_j v$ and $\hat{\Sigma}\hat{v} = \hat{\lambda}_j \hat{v}$, then*

$$\sin \angle(\hat{v}, v) \leq \frac{2\|\hat{\Sigma} - \Sigma\|}{\min(\lambda_{j-1} - \lambda_j, \lambda_j - \lambda_{j+1})}.$$

Moreover, if $\hat{v}^T v \geq 0$, then

$$\|\hat{v} - v\| \leq \frac{2^{3/2}\|\hat{\Sigma} - \Sigma\|}{\min(\lambda_{j-1} - \lambda_j, \lambda_j - \lambda_{j+1})}.$$

Here $\|\cdot\|$ denotes the matrix 2-norm.

Lemma 3 (Wasserstein Distance Between Isotropic Gaussians) *Let* $\alpha_i = \mathcal{N}(\omega_i, c_i I)$ *and* $\alpha_j = \mathcal{N}(\omega_j, c_j I)$ *be two Gaussians supported on* $\mathbb{R}^n$. *The* W_2 *distance between* α_i *and* α_j *is*

$$W_2^2(\alpha_i, \alpha_j) = \|\omega_i - \omega_j\|^2 + n(\sqrt{c_i} - \sqrt{c_j})^2.$$

Proof By [27, Remark 2.31], the Wasserstein distance between two Gaussians $\alpha = \mathcal{N}(m_\alpha, \Sigma_\alpha)$ and $\beta = \mathcal{N}(m_\beta, \Sigma_\beta)$ is given by

$$W_2^2(\alpha, \beta) = \|m_\alpha - m_\beta\|_2^2 + B(\Sigma_\alpha, \Sigma_\beta)^2$$

where

$$B(\Sigma_\alpha, \Sigma_\beta)^2 = \mathrm{Tr}(\Sigma_\alpha + \Sigma_\beta - 2(\Sigma_\alpha^{1/2}\Sigma_\beta\Sigma_\alpha^{1/2})^{1/2})$$

is the Bures distance; see, e.g., [4]. We let $\alpha_i = \mathcal{N}(\omega_i, c_i I)$ and $\alpha_j = \mathcal{N}(\omega_j, c_j I)$, which implies

$$W_2^2(\alpha_i, \alpha_j) = \|\omega_i - \omega_j\|^2 + B(c_i I, c_j I)^2.$$

It is easy to see that $B(c_i I, c_j I)^2 = n(\sqrt{c_i} - \sqrt{c_j})^2$, from which the result follows. □

References

1. Ans, B., Hérault, J., Jutten, C.: Architectures neuromimétiques adaptatives: Détection de primitives. In: Proceedings of Cognitiva 85, pp. 593–597. Paris (1985)
2. Back, A.D., Weigend, A.S.: A first application of independent component analysis to extracting structure from stock returns. Int. J. Neural Syst. **8**(04), 473–484 (1997). https://doi.org/10.1142/S0129065797000458
3. Bruggner, R.V., Bodenmiller, B., Dill, D.L., Tibshirani, R.J., Nolan, G.P.: Automated identification of stratifying signatures in cellular subpopulations. Proc. Natl. Acad. Sci. **111**(26), E2770–E2777 (2014)
4. Bures, D.: An extension of Kakutani's theorem on infinite product measures to the tensor product of semifinite *-algebras. Trans. Am. Math. Soc. **135**, 199–212 (1969). https://api.semanticscholar.org/CorpusID:53529518
5. Chen, Y., Cruz, F.D., Sandhu, R., Kung, A.L., Mundi, P., Deasy, J.O., Tannenbaum, A.: Pediatric sarcoma data forms a unique cluster measured via the earth mover's distance. Sci. Rep. **7**(1), 7035 (2017). https://doi.org/10.1038/s41598-017-07551-8
6. Cloninger, A., Hamm, K., Khurana, V., Moosmüller, C.: Linearized wasserstein dimensionality reduction with approximation guarantees. Appl. Comput. Harmon. Anal. **74**, 101718 (2025). https://doi.org/10.1016/j.acha.2024.101718
7. Comon, P.: Independent component analysis, a new concept? Signal Process. **36**(3), 287–314 (1994)
8. Comon, P., Jutten, C.: Handbook of Blind Source Separation: Independent Component Analysis and Applications, 1st edn. Academic Press, New York (2010)

9. Davis, C., Kahan, W.M.: Some new bounds on perturbation of subspaces. Bull. Am. Math. Soc. **75**(4), 863–868 (1969)
10. Delorme, A., Makeig, S.: EEGLAB: an open source toolbox for analysis of single-trial EEG dynamics including independent component analysis. J. Neurosci. Methods **134**(1), 9–21 (2004). https://doi.org/10.1016/j.jneumeth.2003.10.009
11. Eldridge, J., Belkin, M., Wang, Y.: Unperturbed: spectral analysis beyond Davis-Kahan. In: Janoos, F., Mohri, M., Sridharan, K. (eds.) Proceedings of Algorithmic Learning Theory. Proceedings of Machine Learning Research, vol. 83, pp. 321–358. PMLR (2018)
12. Flamary, R., Courty, N., Gramfort, A., Alaya, M.Z., Boisbunon, A., Chambon, S., Chapel, L., Corenflos, A., Fatras, K., Fournier, N., Gautheron, L., Gayraud, N.T., Janati, H., Rakotomamonjy, A., Redko, I., Rolet, A., Schutz, A., Seguy, V., Sutherland, D.J., Tavenard, R., Tong, A., Vayer, T.: Pot: Python optimal transport. J. Mach. Learn. Res. **22**(78), 1–8 (2021). http://jmlr.org/papers/v22/20-451.html
13. Giannakopoulos, X., Karhunen, J., Oja, E.: An experimental comparison of neural ICA algorithms. In: ICANN 98: Proceedings of the 8th International Conference on Artificial Neural Networks, Skövde, Sweden, pp. 651–656. Springer, Berlin (1998)
14. Hamm, K., Henscheid, N., Kang, S.: Wassmap: Wasserstein isometric mapping for image manifold learning. SIAM J. Math. Data Sci. **5**(2), 475–501 (2023). https://doi.org/10.1137/22M1490053
15. Hamm, K., Moosmüller, C., Schmitzer, B., Thorpe, M.: Manifold learning in Wasserstein space. SIAM J. Math. Anal. **57**(3), 2983–3029 (2025). https://doi.org/10.1137/23M161754
16. Haykin, S., Kosko, B.: Image denoising by sparse code shrinkage. In: Intelligent Signal Processing, pp. 554–568. IEEE, Piscataway (2001). https://doi.org/10.1109/9780470544976.ch15
17. Hyvärinen, A.: New approximations of differential entropy for independent component analysis and projection pursuit. In: Advances in Neural Information Processing Systems, vol. 10 (1997)
18. Hyvärinen, A., Karhunen, J., Oja, E.: Independent Component Analysis. Adaptive and Cognitive Dynamic Systems: Signal Processing, Learning, Communications and Control. Wiley, London (2004). https://books.google.com/books?id=96D0ypDwAkkC
19. Hyvärinen, A., Oja, E.: Independent component analysis: algorithms and applications. Neural Netw. **13**(4–5), 411–430 (2000)
20. Kantorovich, L.: On the transfer of masses. Doklady Akademii Nauk **37**(2), 227–229 (1942)
21. Kileel, J., Moscovich, A., Zelesko, N., Singer, A.: Manifold learning with arbitrary norms. J. Fourier Anal. Appl. **27**(5), 1–56 (2021)
22. Mathews, J., Pouryahya, M., Moosmüller, C., Kevrekidis, I.G., Deasy, J., Tannenbaum, A.: Molecular phenotyping using networks, diffusion, and topology: soft-tissue sarcoma. Sci. Rep. (2019). https://doi.org/10.1038/s41598-019-50300-2
23. Mishne, G., Talmon, R., Meir, R., Schiller, J., Lavzin, M., Dubin, U., Coifman, R.R.: Hierarchical coupled-geometry analysis for neuronal structure and activity pattern discovery. IEEE J. Sel. Topics Signal Process. **10**(7), 1238–1253 (2016). https://doi.org/10.1109/JSTSP.2016.2602061
24. Mohanaprasad, K., Singh, A., Sinha, K., et al.: Noise reduction in speech signals using adaptive independent component analysis (ICA) for hands free communication devices. Int. J. Speech Technol. **22**, 169–177 (2019). https://doi.org/10.1007/s10772-019-09595-9
25. Monge, G.: Mémoire sur la théorie des déblais et des remblais. De l'Imprimerie Royale (1781)
26. Nath, M.K., Sahambi, J.: Independent component analysis of functional MRI data. In: TENCON 2008—2008 IEEE Region 10 Conference, pp. 1–6 (2008). https://doi.org/10.1109/TENCON.2008.4766666
27. Peyré, G., Cuturi, M.: Computational optimal transport. Found. Trends Mach. Learn. **11**(5-6), 355–607 (2019). https://doi.org/10.1561/2200000073
28. Raychaudhuri, S., Sutphin, P., Chang, J., Altman, R.: Basic microarray analysis: grouping and feature reduction. Trends Biotechnol. **19**, 189–193 (2001). https://doi.org/10.1016/S0167-7799(01)01599-2

29. Rubner, Y., Tomasi, C., Guibas, L.J.: The earth mover's distance as a metric for image retrieval. Int. J. Comput. Vis. **40**(2), 99–121 (2000)
30. Santambrogio, F.: Optimal Transport for Applied Mathematicians: Calculus of Variations, PDEs, and Modeling. Progress in Nonlinear Differential Equations and Their Applications. Springer, Berlin (2015). https://books.google.com/books?id=UOHHCgAAQBAJ
31. Singer, A.: Spectral independent component analysis. Appl. Comput. Harmon. Anal. **21**(1), 135–144 (2006). https://doi.org/https://doi.org/10.1016/j.acha.2006.03.003
32. Villani, C.: Topics in Optimal Transportation. Graduate studies in mathematics. American Mathematical Society (2003). https://books.google.com/books?id=idyFAwAAQBAJ
33. Villani, C.: Optimal Transport: Old and New. Grundlehren der mathematischen Wissenschaften, vol. 338. Springer, Berlin, Heidelberg (2009)
34. Yu, Y., Wang, T., Samworth, R.J.: A useful variant of the Davis–Kahan theorem for statisticians. Biometrika **102**(2), 315–323 (2015)
35. Zelesko, N., Moscovich, A., Kileel, J., Singer, A.: Earthmover-based manifold learning for analyzing molecular conformation spaces. In: 2020 IEEE 17th International Symposium on Biomedical Imaging (ISBI), pp. 1715–1719 (2020). https://doi.org/10.1109/ISBI45749.2020.9098723
36. Zhang, Y., Jin, R., Zhou, Z.H.: Understanding bag-of-words model: a statistical framework. Int. J. Mach. Learn. Cybernet. **1**(1–4), 43–52 (2010)
37. Zhao, J., Jaffe, A., Li, H., Lindenbaum, O., Sefik, E., Jackson, R., Cheng, X., Flavell, R.A., Kluger, Y.: Detection of differentially abundant cell subpopulations in scRNA-seq data. Proc. Natl. Acad. Sci. **118**(22) (2021). https://doi.org/10.1073/pnas.2100293118

Faster HodgeRank Approximation Algorithm for Statistical Ranking and User Recommendation Problems

Shelby Ferrier, Junyuan Lin, and Guangpeng Ren

1 Introduction

Statistical ranking and user recommendation problems are central to a wide range of applications, including sports, web search, and literature search. Over the years, researchers have been searching for robust algorithms that can obtain accurate ranking [2, 4–6, 10–12, 16, 18]. Out of those methods, the HodgeRank algorithm, proposed by Jiang et al. [15], is able to derive a global ranking from subjective, incomplete data sets that contain voters (e.g., people who have reviewed some movies) and scored elements (e.g., the ratings each person gives to a movie). The HodgeRank algorithm, distinguished from other ranking methods, analyzes pairwise differences represented as edge flows on a graph using discrete or combinatorial Hodge theory. In Sect. 2, we provide a detailed summary of the HodgeRank algorithm proposed by Jiang et al. [15], particularly how the HodgeRank algorithm formulates the statistical ranking problems into linear least squares problems on graphs.

In recent years, there have been several algorithms developed to approximate the HodgeRank ranking algorithm, with a primary focus on specific applications. Xu

S. Ferrier
Department of Integrative Structural and Computational Biology, The Scripps Research Institute, San Diego, CA, USA
e-mail: sferrier@scripps.edu

J. Lin (✉)
Department of Mathematics, Statistics and Data Science, Loyola Marymount University, Los Angeles, CA, USA
e-mail: Junyuan.Lin@lmu.edu

G. Ren
Institute of Mathematical Sciences, Claremont Graduate University, Claremont, CA, USA
e-mail: guangpeng.ren@cgu.edu

C. Garcia-Cardona, H. Lee (eds.), *Advances in Data Science*, Association for Women in Mathematics Series 37, https://doi.org/10.1007/978-3-031-87804-6_7

et al. [25] propose a robust ranking framework that uses Hodge decomposition and focus on detecting outliers through sparse approximations in computer vision. Wei et al.'s [24] work applies HodgeRank model to characterize structural properties of biomolecules, with a focus on accuracy in identifying cycles, loops, and folding patterns within complex biological data while placing less emphasis on efficiency with large-scale data.

By leveraging the graph properties, there are two main benefits to the problem of ranking data sets: (1) the HodgeRank algorithm can be applied to data sets that are incomplete and imbalanced; that is, not every voter rates every element, and not all elements get an equal amount of rates. This is done by forming each voter's rating items into a fully connected subgraph and merging these subgraphs into a larger item graph with weighted edges. (2) The HodgeRank algorithm can be found particularly useful on data sets where the bias of the voters may need to be considered. For example, some people give most movies top ratings, whereas others may have higher standards for what a good movie is. The pairwise rankings in the HodgeRank algorithm resolves this by focusing on the difference one voter gives between two different elements, as opposed to their individual ratings. These pairwise differences are stored in the graph as edge flows. This feature contributes to the popularity of the HodgeRank algorithm among statistical ranking methods.

Another significance of deriving the statistical ranking problems into linear least squares problems on graphs is that it provides a measurement for the quality of the global ranking. With the relationship to least squares problems on graphs, many mathematical solvers can be applied. A baseline solver to compute the least squares problems is a direct solve on the pseudo-inverse of an $n \times n$ matrix where n is the number of elements to be ranked. The time complexity of this direct solve is $O(n^3)$; therefore, the HodgeRank algorithm becomes computationally limited as the number of items being rated increases. We measured run times of more than an hour as the number increased over 2000, which we show in Sect. 4. As a result, for very large sets of elements (more than 10,000), it is preferable to approximate the ranking rather than use the original algorithm. As mentioned in [3], the unsmoothed aggregation algebraic multigrid (UA-AMG) [21] as a preconditioner for conjugate gradient (CG) yields efficient computation of the least squares problems. In [3], authors used the algebraic multigrid (AMG) method [9] to cut down on run time to $O(n \log n)$ while closely approximating the universal ranking.

To further reduce the computation complexity, in Sect. 3, we present a new method to cut down the time complexity of HodgeRank, which sections the data into groups before computing the ranking. We tested the grouping method on IMDb movie rating data [1] and found the resulting ranking to be a strong approximation for the Hodge ranking. Additionally, we saw that the accuracy of the results was generally dependent on how many groups were used, with fewer groups producing more accurate results. Finally, we analyze the time complexity of the method and discuss the trade-off between run time and accuracy when picking the best group size.

Table 1 Example ratings from patients 1–3 on four symptoms. *The "X" in the above data set represents a missing value

	Fever	Sore throat	Cough	Nausea
Patient 1	3	2	2	5
Patient 2	7	8	9	X*
Patient 3	2	2	1	3

2 Method and Model

The goal of the HodgeRank algorithm is to obtain a relative ranking of elements in a set, based on ratings given to them by individual voters.

To demonstrate the method created by Jiang et al. [15], we take an example through the method in Table 1. Suppose we aim to rate the symptoms of COVID-19 by severity; thus, we survey three patients on a set of symptoms (fever, sore throat, cough, and nausea):

A straightforward approach would be to rank the severity of symptoms by computing the mean of each column. In this data set, the mean of each column is four, which implies that each symptom is equally as severe. However, examining the data reveals that some symptoms should be rated higher than others: nausea, for example, is rated the most severe by every patient who reviewed it. To get a more accurate ranking, we use the HodgeRank method instead.

2.1 Terminology

There is some terminology needed to understand HodgeRank, which we detail here. Following the notation used in Jiang et al. [15] and Colley et al. [3], we define Λ to be the set of voters and V to be the set of elements that are voted on. For $\alpha \in \Lambda$, we denote V_α to be the set of elements rated by voter α. Similarly, we let Λ_{ij} denote the set of voters who rated both elements i and j.

In our example, we have the following:

$$\Lambda = \{\text{Patient 1, Patient 2, Patient 3}\}$$

$$V = \{\text{Fever, Sore Throat, Cough, Nausea}\}$$

$$V_{\text{Patient 2}} = \{\text{Fever, Sore Throat, Cough}\}$$

Also, we define the rankings as $R : \Lambda \times V \to \mathbb{R}$. For example, if voter α gave element i a score of 5, we would say $R(\alpha, i) = 5$.

Using this terminology, we find a universal rating, which is a rating that applies to all elements that have been voted on.

2.2 Graph Building

The HodgeRank method involves building a complete graph with $|V|$ nodes that encodes information from our entire data set. In graph theory, a complete graph is a graph that has an edge connecting every pair of nodes. To build a graph that encodes the entire data set, we create one graph for each voter.

Using the elements in V_α as nodes, we form a complete graph for every voter where each node represents an element (see Fig. 1). Note in the example below that we replace the name of each symptom (fever, sore throat, cough, nausea) with letters (A, B, C, D), respectively.

Let E denote the set of all edges. For every edge, we define an orientation by indiscriminately designating one node to be the sink node and the other to be the source node. To keep things simple, we let nodes that are indexed earlier be the source nodes (Fig. 2).

The relationship between pairs of nodes is described with the pairwise comparison function, $f^\alpha(i, j)$, where α is a voter.

$$f^\alpha(i, j) = R(\alpha, j) - R(\alpha, i) \tag{1}$$

We can now define one graph, G, pertaining to all voters' data. G is a complete graph containing every alternative in V, so long as it has been voted on, as well as $\binom{|V|}{2}$ edges.

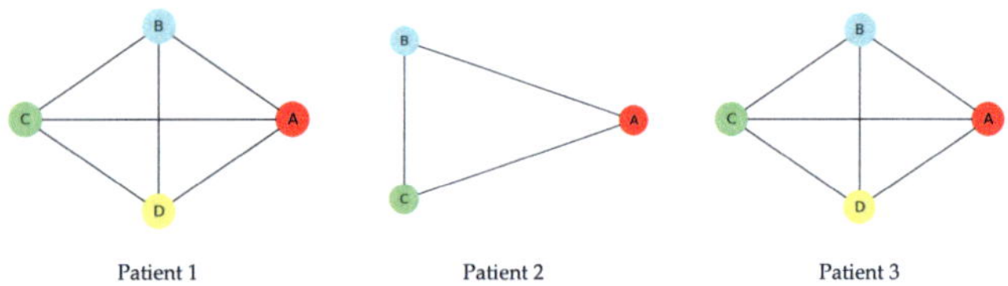

Fig. 1 Complete graphs for patients 1–3. The letters represent symptoms as follows: A (fever), B (sore throat), C (cough), and D (nausea)

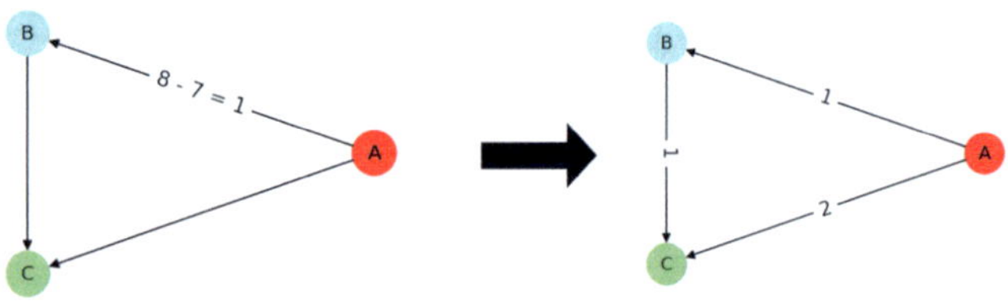

Fig. 2 Patient 2's graph with pairwise comparisons. The orientation and weights are defined by the pairwise comparison function. An example of how the score is calculated

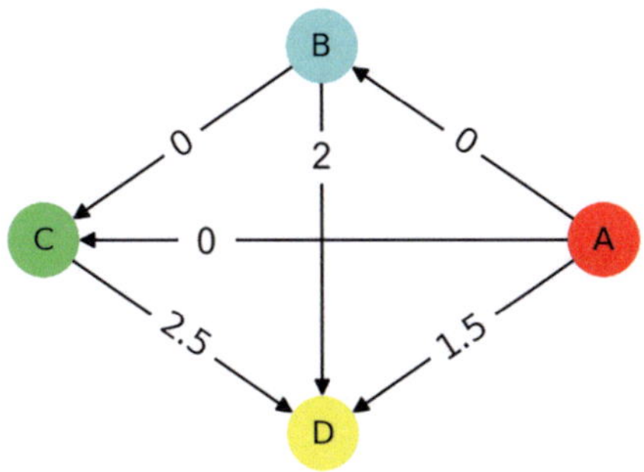

Fig. 3 The edge flow of the entire group's graph based on the ratings in Table 1. The letters are symptoms. The orientation and weights are defined by the pairwise comparison function

We should take into higher consideration pairs of elements that were voted on by many people. With this in mind, we define the weights for each edge as the number of voters who rated both alternatives:

$$\omega_{ij} = |\Lambda_{ij}| \tag{2}$$

where Λ_{ij} is the set of voters who rated both i and j.

The edge flow of the entire group's graph, $f : V \times V \rightarrow R$, represents the average pairwise difference of each edge:

$$f(i, j) = \frac{1}{|\Lambda_{ij}|} \sum_{\alpha \in \Lambda_{ij}} f^{\alpha}(i, j) \tag{3}$$

We show the edge flow of the aforementioned example in Fig. 3.

Later in this paper, we'll refer to the edge flow as the vectorized version of f, indexed by the set of edges E, such that $\mathbf{f} \in R^{|E|}$.

2.3 Least Squares Problem

Our goal is to find a universal rating $r : V \rightarrow R$ that maps every element to its relative rating. By comparing r to the data processed in our graph, we can evaluate the efficacy of our rating. A good choice for r should agree highly with our edge flow, so for each pair of nodes, we aim to minimize

$$f(i, j) - (r(j) - r(i)) \tag{4}$$

We also take into account the number of voters who evaluated both i and j: ω_{ij}. Edges corresponding to item pairs rated by many voters should be given greater weight than those rated by few. This handles the imbalanced nature of our data. Thus, we arrive at the function we use to judge the efficacy of any ranking r:

$$\sum_{i,j \in V} \omega_{ij}(f(i,j) - (r(j) - r(i))^2 \quad (5)$$

Notably, multiplying by ω_{ij} does not boost any element's rating. Instead, it places more importance (or lack thereof) on the balance of $f(i, j)$ and the difference of the universal ratings.

2.4 Assessing Inconsistencies in the Graph

In many data sets, there are contradictions in the graph; paradoxical cycles often arise in G, especially when each voter only reviews some of the elements in V. Let Fig. 4 be a graph processed using Hodge, where the values along the edges are edge flows.

In Fig. 4, the edge flows suggest that element A should be scored lower than element B, which should be scored lower than element C, which should be scored lower than element A. This is paradoxical. Thus, defining a consistent global ranking becomes infeasible in such scenarios. In their paper proposing the HodgeRank algorithm, Jiang et al. [15] propose a method to quantify the extent of local inconsistencies in a vector, which we call **c**. A detailed derivation of **c** is not shown in this work, but for interested readers, we provide the basic algorithm in Sect. 2.6.

For reference later in this paper, **c** is indexed by C where C is the set of all 3-cycles in the graph G. Similar to the set edges, we arbitrarily define an orientation for each 3-cycle in C. Also, we can observe that any paradoxical cycle of arbitrary length can be decomposed into a combination of 3-cycles since 3-cycles are the simplest type of cycle. Therefore, it is sufficient to only consider 3-cycles for any inconsistency in the graph.

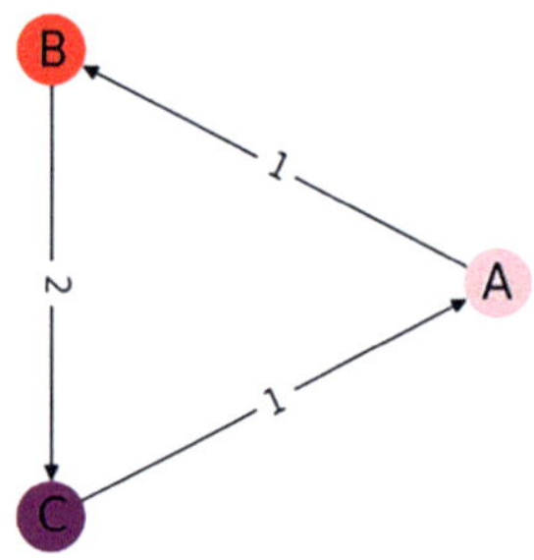

Fig. 4 Example of edge inconsistency. The letters are voted elements. The orientation and weights are defined by the pairwise comparison function

2.5 Hodge Decomposition

In this section, we demonstrate how Hodge decomposition is used to show that it is possible to find a ranking $\mathbf{r} \in R^{|V|}$ (where $\mathbf{r}$ is the vectorized form of r indexed by V) and local consistency $\mathbf{c} \in R^{|C|}$ for any edge flow $\mathbf{f} \in R^{|E|}$. Local consistency means that any 3-cycles in the graph should not create contradictions with each other or the larger structure.

First, we must define the boundary operators of the graph.

The negative divergence, denoted by $\partial_1 : R^{|E|} \to R^{|V|}$, is defined as

$$(\partial_1)_{ij} = \begin{cases} -1, & \text{if } v_i \text{ is the source node in } e_j, \\ 1, & \text{if } v_i \text{ is the sink node in } e_j, \\ 0, & \text{else.} \end{cases}$$

Note that the divergence is simply ∂_1^T.

The curl, $\partial_2 : R^{|C|} \to R^{|E|}$, is defined as

$$(\partial_2)_{ij} = \begin{cases} 1, & \text{if } e_i \in C_j \text{with same orientation as } C_j, \\ -1, & \text{if } e_i \in C_j \text{with same orientation as } C_j, \\ 0, & \text{else.} \end{cases}$$

The 1-Hodge Laplacian is $L_1 = \partial_1^T \partial_1 + \partial_2 \partial_2^T$. Due to Hodge decomposition [3, 15, 19], we can show the following:

$$R^{|E|} = im(\partial_1^T) \oplus ker(L_1) \oplus im(\partial_2^T)$$

$R^{|E|}$ denotes the vector space of all edge flows in the graph. $im(\partial_1^T)$ is the subspace of edge flows that are gradient flows of the score function. $ker(L_1)$ is the kernel of the Laplacian operator L_1 on edges. It corresponds to the space of harmonic edge flows that are both curl-free and divergence-free, representing equilibrium conditions in the graph. $im(\partial_2^T)$ corresponds to the curl operator, the subspace of locally cyclic pairwise rankings with nonzero curls.

Therefore, for any $\mathbf{f} \in R^{|E|}$, we can find $\mathbf{r} \in R^{|V|}$, $\mathbf{c} \in R^{|C|}$, and $\mathbf{x_h} \in ker(L_1)$ such that

$$\mathbf{f} = \partial_1^T \mathbf{r} + \partial_2 \mathbf{c} + \mathbf{x}_h$$

This shows that for any **f**, one is able to find a ranking **r** and a local consistency **c**. An extensive explanation of Hodge decomposition can be found in Lim et al.'s work [19], with implementations demonstrated in [3, 15].

2.6 Solving with Linear Algebra

Using linear algebra, the minimization problem from Sect. 2.3 can be rewritten. First, let $\boldsymbol{\omega}$ be the vectorized version of ω indexed by E. Using the negative divergence of the system, the minimization becomes

$$\min_{\mathbf{r}\in R^{|\Lambda|}} ||\mathbf{f} - \partial_1^T \mathbf{r}||_W^2 \tag{6}$$

where $W \in \mathbb{R}^{|E|\times|E|}$ is the diagonal matrix whose entries are $\boldsymbol{\omega}$.

Using some basic calculus, the minimization reduces to the following:

$$\partial_1 W \partial_1^T \mathbf{r} = \partial_1 W \mathbf{f} \tag{7}$$

The only unknown in this equation is $\mathbf{r}$, so this is an $A\mathbf{x} = \mathbf{b}$ problem. The matrix $\partial_1 W \partial_1^T$ is a well-studied matrix called the graph Laplacian, which has no inverse, so when solving for $\mathbf{r}$, the pseudo-inverse must be taken. The time complexity of this operation is $O(n^3)$ where $n = |V|$.

Following a similar derivation, the solution of $\mathbf{c}$ follows the same pattern. The minimization problem reduces to

$$\min_{\mathbf{c}\in R^{|E|}} ||\mathbf{f} - \partial_2 \mathbf{c}||_W^2 \tag{8}$$

Similarly, this reduces to

$$\partial_2^T W \partial_2 \mathbf{c} = \partial_2^T W \mathbf{f}, \tag{9}$$

This equation has one unknown, $\mathbf{c}$, which can be solved with complexity $O(n^3)$, where $n = |E|$.

It should be noted that while $\mathbf{r}$ represents the universal ranking on the set of rated elements, $\mathbf{c}$ represents the consistency of the graphical model created from the raw data. Thus, the two are calculated independently.

2.7 Methods to Reduce Run Time

The usability of this method is impacted by the potentially great computational run time. The most computationally expensive step of the method is taking the pseudo-inverse of the graph Laplacian, which is an operation of order $O(n^3)$ where n is $|V|$.

There are a few established methods to reduce the cost of solving for $\mathbf{r}$ in $\partial_1 W \partial_1^T \mathbf{r} = \partial_1 W \mathbf{f}$.

One is the algebraic multigrid (AMG) method, which lets $\mathbf{x} \in R^n$ be approximated with linear complexity, $O(n)$, where $\mathbf{x}$ is the only unknown in $A\mathbf{x} = \mathbf{b}$. Furthermore, the work can be done in parallel across multiple machines, making it an ideal choice for implementing HodgeRank when the number of elements to be ranked is very large. In short, the method is a successive subspace correction method that recursively partitions the solution space to approximate the best solution. More information can be found in the article by Falgout et al. [9], which introduces the method, as well as in more recent advancements of the adaptive AMG [14] and spanning tree-based AMG [13] for better convergence results.

Tai et al. [22] detail a successive subspace correction method (SSC), which is a general convex optimization algorithm that decomposes the original problem into a number of smaller optimization problems.

Additionally, there is a method created by Drineas et al. [7, 8], which employs a row sampling method to approximate large-scale matrix multiplication and reduce graph size.

We introduce a new method in Sect. 3 that falls under the umbrella of dimensional reduction and is specifically suited to reduce the run time of the least squares solver on universal ranking problems.

3 Grouping Method

In this section, we propose an algorithm that reduces the computational cost of the method by reducing the size of the matrix that we take the pseudo inverse of.

The key idea is that partitioning the data into smaller subsets can significantly reduce overall run time. An in-depth description of the method is given in the next section.

3.1 *Naive Ranking*

The first step in the grouping method is to obtain a naive ranking of elements, which we denote $r_0 : V \to R$. Several strategies can be employed for this initial step.

- **Arithmetic mean of rating:** In this ranking, elements are ordered by their average rating. This is similar to sorting search results by "top rated":

$$r_0(i) = \frac{\sum_{\alpha \in \Lambda_i} R(\alpha, i)}{|\Lambda_i|} \tag{10}$$

- **Arithmetic mean of edge flow:** Here, r_0 is the result of averaging the edge flow between one node and every other node. The edge flow refers to f, which we defined in Eq. 3:

$$r_0(i) = \frac{1}{|V|} \sum_{j \in V} f(j, i) \tag{11}$$

- **Weighted arithmetic mean of edge flow:** Here, r_0 is the average edge flow between one node and every other node weighted by ω:

$$r_0(i) = \frac{\sum_{j \in V} \omega_{ij} f(j, i)}{\sum_{j \in V} \omega_{ij}} \tag{12}$$

Defining the naive rank as the weighted arithmetic mean of the edge flow yielded better empirical results, as we include in the later section. We assume this is the case because it incorporates the edge flow and the edge weight, which are both key components of the final step of the HodgeRank.

3.2 Splitting into Groups

Next, we evenly split the group into k subgroups by their naive rank; that is, the highest-scoring elements and lowest-scoring elements are kept together. See expression (13) for an example where the elements in V, which are indexed by their naive ranking, are split into three subgroups:

$$V = [v_1, v_2, v_3, v_4, v_5, v_6, v_7, v_8, v_9] \Longrightarrow \begin{aligned} V_1 &= \{v_1, v_2, v_3\} \\ V_2 &= \{v_4, v_5, v_6\} \\ V_3 &= \{v_7, v_8, v_9\} \end{aligned} \tag{13}$$

It is possible to run the HodgeRank on each of the groupings to achieve a universal rating, but doing so omits much data (see Fig. 5).

To demonstrate this, we count the number of edges omitted. The graph that might be formed in the normal HodgeRank algorithm has n nodes and $\binom{n}{2}$ edges. Splitting the elements into k groupings and building graphs for each of the groupings with $\lceil \frac{n}{k} \rceil$ nodes results in a total of at most $k\binom{\lceil \frac{n}{k} \rceil}{2}$ edges. The number of edges that would be dropped is

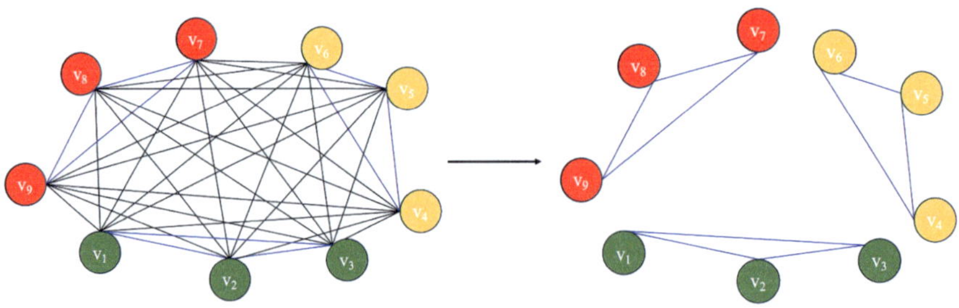

Fig. 5 Running the HodgeRank on smaller groups; all omitted edges in black

$$\binom{n}{k} - k\binom{\lceil \frac{n}{k} \rceil}{2} \approx \frac{n(n-1)}{2} - \frac{n(\frac{n}{k}-1)}{2} = \frac{n(n-\frac{n}{k})}{2}$$

Ideally, all edges should contribute to the final ranking. To resolve this issue, we introduce pseudo-nodes into each subgroup, which will be placeholders for subgroups connections to other subgroups.

3.3 Adding Pseudo-nodes

Let V be the set of elements and $\{V_1, V_2, \ldots, V_k\}$ be the set of subgroups. Then, we let W_n for $n \in 1, 2, \ldots, k$ denote the set of nodes that HodgeRank will run on such that $W_n = \{v, V_m | v \in V_n, m \neq n\}$. We modify the definitions of edge flow and edge weight to suit the introduction of subgroups as pseudo-nodes (see example in Fig. 6):

$$f(i,j) = \begin{cases} \frac{1}{|\Lambda_{ij}|} \sum_{\alpha \in \Lambda_{ij}} f^{\alpha}(i,j) \\ \sum_{v \in i} \left[\frac{1}{|\Lambda_{vj}|} \sum_{\alpha \in \Lambda_{vj}} f^{\alpha}(v,j) \right] \\ \sum_{v \in i, u \in j} \left[\frac{1}{|\Lambda_{vu}|} \sum_{\alpha \in \Lambda_{vu}} f^{\alpha}(v,u) \right] \end{cases} \tag{14}$$

$$w(i,j) = \begin{cases} |\Lambda_{ij}| \\ \sum_{v \in i} |\Lambda_{vj}| \\ \sum_{v \in i, u \in j} |\Lambda_{vu}| \end{cases} \tag{15}$$

where the first situation for both Eqs. 14 and 15 is when i and j represent single nodes, the second situation is when i represents a subgroup and not j, and the third situation is when i and j represent subgroups.

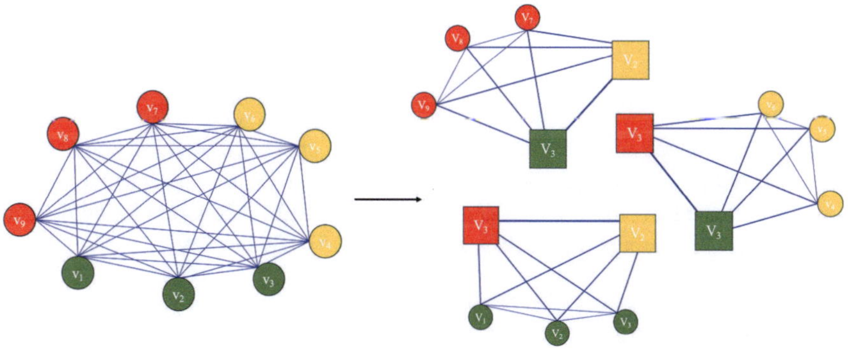

Fig. 6 Groupings with pseudo-nodes; V_1 represents the elements in the first group $\{v_1, v_2, v_3\}$, and similarly for V_2 and V_3. Edges on the right represent all edges between respective nodes, with thicker edges representing more edges from the original graph. Pseudo-nodes represented by squares

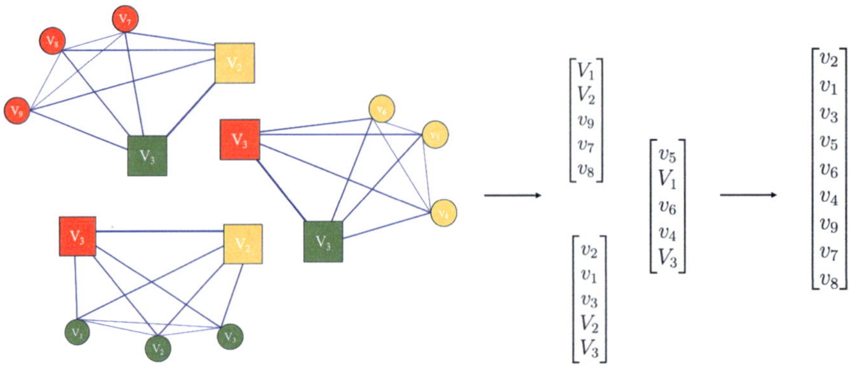

Fig. 7 Universal ranking from grouping method. Note that pseudo-nodes are omitted from the final ranking

Finally, HodgeRank is executed independently on each W_n, yielding k rankings. We achieve our final ranking by stacking the groupings' rankings on top of each other according to their order from the naive rank (see example in Fig. 7).

4 Results

To test the grouping method, we used the IMDb movie data set [1]. The accuracy of a rating found using HodgeRank with grouping is measured by its similarity to the ranking found using the original HodgeRank algorithm. We employ the rank-biased overlap (RBO) [23], which we detail later in the paper. Finally, from our experimentation, we discuss how to pick the best group size by balancing accuracy and run time.

4.1 Data

For our experimentation, we used the IMDb movie data set, which is a collection of IMDb movies updated on 27 December 2020 including their UserID, MovieID, and ratings collected by Vahid Baghi and uploaded to IEEE Dataport [1]. The data set offers 4,669,820 ratings from 1,499,238 users to 351,109 movies.

4.1.1 Preprocessing

Since the HodgeRank algorithm uses pairwise differences to build edge flows in the graphs, inspired by how Page et al. [20] handled dangling links by removing them, we first filter out users who have only voted for one movie, since the user subgraph would only be one node. To ensure all the movie ratings are statistically meaningful, we also filter the movies by requiring they get ratings from at least two different users.

Then, we derive the diagonal weighted matrix W and the negative divergence matrix ∂_1 based on the HodgeRank with the selected movies and the users who rated them. We define the edge flows $\mathbf{f}$ according to the ratings that are related to the filtered movies. We formulate the matrix $L = \partial_1 W \partial_1^T$, which is the graph Laplacian, and $\mathbf{b} = \partial_1 W\mathbf{f}$, which is the right-hand side in Eq. 7. The largest connected component of the graph Laplacian in the IMDb movie data set is taken out and analyzed. After filtering, the set contains a total of 62,917 movies, including 29,945 movies that were rated by at least two people in this set. To test the grouping method outlined in Sect. 3, we also select different sizes of graphs corresponding to different divides of MovieID ranging from 1000 to 20,000 and only maintain the largest connected component of the subgraph. Thus, we can test the performance of the proposed algorithm on various sizes n for its robustness and accuracy. Finally, we produce the ranking $\mathbf{r}$ by solving each small HodgeRank linear system $L\mathbf{r} = \mathbf{b}$ using AMG, as suggested in the paper by Colley et al. [3], which implements AMG directly with HodgeRank.

4.1.2 Organizing by Popularity

Figures 8 and 9 present breakdowns of the number of voters per the n'th most popular movie. Upon running the grouping method on the IMDb data set, we

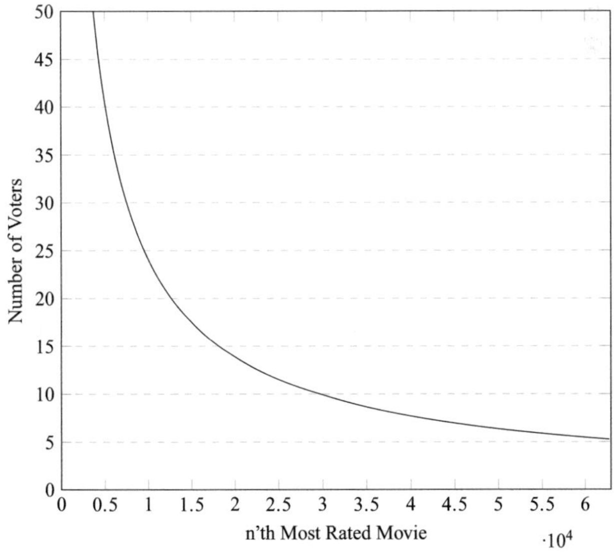

Fig. 8 Average number of voters for first n most-rated movies (all movies)

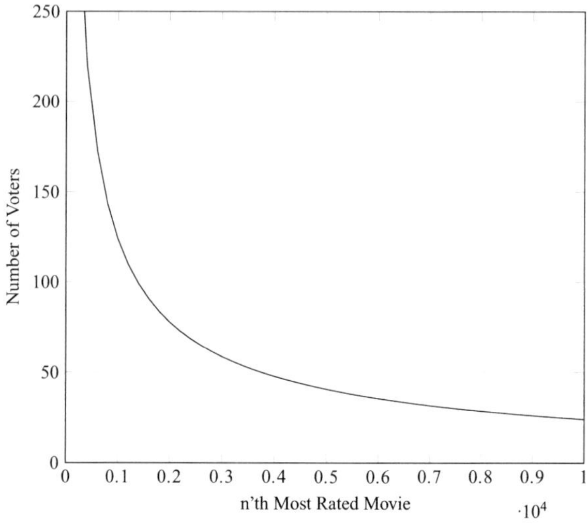

Fig. 9 Average number of voters for first n most-rated movies (top 10,000 movies)

observe that the average number of voters varies significantly across the data set. The most-rated movie has 1441 total reviews, while the least-rated movie receives around 5 reviews. This imbalanced data set motivates us to organize the data set by popularity before truncating them into different sizes. We believe that the most-rated movies contain more information across the different users and potentially help preserve the accuracy of the approximated rankings.

4.2 Computing Environment

The numerical tests are conducted with a 1.80 GHz Intel Core i7-8550U CPU, a quad-core processor, and 16 GB of RAM.

4.3 Accuracy

To evaluate ranking accuracy, we employ the rank-biased overlap (RBO) statistical method [23] to compare our approximated rankings resulting from the grouping method with the traditional HodgeRank ranking. RBO is a similarity measure for ranked lists, which takes into account the position of items in ranked lists. Therefore, it is a more appropriate measure of ranking similarity than Kendall rank correlation coefficient (Kendall's tau) [17], which scores similarity of two rankings through pair-wise concordance of elements found in both lists. RBO uses weights for each rank position, which are derived from a convergent series. The goals of RBO are to handle non-conjointness, weight high ranks more heavily than low, and be monotonic with increasing depth of evaluation:

$$RBO(S, T, p) = (1 - p) \sum_{d=1}^{\infty} p^{d-1} A_d$$

where S and T are two distinct ranked lists and d is the depth of the list. A_d is the agreement between S and T by the proportion of the overlap size over the depth, essentially $\frac{X_{S,T,d}}{d}$, where $X_d = |I_{S,T,d}|$ such that I is the intersection between the subset of lists S and T such that each subsetted list contains the top d ranked elements.

The parameter p represents the steepness of decline in weights. For example, a smaller p indicates a more top-weighted metric.

The RBO score falls into the interval of [0, 1], where 0 means disjoint (i.e., no correlation) and 1 means identical (i.e., perfect correlation).

The following is our derivation of RBO, which changes the infinite summation into a finite one. Starting from the original RBO, we have

$$RBO(S, T, p) = (1 - p) \sum_{d=1}^{\infty} p^{d-1} A_d$$

Taking data up to some finite depth of k, we have

$$RBO_{\text{truncated}}(S, T, p, k) = (1 - p) \sum_{d=1}^{k} p^{d-1} A_d$$

To account for the ranks beyond k, we assume that the pattern observed up to depth k continues. Let $X_k = \sum_{d=1}^{k} A_d$ be the cumulative agreement up to depth k. Then, the average agreement up to depth k is $\frac{X_k}{k}$, which we assume that it holds for all depths beyond k as well. Therefore, the agreement beyond k can be written as

$$RBO_{beyond}(S, T, p, k) = (1 - p) \sum_{d=k+1}^{\infty} p^{d-1} \cdot \frac{X_k}{k}$$

Using geometric series, we obtain

$$\sum_{d=k+1}^{\infty} p^{d-1} = p^k \sum_{d=0}^{\infty} p^d = \frac{p^k}{1 - p}$$

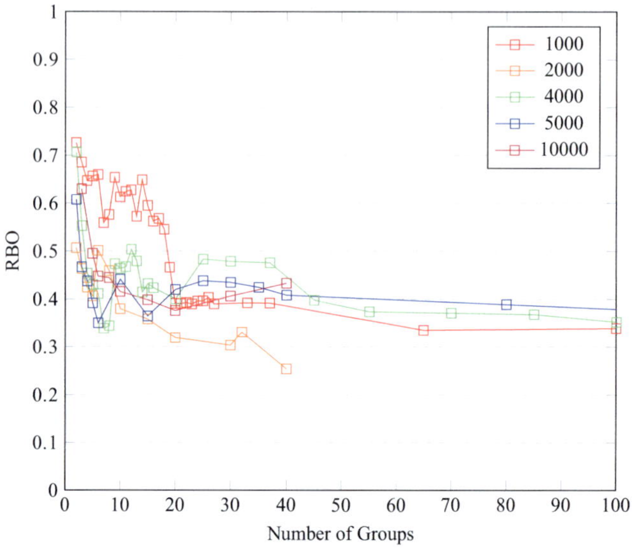

Fig. 10 The effect of the number of groups on ranking accuracy

We can rewrite the agreement beyond k as the following:

$$RBO_{beyond}(S, T, p, k) = \frac{X_k}{k} \cdot p^k$$

We can combine with the truncated RBO:

$$RBO(S, T, p, k) = \frac{X_k}{k} \cdot p^k + (1 - p) \sum_{d=1}^{k} p^{d-1} A_d$$

Letting $A_k = \frac{X_k}{k}$, we have

$$RBO(S, T, p, k) = A_k p^k + (1 - p) \sum_{d=1}^{k} p^{d-1} A_d$$

Figure 10 shows higher RBOs when very few groups are assigned for the trials with 1000, 2000, 4000, 5000, and 10,000 movies. This is expected, as when group size $k = 1$, it is essentially applying the HodgeRank algorithm on some most-rated movie subsets, and the ranking should be highly correlated to the HodgeRank ranking on the entire movie set. As the number of groups k grows, the RBO scores for all sets drop and most of them converge to around 0.4. This is because, as k grows to the size of the movie set, the ranking would reflect the naive rank instead of the HodgeRank, resulting in a low RBO score with the HodgeRank.

Another observation is that when the movie set size is smaller (e.g.,$n = 1000$), increasing the group number k does not seem to drop the RBO by a lot. This likely occurs because the group with 1000 movies has more movies that were rated by more people than the other sets. The accuracy of the proposed grouping method hinges on the comparison of ratings among distinct movie groups. Consequently, when a movie has ratings from users spanning diverse groups, the edge flows across these groups become more representative, culminating in a higher overall rating accuracy.

Overall, the rankings found using the grouping method show a high correlation with the ranking found using just the HodgeRank. As we see in the next section, the run time is greatly reduced by using groupings.

4.4 Run Time

First, we calculate the theoretical time complexity for this method and recommend a general formula for selecting the optimal number of groups.

Since the most computationally expensive step in both algorithms is taking the pseudo-inverse, we focus on this operation. The complexity of directly computing the pseudo-inverses during HodgeRank with grouping is

$$O(k(\frac{n}{k} + k - 1)^3) \tag{16}$$

Here, $\frac{n}{k} + k - 1$ is the number of nodes including the pseudo-nodes in each group, and directly solving this group's pseudo-inverse is $O(\frac{n}{k} + k - 1)^3$. Since we have to solve it for all k even groups, $O(k(\frac{n}{k} + k - 1)^3)$ is the total complexity.

The selection of k gives us freedom in balancing accuracy and efficiency. As mentioned before, when k is small, the ranking algorithm is more similar to HodgeRank and takes more time. However, a large k results in a fast naive ranking. When the item set is relatively small and accuracy is the priority, we recommend using as few groups as possible, with two groups being ideal, while maintaining a sufficiently small run time. As the number of items being ranked increases, it may not be feasible to compute the universal ranking with two groups; thus, we generally recommend using $k = log_{21}(n)$, where k is the number of groups. This recommendation comes from our empirical observation of the group size, which will balance high conformance to the original ranking with feasible compute times; however, it is recommended to use the least number of groups as their computational resources allow.

Plugging our recommended k into Eq. 16, the run time of approximating the ranking drops from $O(n^3)$ to $O(\frac{n^3}{log_{21}^2(n)})$, even when using a direct solve.

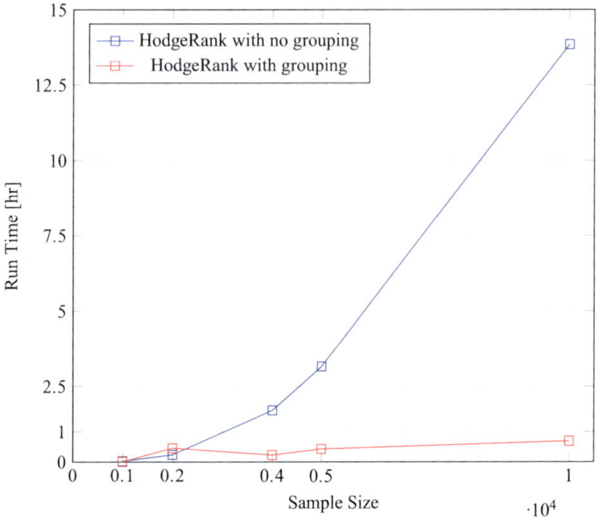

Fig. 11 Sample size versus time

Using the IMDb movie rating data set, we obtained the following run times. All run times are measured on trials where $k = 2$, which gave the most accurate results. One can increase the group size k for more robust approximations.

As Fig. 11 shows, the run times of all trials, which used the grouping method, took less than an hour to run, whereas the longest run time for the original HodgeRank algorithm is 13.85 hours, for 10,000 movies.

In our pre-processing, we used the n movies that were voted on by the greatest number of people for evaluation of HodgeRank with grouping, where n is the sample size. Thus, as sample sizes become bigger, the resulting graphs are more and more sparse. Although the sparsity of our graph may be negatively correlated with run time improvements, we clearly see that the most sparse data set we tested on (sample size $=$ 10,000) saw major improvements in run time.

5 Conclusion

HodgeRank is an algorithm that provides a ranking on data sets that may feature bias and incompleteness. It also offers a metric for judging the correctness of the ranking as well as a quantization of inconsistencies in the graph.

We propose a new group-based method to decrease time complexity while upholding ranking integrity. This approach entails segmenting the set into distinct groups to diminish the matrix inversion's dimensionality. Concretely, we achieve fast ranking through naive ranking methods, partition the item sets into groups guided by the naive ranking, and introduce pseudo-nodes to each subgraph to retain cross-group connections. This strategy effectively dissects the extensive Laplacian system generated by the HodgeRank algorithm into more manageable components, each amenable to efficient processing.

The proposed grouping method introduces a group size parameter k, which controls whether the resulting ranking is more similar to the HodgeRank or the naive rank. When choosing a smaller group size k, the grouping method yields a ranking that is highly correlated to the original HodgeRank while decreasing its run time. When setting k at a higher value, the resulting ranking is closer to the naive rank, which drifts away from the HodgeRank results. The choice of k represents the balance between accuracy and efficiency.

Since the grouping method is meant to address issues with data sets with many elements, we theoretically analyzed its run time and showed that the complexity of direct solving the least squares is reduced from $O(n)$ to $O(k(\frac{n}{k} + k - 1)^3)$, where n is the number of rated items and k is the number of groups. Picking a good number of groups is important to reduce runtime while maintaining accuracy; for n nodes, we found that a practical number is $k = \log_{21} n$ to yield better numerical results.

Further work on the grouping method might address the issue of edge cases, where nodes that are sorted into the wrong tier at the start cause errors in the final ranking. Splitting up the groups in a more sophisticated way, such as adaptive splitting and groups with overlapping elements instead of evenly splitting, may help address this issue. Additionally, it would be interesting to look into the performance and run time of embedded groupings for data sets with a great number of elements.

Competing Interests The work of Junyuan Lin was partially supported by the National Science Foundation under grant DMS-2418877.

References

1. Baghi, V.: Imdb users' ratings dataset (2020). https://doi.org/10.21227/br41-bd49
2. Cohen, W.W., Schapire, R.E., Singer, Y.: Learning to order things. In Advances in Neural Information Processing Systems, vol. 10 (1997)
3. Colley, C., Lin, J., Hu, X., Aeron, S.: Algebraic multigrid for least squares problems on graphs with applications to HodgeRank. In: 2017 IEEE International Parallel and Distributed Processing Symposium Workshops (IPDPSW), pp. 627–636. IEEE, Piscataway (2017)
4. Colley, W.: Colley's Bias Free College Football Ranking Method. Princeton University, Princeton (2002)
5. Crammer, K., Singer, Y.: Pranking with ranking. In: Advances in Neural Information Processing Systems, vol. 14 (2001)
6. David, H.A.: The Method of Paired Comparisons, vol. 12. London (1963)
7. Drineas, P., Kannan, R., Mahoney, M.W.: Fast monte carlo algorithms for matrices I: approximating matrix multiplication. SIAM J. Comput. **36**(1), 132–157 (2006)
8. Drineas, P., Mahoney, M.W.: Effective resistances, statistical leverage, and applications to linear equation solving (2010). arXiv preprint arXiv:1005.3097
9. Falgout, R.D.: An introduction to algebraic multigrid. Tech. rep., Lawrence Livermore National Lab.(LLNL), Livermore (2006)

10. Ford, Jr., L.R.: Solution of a ranking problem from binary comparisons. Am. Math. Monthly **64**(8P2), 28–33 (1957)
11. Freund, Y., Iyer, R., Schapire, R.E., Singer, Y.: An efficient boosting algorithm for combining preferences. J. Mach. Learn. Res. **4**(Nov), 933–969 (2003)
12. Guttman, L.: An approach for quantifying paired comparisons and rank order. Ann. Math. Stat. **17**(2), 144–163 (1946)
13. Hu, X., Lin, J.: Solving graph laplacians via multilevel sparsifiers. SIAM J. Sci. Comput. **46**(2), S378–S400 (2024)
14. Hu, X., Lin, J., Zikatanov, L.T.: An adaptive multigrid method based on path cover. SIAM J. Sci. Comput. **41**(5), S220–S241 (2019)
15. Jiang, X., Lim, L.H., Yao, Y., Ye, Y.: Statistical ranking and combinatorial Hodge theory. Math. Program. **127**(1), 203–244 (2011)
16. Kano, M., Sakamoto, A.: Ranking the vertices of a paired comparison digraph. SIAM J. Algebraic Discrete Methods **6**(1), 79–92 (1985)
17. Kendall, M.G.: A new measure of rank correlation. Biometrika **30**(1/2), 81–93 (1938)
18. Kendall, M.G., Smith, B.B.: On the method of paired comparisons. Biometrika **31**(3/4), 324–345 (1940)
19. Lim, L.H.: Hodge Laplacians on graphs. SIAM Rev. **62**(3), 685–715 (2020)
20. Page, L., Brin, S., Motwani, R., Winograd, T.: The pagerank citation ranking : Bringing order to the web. In: The Web Conference (1999). https://api.semanticscholar.org/CorpusID:1508503
21. Stüben, K.: A review of algebraic multigrid. Numerical Analysis: Historical Developments in the 20th Century, pp. 331–359 (2001)
22. Tai, X.C., Xu, J.: Global and uniform convergence of subspace correction methods for some convex optimization problems. Math. Comput. **71**(237), 105–124 (2002)
23. Webber, W., Moffat, A., Zobel, J.: A similarity measure for indefinite rankings. ACM Trans. Inform. Syst. **28**(4), 1–38 (2010)
24. Wei, R.K.J., Wee, J., Laurent, V.E., Xia, K.: Hodge theory-based biomolecular data analysis. Sci. Rep. **12**, 9699 (2022)
25. Xu, Q., Xiong, J., Cao, X., Huang, Q., Yao, Y.: Evaluating visual properties via robust HodgeRank. Int. J. Comput. Vis. **129**, 1732–1753 (2021)

A Comparison Study of Graph Laplacian Computation

Michela Marini, Haiyan Cheng, Cristina Garcia-Cardona, Weihong Guo, Sara Hahner, Yuan Liu, Yifei Lou, and Sui Tang

1 Introduction

In recent years, graph signal processing has become popular in many data-driven applications [4, 8, 15, 18, 22, 23], offering a versatile framework for representing and analyzing relationships within complex datasets. By using nodes to signify entities

M. Marini
Department of Mathematics, University of Houston, Houston, TX, USA
e-mail: mmarini2@uh.edu

H. Cheng
School of Computing and Information Sciences, Willamette University, Salem, OR, USA
e-mail: hcheng@willamette.edu

C. Garcia-Cardona
Los Alamos National Laboratory, Los Alamos, NM, USA
e-mail: cgarciac@lanl.gov

W. Guo
Department of Mathematics, Applied Mathematics and Statistics, Case Western Reserve University, Cleveland, OH, USA
e-mail: wxg49@case.edu

S. Hahner
Fraunhofer Institute for Scientific Computing and Algorithms (SCAI), Sankt Augustin, Germany

Y. Liu
Department of Mathematics, Statistics and Physics, Wichita State University, Wichita, KS, USA
e-mail: yuan.liu@wichita.edu

Y. Lou (✉)
Department of Mathematics & School of Data Science and Society, University of North Carolina at Chapel Hill,, Chapel Hill, NC, USA
e-mail: yflou@unc.edu

S. Tang
Department of Mathematics, University of California Santa Barbara, Santa Barbara, CA, USA
e-mail: suitang@ucsb.edu

C. Garcia-Cardona, H. Lee (eds.), *Advances in Data Science*, Association for Women in Mathematics Series 37, https://doi.org/10.1007/978-3-031-87804-6_8

and edges to denote connections between them, graphs can model a wide array of structures, from social networks and biological systems to transportation grids and recommendation engines.

Consider a collection of data points $\{\mathbf{x}_i\}_{i=1}^n \subseteq \mathbb{R}^\nu$, where n is the number of points and ν is the dimension of each feature vector. One constructs a graph $G(V, E)$ by treating each point as a vertex $v_i \in V$, $i = 1, \ldots, n$, and E an edge connectivity representing specific relations between vertices. E can be represented by a matrix, called an adjacency matrix. Specifically, for a graph with n nodes, the adjacency matrix, denoted by A, is an $n \times n$ matrix where each element a_{ij} indicates whether there is an edge from node i to node j. The value of a_{ij} is typically either 1 (indicating the connection) or 0 (no connection). A generalization of the adjacency matrix is a similarity matrix W associated with a weighted graph where each edge is characterized by a real weight w_{ij} representing application-specific meanings, usually a measure of how similar nodes i and j are. This paper focuses on an undirected and unsigned graph corresponding to a symmetric and nonnegative weight function, i.e., $w_{ij} = w_{ji} \geq 0, \forall 1 \leq i, j \leq n$.

The graph Laplacian, derived from the similarity matrix of a weighted graph, is a fundamental tool in spectral graph theory [11]. Let the degree matrix D be a diagonal matrix where each diagonal element is defined by $d_{ii} = \sum_j w_{ij}$. The unnormalized graph Laplacian L, defined as $L = D - W$, encapsulates important structural properties of the graph, such as connectivity and the presence of clusters. For data science applications, it is widely recognized [4, 18] the computational and performance advantages of deploying the symmetric normalized Laplacian, which is defined as

$$L_s = I - D^{-1/2} W D^{-1/2} . \tag{1}$$

The eigenvalues and eigenvectors of L_s are particularly useful, providing insights into graph partitioning [9], clustering [4, 19, 22, 26], machine learning [8, 13], and the behavior of diffusion processes on the graph [10].

However, it is computationally intensive to obtain the similarity matrix and the graph Laplacian, often becoming a bottleneck in dealing with "big data." Specifically, the computational complexity of constructing a graph Laplacian is of the order $O(n^2)$, making it intractable when n is extremely large. In addition, when the graph Laplacian is used in certain applications [4, 23], the eigendecomposition and/or singular value decomposition (SVD) is often required, which is in the computational complexity of $O(n^3)$. Consequently, accelerating the construction of the graph Laplacian together with its decompositions is essential for handling large-scale graph-based applications.

This paper studies three methods to approximate L_s. The first method, called K-nearest neighbors (KNN), involves creating a sparse approximation by computing a small number of pairwise weight functions for each node, resulting in a sparse matrix. The other two methods focus on low-rank approximations and are called Nyström method [14] and its variant using the QR decomposition [6]. Our empirical evaluation of the methods yields the following observations:

- Nyström methods (the original one and its QR-based variant) provide good approximations to the eigendecomposition of the Laplacian for the fully connected graph while considerably reducing computation times since they require computations for only a handful of samples in the dataset. This is observed in both benchmarks and high-dimensional datasets.
- Both Nyström-based methods are particularly advantageous when an eigendecomposition is required for downstream tasks, as they provide efficient algorithms for computing accurate approximations without increasing time demands.
- The KNN method provides an excellent approximation to the Laplacian of the fully connected graph (given that the similarity metric is sufficiently smooth), but requires computations over the entire dataset, which can become intractable for datasets with a large number of nodes.

The rest of the paper is organized as follows. Section 2 provides a brief review of the methods: KNN, Nyström, and QR-based Nyström. We then investigate the performance of these approximations in Sect. 3 in terms of accuracy to approximate the fully connected graph, computational time, and efficiency in applications of classification, clustering, and CT reconstruction. Finally, the conclusions are given in Sect. 4.

2 Method Review

A fully connected weighted graph can be represented via a dense weight matrix W of dimensions $n \times n$, where every pair of nodes is connected with an assigned similarity value. In this work, we use the Gaussian similarity metric, where each weight entry is defined as

$$w_{ij} = \exp\left\{\frac{-d(\mathbf{x}_i, \mathbf{x}_j)^2}{2\sigma^2}\right\}, \quad i, j = 1, \ldots, n, \tag{2}$$

with $d(\mathbf{x}_i, \mathbf{x}_j)$ being the Euclidean distance between the two samples (i.e., vertices) $\mathbf{x}_i$ and $\mathbf{x}_j$, which can be computed as $d_E(\mathbf{x}_i, \mathbf{x}_j) = \|\mathbf{x}_i - \mathbf{x}_j\|_2$, i.e., the conventional measure for calculating the distance between two points in the Euclidean space. Note that $\sigma > 0$ controls the smoothness of the similarity metric, providing more drastic differences when its value is small and more gradual transitions when its value is large. Note that the diagonal element $w_{ii} = 1$ follows the definition in Eq. (2), which is reasonable due to self-similarity.

When dealing with big data, e.g., hyperspectral data where the number of pixels in the image could be in the order of 10^6, the weight matrix presents computational challenges and requires significant storage space. We review three ways to approximate the weight matrix, namely, K-nearest neighbor [12], Nyström method [14], and QR-based Nyström decomposition [6]. In the experimental section, we compare their performance in terms of accuracy and efficiency.

2.1 K-Nearest Neighbor Graph

The K-nearest neighbor (KNN) graph is frequently used in machine learning and data analysis, particularly in pattern recognition, classification, and clustering tasks [24, 27, 29]. As the name suggests, KNN constructs a graph by connecting each node to its K-nearest neighbors based on a chosen distance metric. To do this, one must first determine an appropriate distance metric and select a value for K.

For each data point, a distance metric is computed between this point and the other points, followed by Eq. (2) to obtain the similarity measures between any pair. Subsequently, weights are only stored for the K-nearest neighbors, corresponding to the K largest similarity values. This process results in a sparse weight matrix W with each row having at most $K (\ll n)$ nonzero elements.

The naive KNN does not guarantee a symmetric matrix, since the node i being in the top K neighbor of j does not entail j being in the top K neighbor of i. To make the weight symmetric, we adopt a simple approach by taking the average of the weight and its transpose, i.e., $W \leftarrow \frac{1}{2}(W + W^\top)$. Another alternative is the mutual KNN [20], which is out of the scope of this paper.

2.2 Nyström Method

To reduce the time/space complexity, Fowlkes et al. [14] proposed the Nyström method to approximate the eigenvalues and eigenvectors of $W \in \mathbb{R}^{n\times n}$ by using only p sampled data points with $p \ll n$. Up to permutations, we adopt a block-matrix form to represent the weight matrix W as follows:

$$W = \begin{bmatrix} W_{11} & W_{12} \\ W_{21} & W_{22} \end{bmatrix}, \tag{3}$$

where $W_{11} \in \mathbb{R}^{p\times p}$ is the weight (similarity) matrix between the sampled data points, $W_{12} = W_{21}^\top$ is the one between the sampled points and the unsampled points, and W_{22} is the one between the unsampled points. The idea of Nyström extension is to approximate the matrix W and its corresponding normalized graph Laplacian, L_s defined in Eq. (1), using W_{11} and W_{12}, thereby avoiding the computation of the relatively large matrix W_{22}. Since the matrix L_s involves the degree matrix, we begin by normalizing W so that its degree matrix becomes the identity. In particular, we define a matrix

$$\overline{W} = \begin{bmatrix} W_{11} & W_{21}^\top \\ W_{21} & W_{21} W_{11}^{-1} W_{21}^\top \end{bmatrix}, \tag{4}$$

and its row-sum vector in a block form:

$$\begin{bmatrix} \mathbf{d}_1 \\ \mathbf{d}_2 \end{bmatrix} = \overline{W}\mathbf{1}_n = \begin{bmatrix} W_{11} & W_{21}^\top \\ W_{21} & W_{21}W_{11}^{-1}W_{21}^\top \end{bmatrix} \begin{bmatrix} \mathbf{1}_p \\ \mathbf{1}_{n-p} \end{bmatrix},$$

where $\mathbf{1}_k$ denotes the k-dimensional all one vector. Denoting $\mathbf{s}_1 = \sqrt{\mathbf{d}_1}$ and $\mathbf{s}_2 = \sqrt{\mathbf{d}_2}$, we can normalize the matrices W_{11} and W_{21} by

$$\widetilde{W}_{11} = W_{11} \oslash (\mathbf{s}_1\mathbf{s}_1^\top) \quad \widetilde{W}_{21} = W_{21} \oslash (\mathbf{s}_1\mathbf{s}_2^\top), \tag{5}$$

where $\oslash$ denotes the componentwise division. In the same block format as $\overline{W}$, we define

$$\widetilde{W} = \begin{bmatrix} \widetilde{W}_{11} & \widetilde{W}_{21}^\top \\ \widetilde{W}_{21} & \widetilde{W}_{21}\widetilde{W}_{11}^{-1}\widetilde{W}_{21}^\top \end{bmatrix}. \tag{6}$$

By definition, the degree matrix corresponding to $\widetilde{W}$ becomes the identity, and the symmetric normalized graph Laplacian becomes

$$\widetilde{L}_s = I - \widetilde{W}. \tag{7}$$

Next, we describe the SVD of $\overline{W}$ and use it to represent the symmetric normalized graph Laplacian $\widetilde{L}_s$. We assume $\widetilde{W}_{11}$ is positive definite (by choosing a proper value of σ in Eq. (2)); then it is invertible and we further denote $\widetilde{W}_{11}^{1/2}$ as its square root. We can express $\widetilde{W}$ in the following way:

$$\begin{aligned} \widetilde{W} &= \begin{bmatrix} \widetilde{W}_{11} \\ \widetilde{W}_{21} \end{bmatrix} \widetilde{W}_{11}^{-1} [\widetilde{W}_{11} \quad \widetilde{W}_{21}^\top] \\ &= \left\{ \begin{bmatrix} \widetilde{W}_{11} \\ \widetilde{W}_{21} \end{bmatrix} \widetilde{W}_{11}^{-1/2} U \Sigma^{-1/2} \right\} \Sigma \left\{ \Sigma^{-1/2} U^\top \widetilde{W}_{11}^{-1/2} [\widetilde{W}_{11} \quad \widetilde{W}_{21}^\top] \right\}, \end{aligned} \tag{8}$$

for any diagonal matrix Σ and unitary matrix U, both of which can be determined by the requirement that $V^\top V = I$ with

$$V := \begin{bmatrix} \widetilde{W}_{11} \\ \widetilde{W}_{21} \end{bmatrix} \widetilde{W}_{11}^{-1/2} U \Sigma^{-1/2}.$$

We elaborate on this requirement by expressing it into

$$I = V^\top V = \left\{ \Sigma^{-1/2} U^\top \widetilde{W}_{11}^{-1/2} [\widetilde{W}_{11} \quad \widetilde{W}_{21}^\top] \right\} \left\{ \begin{bmatrix} \widetilde{W}_{11} \\ \widetilde{W}_{21} \end{bmatrix} \widetilde{W}_{11}^{-1/2} U \Sigma^{-1/2} \right\}.$$

Multiplying the above equation from the left by $U\Sigma^{1/2}$ and from the right by $\Sigma^{1/2}U^\top$ yields

$$U\Sigma U^\top = \widetilde{W}_{11} + \widetilde{W}_{11}^{-1/2}\,\widetilde{W}_{21}^\top\,\widetilde{W}_{21}\,\widetilde{W}_{11}^{-1/2},$$

which implies that U and Σ can be obtained by the SVD of the matrix $\widetilde{W}_{11} + \widetilde{W}_{11}^{-1/2}\widetilde{W}_{21}^\top\widetilde{W}_{21}\widetilde{W}_{11}^{-1/2}$. In summary, we have $\widetilde{W} = V\Sigma V^\top$ with $V^\top V = I$.

Using the SVD of $\widetilde{W}$, we further approximate $\widetilde{L}_s$, defined in Eq. (7), by

$$\widetilde{L}_s \approx V(I - \Sigma)V^\top = V\Lambda V^\top, \tag{9}$$

with $\Lambda = I - \Sigma$. This is an approximation, as $VV^\top$ is generally not the identity matrix. An improvement, originally suggested in [7], is to use the decomposition to approximate $I - L_s$ instead, i.e., $\widetilde{L}_s \approx I - V\Sigma V^\top$. We denote this alternative approximation as Nyström ($I - L_s$). In Sect. 3, we compare the performance of the two Nyström-based alternatives to compute $\widetilde{L}_s$ through numerical experiments.

Overall, the Nyström approach significantly reduces the computational costs by computing pairwise similarities only for a subset of the dataset, resulting in the computational complexity and storage requirements of $O(n)$ instead of $O(n^2)$, as p is negligible compared to n.

2.3 QR-Based Nyström Decomposition

The Nyström method requires $\widetilde{W}_{11}$ to be positive definite so that its square root is well-defined in Eq. (8) to calculate the SVD of the corresponding normalized graph Laplacian. If $\widetilde{W}_{11}$ is indefinite, Fowles et al. [14] provided a feasible solution based on [3], but unfortunately, this approach incurs additional computational cost and is prone to numerical errors.

Inspired by the work of [1] that used a recompression technique in [2] for computing a fully connected graph Laplacian, Budd et al. [6] employed the QR decomposition instead of SVD when approximating the normalized graph Laplacian. Specifically, we consider the thin QR decomposition of

$$\begin{bmatrix} \widetilde{W}_{11} \\ \widetilde{W}_{21} \end{bmatrix} = QR, \tag{10}$$

where $\widetilde{W}_{11}$ and $\widetilde{W}_{12}$ are obtained in Eq. (5), $Q \in \mathbb{R}^{n\times p}$ is orthonormal, and $R \in \mathbb{R}^{p\times p}$ is upper triangular. Then, we have the eigendecomposition:

$$R\widetilde{W}_{11}^{-1}R^\top = \Phi\Sigma\Phi^\top, \tag{11}$$

where $\Phi \in \mathbb{R}^{p\times p}$ is orthonormal and $\Sigma \in \mathbb{R}^{p\times p}$ is diagonal. We define $\Psi = Q\Phi$, which is orthonormal and adopt the following eigendecomposition of the symmetric normalized Laplacian:

Table 1 The computational complexity for KNN and Nyström methods for obtaining a normalized graph Laplacian of size $n \times n$, with K as the internal parameter for KNN and p for both Nyström methods

Method	Complexity
KNN	$O(Kn)$
Nyström	$O(np^2 + p^3)$
QR	$O(n^2 p + p^3)$

$$L_s \approx \Psi(I - \Sigma)\Psi^\top = \Psi \Lambda \Psi^\top. \tag{12}$$

Please refer to [2, 6] for more details.

Similar to the Nyström case, the decomposition can be used to approximate $I - L_s$ instead, i.e., $L_s \approx I - \Psi \Sigma \Psi^\top$. We denote this alternative approximation as QR ($I - L_s$). In Sect. 3, we compare the performance of the two QR-based alternatives to compute $\widetilde{L}_s$ through numerical experiments.

2.4 Summary

The choice of method depends on the specific requirements of the task, such as the size of the dataset, the desired accuracy, and the available computational resources. KNN is a simple and intuitive method for computing the weight matrix. It is effective for processing data with a clear local structure, but it can be sensitive to the choice of K and less effective for large, nonuniform datasets. Both Nyström methods can achieve good approximations for the symmetric normalized graph Laplacian with a relatively small number of columns, though random selection can sometimes lead to poor performance. The QR variant of the Nyström method enhances numerical robustness in the approximation but comes with higher computational costs compared to the standard Nyström method. The computational complexity of each method is provided in Table 1.

3 Numerical Experiments

We conduct numerical experiments on two benchmark datasets and one high-dimensional dataset to evaluate the efficacy of three graph Laplacian computation approaches, including KNN, Nyström, and QR-based Nyström (QR in short). The two benchmark datasets are obtained from the Scikit-learn library [21], while a high-dimensional dataset is the low-dose CT dataset [17] as processed for CT reconstruction in [28].

3.1 Benchmark Datasets

We use two benchmark datasets from Scikit-learn, namely, the two-moon and digits datasets. For each dataset, we compute (i) a fully connected graph with Gaussian similarity for the weight matrix W defined in Eq. (2), (ii) the symmetric normalized Laplacian L_s defined in Eq. (1), and (iii) the corresponding eigendecomposition via the `eigh` function of the `linalg` utilities of the NumPy Python package. This matrix L_s and its eigendecomposition become the *ground truth* with respect to which the performance of the methods is evaluated. We compare the performance of the three aforementioned methods, including the variant of approximating $(I - L_s)$. Note that KNN computes a sparse approximation to the weight matrix W, followed by the symmetric normalization to obtain the graph Laplacian L_s. In this case, we again compute the corresponding eigendecomposition via the `eigh` function of the `linalg` utilities of NumPy. In contrast, Nyström and QR-based Nyström directly compute an eigendecomposition of L_s.

The results reported include a comparison of the eigendecomposition obtained for each method, approximation errors, computation times, and accuracy obtained for unsupervised (clustering) and supervised (classification) tasks using the eigendecomposition as a pre-processing step. For the eigendecomposition, we report results obtained under different σ^2 values in Eq. (2) to reveal a stability issue in the original Nyström method. For the remaining comparisons, we examine two values of σ^2, and for each value, we vary the number of neighbors (K) in KNN and the number of sample data points (p) in Nyström methods. For each combination of parameters, we report mean and standard deviations over 30 repetitions of the whole processing pipeline, consisting of the following steps:

1. Generate data.
2. Split into training (70%) and testing (30%) partitions.
3. Construct Laplacians and their eigendecompositions using the training partition.
4. Evaluate clustering accuracy (over training partition).
5. Evaluate classification accuracy (over testing partition).

Approximation Error The approximation error is computed in terms of the relative Frobenius distance:

$$E_r = \frac{\|\widehat{L}_s - L_s\|_F}{\|L_s\|_F}, \tag{13}$$

where L_s is the ground truth, i.e., symmetric normalized Laplacian for the fully connected graph, and $\widehat{L}_s$ is the approximation, which, as a reminder, corresponds to

- KNN: $\widehat{L}_s = I - \widetilde{D}^{1/2}\widetilde{W}\widetilde{D}^{1/2}$, with $\widetilde{W}$ the similarity matrix including only K-nearest neighbors.
- Nyström: $\widehat{L}_s = V\Lambda V^\top$, computed using p sampled data points.
- Nyström $(I - L_s)$: $\widehat{L}_s = I - V\Sigma V^\top$, computed using p sampled data points.

- QR-based Nyström: $\widehat{L}_s = \Psi \Lambda \Psi^\top$, computed using p sampled data points.
- QR-based Nyström $(I - L_s)$: $\widehat{L}_s = I - \Psi \Sigma \Psi^\top$, computed using p sampled data points.

Computation Time Computation times reported were obtained on a 2.4 GHz 8-Core Intel Core i9 MacBook Pro.

Clustering Accuracy We use spectral clustering [26], i.e., K-means over the eigenvectors of L_s, as an unsupervised graph-based method to partition data into clusters. In each case, we only select a handful (5–25) of the top eigenvectors (i.e., the eigenvectors associated with the 5–25 smallest eigenvalues of matrix L_s). Since this is an unsupervised method, we do not make use of the class labels. To evaluate the accuracy, we only use the training data (i.e., the data used to build the graph Laplacian) and make use of the Scikit-learn [21] `rand_score` metric, which computes the *rand index*, a similarity measure between two clusterings based on "considering all pairs of samples and counting pairs that are assigned in the same or different clusters in the predicted and true clusterings."

Classification Accuracy We apply the support vector machine (SVM) technique for classification [16] using the Scikit-learn [21] functionality. SVM classification is a supervised learning algorithm that tries to find a maximum margin separating hyperplane, i.e., a hyperplane that separates the classes and has the maximum distance between data points in disparate classes. Instead of using the data points in the original domain, we project them onto a subspace defined by the eigenvectors of a Laplacian matrix, i.e., $\widetilde{X} = XU^\top$, where X is a matrix with rows corresponding to the data points and U is the matrix that is composed of eigenvectors of L_s. We only use a subset of eigenvectors corresponding to the dimensionality of the data. In this way, we can project both training and testing partitions. We also use a linear kernel, to evaluate the usefulness of the eigendecomposition as a pre-processing mechanism. To evaluate the accuracy, given that we know the true labels, we use the testing data and make use of the Scikit-learn `accuracy_score` metric, which computes the fraction of correctly classified samples.

3.1.1 Two-Moons Dataset

The *two-moons* dataset comprises a total of 2000 samples. Each sample is a point in a 2D plane, following the arch of a moon. As shown in Fig. 1, the dataset is divided into two classes, purple and yellow points, each containing 1000 samples. Additionally, each class comprises 500 points where the true moon samples have been perturbed with a 10% noise level, and another 500 points where the true moon samples have been perturbed with a 20% noise level.

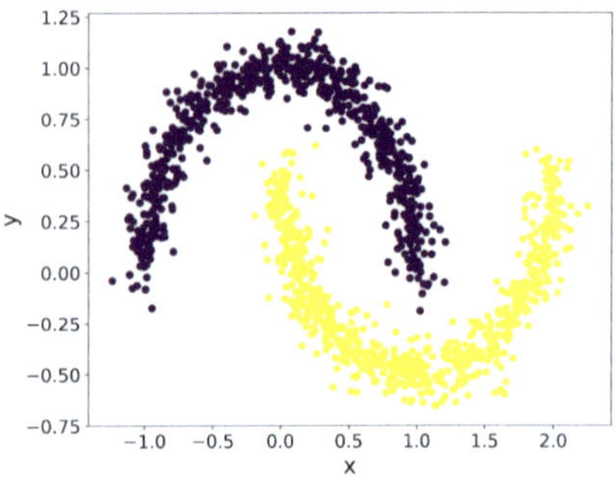

Fig. 1 Two-moons dataset from one random realization of the noise distribution. Each sample is a 2D vector belonging to one of two classes: either purple or yellow

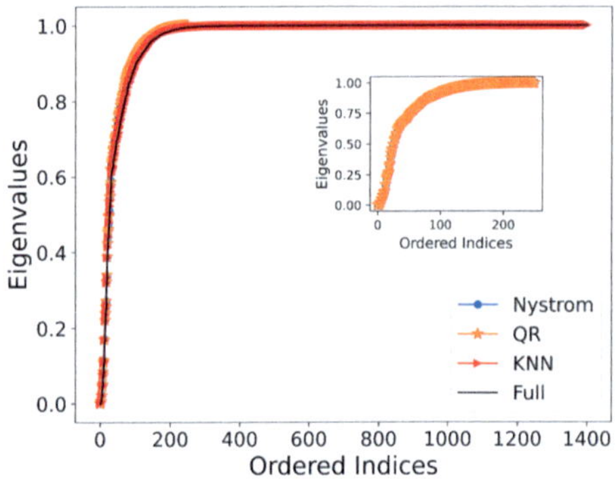

Fig. 2 Eigenvalues of the symmetric normalized Laplacian obtained by KNN ($K = 10$) and Nyström methods ($p = 250$) on the two-moons dataset with $\sigma^2 = 0.01$. Note that Nyström methods completely overlap and only QR, which lays on top of original Nyström, is visible in the plots

Eigendecomposition

Figure 2 compares the eigenvalues obtained by the three methods with $\sigma^2 = 0.01$, $K = 10$ nearest neighbors for KNN and $p = 250$ sampled points for both Nyström methods. It is clear that all the eigenvalues approximate the ones for the fully connected graph (labeled by "Full" in Fig. 2). The inset is included to remark that Nyström methods produce a rank p approximation to the eigendecomposition, thereby making only p eigenvalues available for these methods. Similarly, Fig. 3 compares the Nyström and Nyström $(I - L_s)$ approximations (left) as well as QR and QR $(I - L_s)$ approximations (right). Both methods with two approximation variants display a good agreement with the eigenvalues of the fully connected graph.

We then examine the top three eigenvectors (i.e., the eigenvectors associated with the smallest eigenvalues) of L_s obtained by all the methods in Fig. 4. As the original two-moons data is in 2D, we can plot the distribution of the training set in the x-y plane and color each point according to the value of a specific eigenvector. The row ordering of the input data X establishes the row correspondence to the eigenvector components. Note that the first eigenvector (first row), in which L_s is related to the normalized degree [26], remains consistent between fully connected graph and Nyström approximations. In contrast, it remains almost constant for KNN, as expected, since the normalized degree should be similar for graphs with the same number of nearest neighbors. Likewise, Fig. 5 compares the Nyström , Nyström

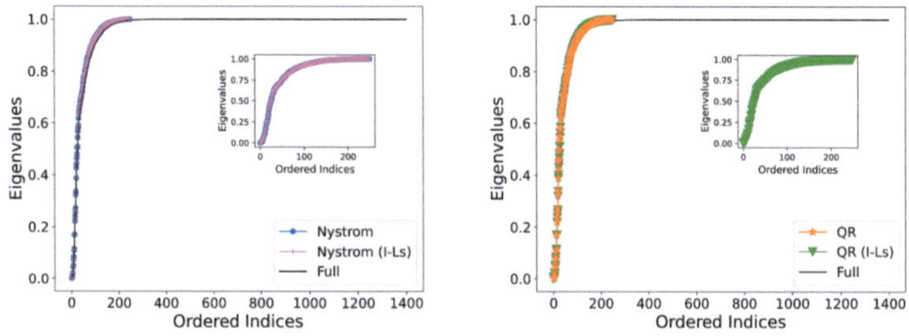

Fig. 3 Eigenvalues of the symmetric normalized Laplacian obtained by Nyström methods ($p = 250$) on the two-moons dataset with $\sigma^2 = 0.01$. Note that the methods completely overlap and only $(I - L_s)$ variants, which lay on top of the direct L_s approximation, are visible in the plots

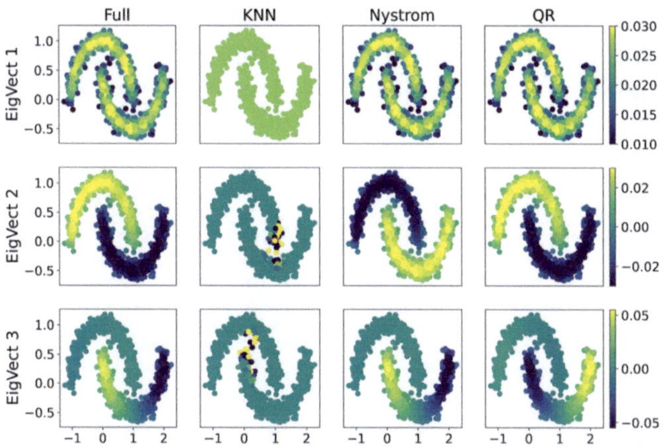

Fig. 4 Top three eigenvectors of the symmetric normalized Laplacian obtained by KNN ($K = 10$) and Nyström methods ($p = 250$) on the two-moons dataset with $\sigma^2 = 0.01$

$(I - L_s)$, QR, and QR $(I - L_s)$ approximations, showing a good agreement among these methods. In summary, Figs. 4 and 5 illustrate that, aside from sign differences in the eigenvectors, all the Nyström variants produce a good approximation to the first eigenvectors. The KNN method, on the other hand, produces much more localized patterns. The errors in the approximations given by Eq. (13) are 0.127935 for KNN, 0.927003 for Nyström , 0.022307 for Nyström $(I - L_s)$, 0.926823 for QR, and 0.021647 for QR $(I - L_s)$. From these error estimations, it is clear that the $(I - L_s)$ variant of the Nyström methods produces much better approximations to the full symmetric normalized Laplacian than the direct L_s approximations.

We investigate the eigenvalues obtained by the competing methods under different values of σ^2; specifically, $\sigma^2 = 0.005, 0.01, 0.07$ and 0.1 are considered in Fig. 6, showing that the smaller σ^2 is, the larger error to the fully connected graph is made by the Nyström approximations. For simplicity, we omit the $(I - L_s)$ approximation variants from Fig. 6, because they fall on top of the graphs for the

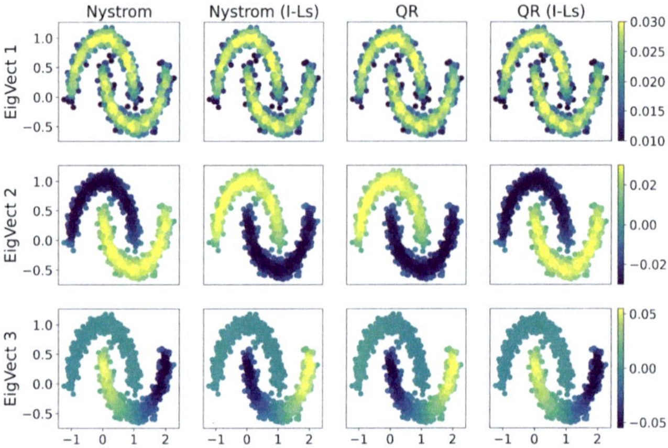

Fig. 5 Top three eigenvectors of the symmetric normalized Laplacian obtained by Nyström methods ($p = 250$) with the two approximation variants on the two-moons dataset with $\sigma^2 = 0.01$

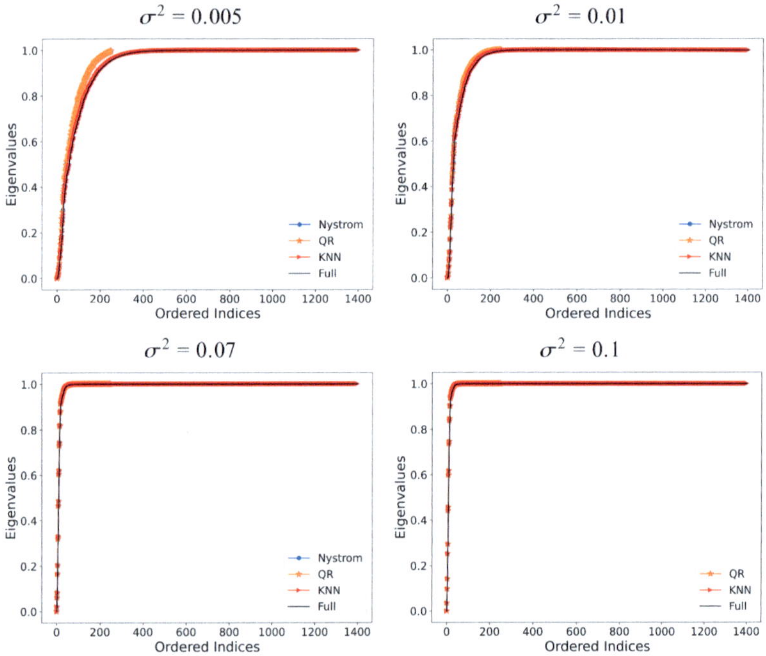

Fig. 6 Eigenvalues of the symmetric normalized Laplacian obtained by KNN ($K = 10$) and Nyström methods ($p = 250$) on the two-moons dataset under different values of σ^2. Note that Nyström methods completely overlap and only QR, which lays on top of original Nyström , is visible in the plots

Table 2 Comparison of the approximations errors, i.e., E_r defined in (13), made by KNN ($K = 10$) and Nyström method ($p = 250$) on the two-moons dataset under different values of σ^2. NaN indicates that the original Nyström method fails at $\sigma^2 = 0.1$ when the submatrix W_{11} is not positive definite

	σ^2			
Method	0.005	0.01	0.07	0.1
KNN	0.180776	0.127935	0.066262	0.058065
Nyström	0.940897	0.927003	0.911308	NaN
Nyström $(I - L_s)$	0.068371	0.022307	0.000006	NaN
QR	0.941037	0.926823	0.911150	0.910177
QR $(I - L_s)$	0.069773	0.021647	0.000008	0.000005

direct L_s approximation when plotted. On the other hand, both variants of the original Nyström method fail for larger values of σ^2, e.g., $\sigma^2 = 0.1$ and $p = 250$ used here, as the submatrix W_{11} is not positive definite. Note that both QR-based variants succeed in this case. Table 2 records the approximation errors for these four values of σ^2. Note that, in general, the $(I - L_s)$ variants yield better approximation results.

Approximation Errors

The approximation errors with respect to a range of K-nearest neighbors in KNN and p sampled data points in both Nyström methods, using both approximation variants, are plotted in Fig. 7 for $\sigma^2 = 0.01$ and $\sigma^2 = 0.07$. The results are averaged over 30 random trials. Since the ranges of K and p are different, the plots include two x-axis: the top one in red corresponds to the K values for KNN, while the bottom one in black corresponds to the p values for Nyström methods. Figure 7 clearly illustrates that the approximation given by the Nyström methods improves as the number of sample points p increases. It also shows that the Nyström method does not converge for larger values of p, where only results for $p \leq 250$ can be computed. Since the original Nyström and QR mostly overlap, it is difficult to observe the lack of convergence of the original Nyström from these error plots. However, the other plots, especially Fig. 9, make this observation more apparent. The approximation errors for the KNN method are generally smaller than the Nyström methods (for the direct L_s approximation) and are relatively independent of K. The Nyström methods that approximate $(I - L_s)$ produce smaller errors, compared to KNN. The performance of Nyström methods on downstream tasks involving the eigendecomposition is better than the KNN method as shown in Figs. 9 and 10.

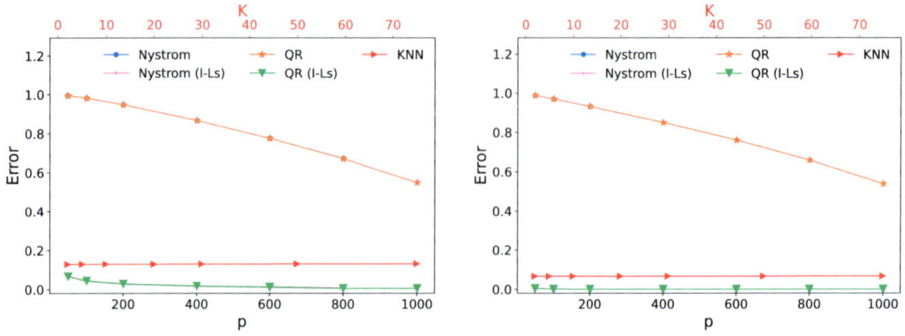

Fig. 7 Error in L_s approximation for KNN as a function of K (top axis) and Nyström methods as a function of p, for $\sigma^2 = 0.01$ (left) and $\sigma^2 = 0.07$ (right), on the two-moons dataset. The results are averaged over 30 random trials and computed means are reported. The standard deviation computed is very small, with practically no-shaded region distinguishable. Note that Nyström methods completely overlap (up to where the original Nyström is stable, i.e., $p \leq 800$ (left) and $p \leq 200$ (right)), and only QR results, which fall on top of the original Nyström (same phenomenon for the QR methods), are visible in the plots

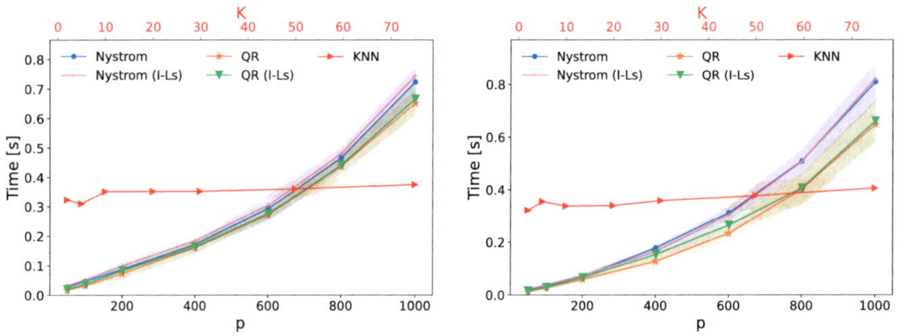

Fig. 8 Computation times for KNN as a function of K (top axis) and Nyström methods as a function of p, for $\sigma^2 = 0.01$ (left) and $\sigma^2 = 0.07$ (right), on the two-moons dataset. The results are averaged over 30 random trials and computed means are reported. The shaded region in the plots represents the standard deviation calculated over the random trials

Computation Time

Under the same setup as the approximation error, the computation times are plotted in Fig. 8, where the standard deviations calculated over 30 random trials are depicted as a shaded region. Note that the times reported for KNN include the eigendecomposition stage, which is naturally included in the Nyström class. Figure 8 shows that the QR-based Nyström is slightly faster than the original Nyström method, and their difference becomes larger as p or σ^2 increases. In addition, the KNN method, utilizing the nearest neighbors routine from the `giotto-tda` Python package [25], ensures stable computation times, remaining almost constant across the range of $K \in [2, 75]$ most probably due to its exploitation of multi-core parallelism. Figure 8

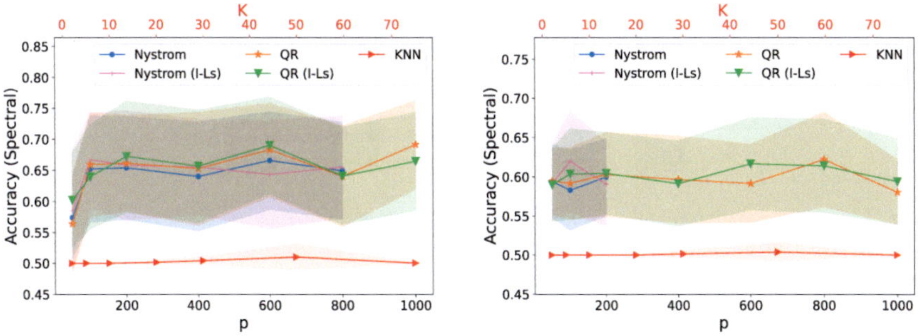

Fig. 9 Accuracy of spectral clustering for KNN as a function of K (top axis) and Nyström methods as a function of p, for $\sigma^2 = 0.01$ (left) and $\sigma^2 = 0.07$ (right), on the two-moons dataset. The results are averaged over 30 random trials and computed means are reported. The shaded region in the plots represents the standard deviation calculated over the random trials

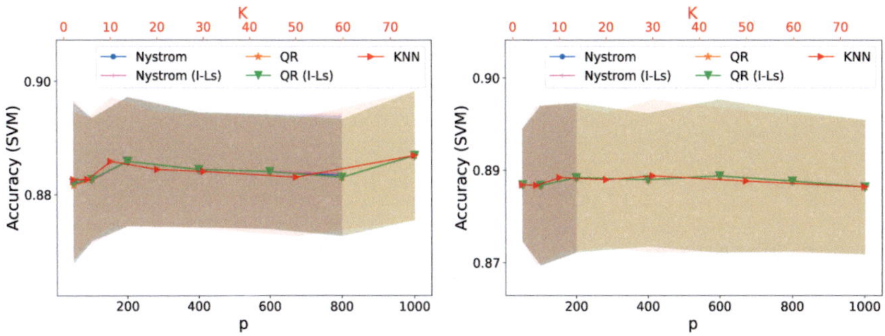

Fig. 10 Accuracy of SVM classification for KNN as a function of K (top axis) and Nyström methods as a function of p, for $\sigma^2 = 0.01$ (left) and $\sigma^2 = 0.07$ (right), for the two-moons dataset. The results are averaged over 30 random trials and computed means are reported. The shaded region in the plots represents the standard deviation calculated over the random trials. Note that Nyström methods completely overlap (up to where the original Nyström is stable, i.e., $p \leq 800$ (left) and $p \leq 200$ (right)), and only QR, which lays on top of original Nyström, is visible in the plots

reveals that there is a range where substantial computation savings can be obtained by using the Nyström approximation methods, without a significant sacrifice in performance (see accuracy plots, e.g., Figs. 9 and 10).

Unsupervised Task

We report the performance of the weight approximation methods in a downstream task: unsupervised clustering. Specifically, averaged accuracy results obtained by spectral clustering over 30 random trials are plotted in Fig. 9 for $\sigma^2 = 0.01$ and $\sigma^2 = 0.07$. The standard deviations calculated over the random trials are depicted

Fig. 11 Representative samples from each of the ten-class digits dataset. Each sample is an 8×8 pattern that can be flattened to a 64-dimensional vector. The training set used has about 1250 samples

as a shaded region in the plots. Given that the eigenvectors tend to be more localized in KNN, 25 eigenvectors are used for the spectral clustering, while only five eigenvectors are used for Nyström methods. It is clear in Fig. 9 that projecting on the eigendecomposition of the Nyström methods produces better results than KNN, but no major improvements are observed for approximations using larger K or p. These plots also make more evident that no results are reported for Nyström $p > 800$ (left plot) and for $p > 250$ (right plot) due to the invalid partial computations (i.e., submatrix W_{11} not positive definite or unstable inversion).

Supervised Task

Another downstream task given by the SVM classification is examined in Fig. 10, showing that supervised learning contributes to a large improvement in the classification results compared to unsupervised clustering. It is also interesting to note that although the Nyström methods that directly approximate L_s yield larger approximation errors than KNN (see Fig. 7), the classification accuracy is similar and relatively high for all the weight approximation methods, probably due to the supervised nature of this task.

3.1.2 Digits Dataset

The *digits* dataset comprises a total of 1797 images of handwritten digits ranging from 0 to 9. Each image is of dimension 8×8 and hence can be represented by a 64-dimensional array of gray-scale intensity values, vectorized from a 2D image. This dataset is a copy of the test set of the UCI ML handwritten digits datasets.[1] An illustration of the images in each class can be found in Fig. 11.

[1] https://archive.ics.uci.edu/ml/datasets/Optical+Recognition+of+Handwritten+Digits

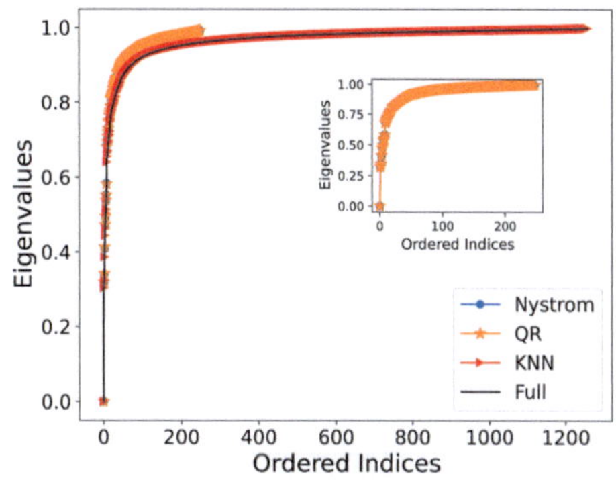

Fig. 12 Eigenvalues of the symmetric normalized Laplacian for the digits dataset for each of the methods with $\sigma^2 = 1.0$, $K = 10$ for KNN, and $p = 250$ for Nyström methods. Note that Nyström methods completely overlap and only QR, which lays on top of original Nyström, is visible in the plots

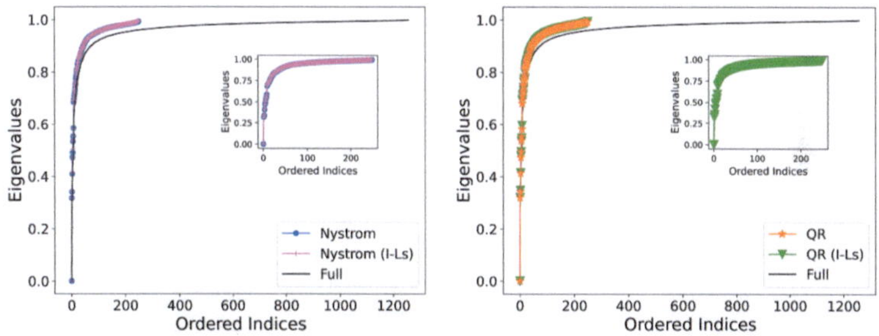

Fig. 13 Eigenvalues of the symmetric normalized Laplacian obtained by Nyström methods ($p = 250$) on the digits dataset with $\sigma^2 = 1.0$. Note that the methods completely overlap and only $(I - L_s)$ variants, which lay on top of the direct L_s approximation, are visible in the plots

Eigendecomposition

Figure 12 compares the eigenvalues obtained by three methods with $\sigma^2 = 1.0$, $K = 10$ nearest neighbors for KNN, and $p = 250$ sampled points for both Nyström methods. All the eigenvalues approximate the ones for the fully connected graph, except that the Nyström methods start to show a slight deviation from the ground truth. Figure 13 compares the Nyström, Nyström $(I - L_s)$, QR, and QR $(I - L_s)$ approximations, showing a very good agreement between them.

Following the two-moons example, we examine the top three eigenvectors of L_s obtained by all the methods in Figs. 14 and 15. As it is difficult to directly visualize the distribution of the original 64-dimensional data in the x-y plane, we plot each eigenvector as a function of the row index and color each component according to the value of such index. Again, the row ordering of the input data X establishes the row correspondence to the eigenvector components. Similar to the two-moons case, the first eigenvector (first row), which is related to the normalized degree [26], is consistent between fully connected graph and all the Nyström approximations, while it is almost constant for KNN. Briefly, Figs. 14 and 15 illustrate that, aside from sign differences in the eigenvectors, both Nyström variants produce a good approximation to the first eigenvectors, while the KNN method produces different

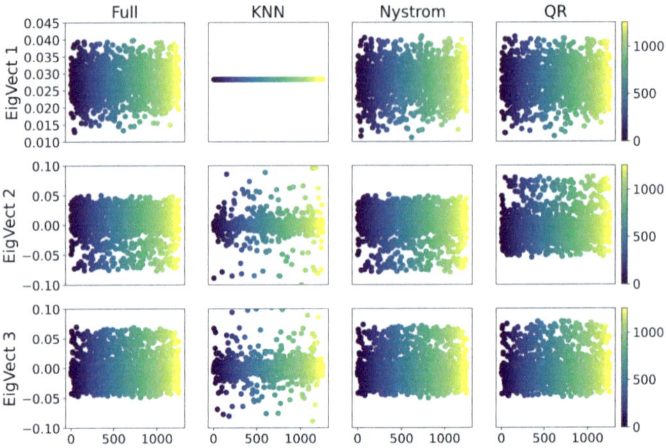

Fig. 14 Eigendecomposition of the symmetric normalized Laplacian for the digits dataset for each of the methods with $\sigma^2 = 1.0$, $K = 10$ for KNN, and $p = 250$ for Nyström methods

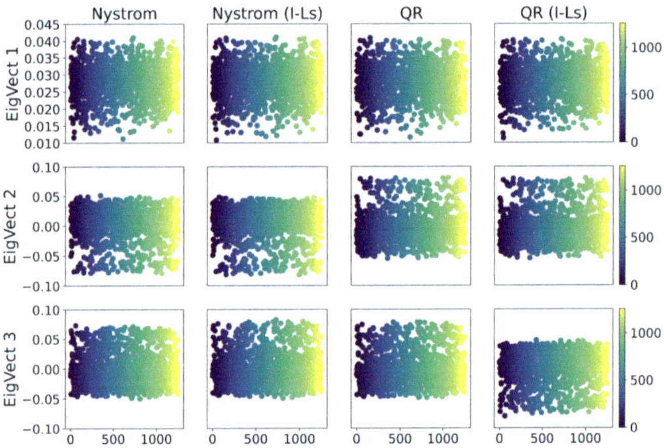

Fig. 15 Eigendecomposition of the symmetric normalized Laplacian obtained by Nyström methods ($p = 250$) with the two approximation variants on the digits dataset with $\sigma^2 = 1.0$

patterns. The errors in the approximations given by Eq. (13) are 0.071614 for KNN, 0.906414 for Nyström, 0.031120 for Nyström $(I - L_s)$, 0.906467 for QR, and 0.030089 for QR $(I - L_s)$. Similar to the two-moons case, from these error estimations, it is clear that the $(I - L_s)$ variants of the Nyström methods produce much better approximations to the full symmetric normalized Laplacian than the direct L_s approximations.

We investigate the eigenvalues obtained by the competing methods under different values of σ^2; specifically, $\sigma^2 = 0.5$, 1.0, 5.3 and 10.3 are considered in

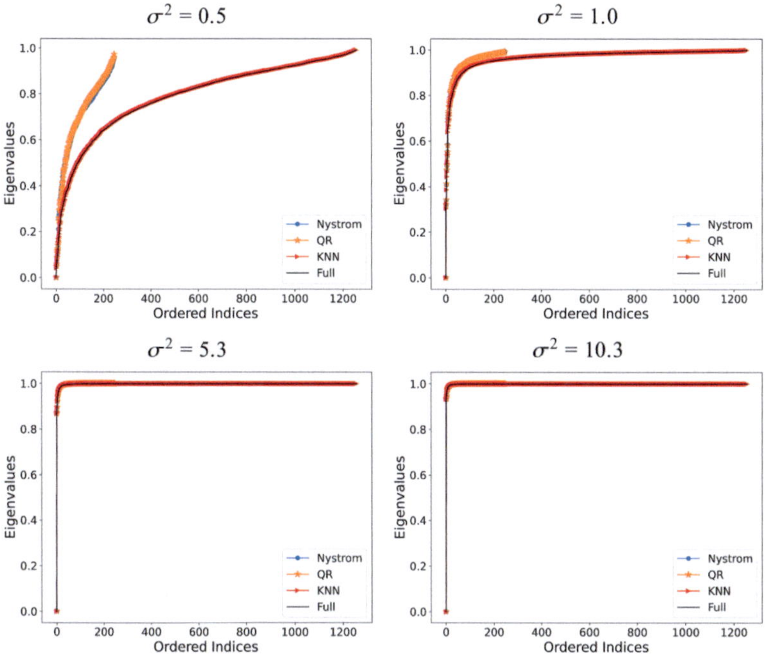

Fig. 16 Eigenvalues of the symmetric normalized Laplacian obtained by KNN ($K = 10$) and Nyström methods ($p = 250$) on the digits dataset under different values of σ^2. Note that Nyström methods completely overlap and only QR, which lays on top of original Nyström, is visible in the plots

Table 3 Comparison of the error E_r (13) of the approximation methods for the digits dataset for different σ^2 and for $K = 10$ (KNN) and $p = 250$ (Nyström methods)

	σ^2			
Method	0.5	1.0	5.3	10.3
KNN	0.337764	0.071614	0.007724	0.003838
Nyström	0.925103	0.906414	0.896262	0.895787
Nyström $(I - L_s)$	0.264682	0.031120	0.000317	0.000063
QR	0.922371	0.906467	0.896263	0.895788
QR$(I - L_s)$	0.270744	0.030089	0.000296	0.000068

Fig. 16, showing that the smaller σ^2 is, the larger error to the fully connected graph is made by the Nyström approximations. However, in contrast with the two-moons case, for all these σ^2 values used, both the original Nyström and QR-based Nyström succeed. Table 3 records the approximation errors for these four values of σ^2. Note again that the $(I - L_s)$ variants yield small approximation errors.

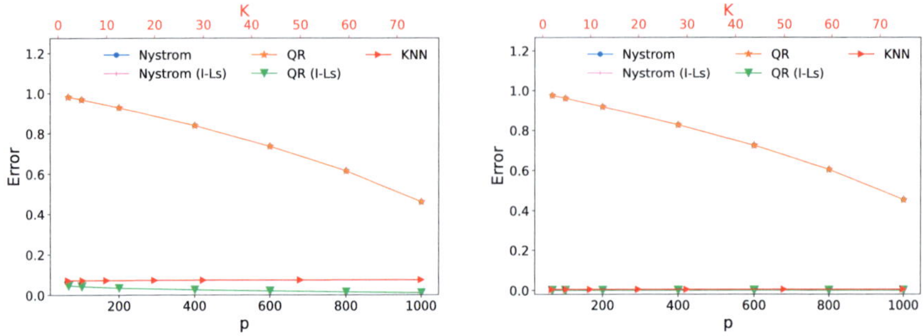

Fig. 17 Error in L_s approximation for KNN as a function of K (top axis) and Nyström methods as a function of p, for $\sigma^2 = 1.0$ (left) and $\sigma^2 = 10.3$ (right), on the digits dataset. The results are averaged over 30 random trials and computed means are reported. The standard deviation computed is very small, with practically no-shaded region distinguishable. Note that Nyström methods completely overlap and only QR results, which fall on top of original Nyström , (or QR($I - L_s$) which fall on top of Nyström ($I - L_s$)), are visible in the plots

Approximation Errors

The approximation errors with respect to a range of K-nearest neighbors in KNN and p sampled data points in both Nyström methods are plotted in Fig. 17 for $\sigma^2 = 1.0$ and $\sigma^2 = 10.3$. The results are averaged over 30 random trials. Since the ranges of K and p are different, the plots include two x-axis: the top one in red corresponds to the K values for KNN, while the bottom one in black corresponds to the p values for Nyström methods. For this dataset, the Nyström method produces valid results across all the parameters tested. Figure 17 agrees with the observations made for the two-moons datasets, showing again that the approximations obtained via Nyström methods improve as the number of sample points p increases and that the error of the KNN method is smaller than the Nyström methods that directly approximate L_s and is relatively independent of K. Nyström methods that approximate $(I - L_s)$ produce smaller errors. The performance of Nyström methods on downstream tasks involving the eigendecomposition is better (see Fig. 19) or matches (see Fig. 20) the performance of the KNN method.

Computation Time

Under the same setup as the approximation error, the computation times are plotted in Fig. 18, where the standard deviations calculated over 30 random trials are depicted as a shaded region. As before, the times reported for KNN include the eigendecomposition stage. Figure 18 shows that the QR-based Nyström is slightly faster than the original Nyström method and that the KNN computation (via `giotto-tda` routine [25]) ensures stable computation times, remaining almost constant across the range $K \in [2, 75]$. Figure 18 reveals that there is a range

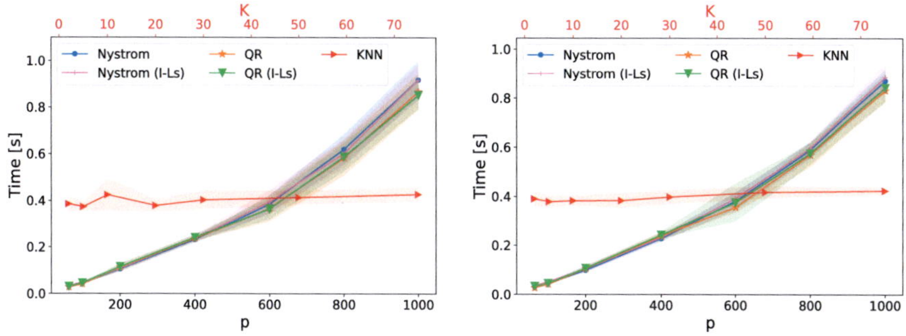

Fig. 18 Computation times for KNN as a function of K (top axis) and Nyström methods as a function of p, for $\sigma^2 = 1.0$ (left) and $\sigma^2 = 10.3$ (right), on the digits dataset. The results are averaged over 30 random trials and computed means are reported. The shaded region in the plots represents the standard deviation calculated over the random trials

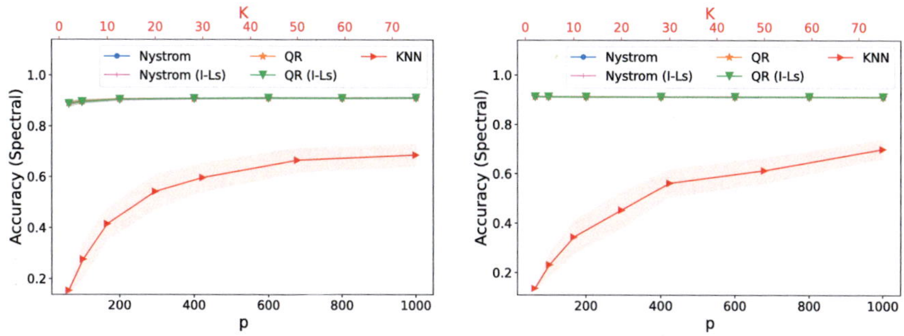

Fig. 19 Accuracy of spectral clustering for KNN as a function of K (top axis) and Nyström methods as a function of p, for $\sigma^2 = 1.0$ (left) and $\sigma^2 = 10.3$ (right), on the digits dataset. The results are averaged over 30 random trials and computed means are reported. The shaded region in the plots represents the standard deviation calculated over the random trials. Note that Nyström methods completely overlap and practically only QR($I - L_s$), which falls on top of the other Nyström variants, is visible in the plots

when substantial computation savings can be obtained by using the Nyström approximation methods, without a significant sacrifice in performance (see accuracy plots, e.g., Figs. 19 and 20).

Unsupervised Task

We report the performance of the weight approximation methods in the downstream task of unsupervised clustering. Averaged accuracy results obtained by spectral clustering over 30 random trials are plotted in Fig. 19 for $\sigma^2 = 1.0$ and $\sigma^2 = 10.3$. The standard deviations calculated over the random trials are depicted as a shaded region in the plots. Given that the eigenvectors tend to be more localized in KNN-

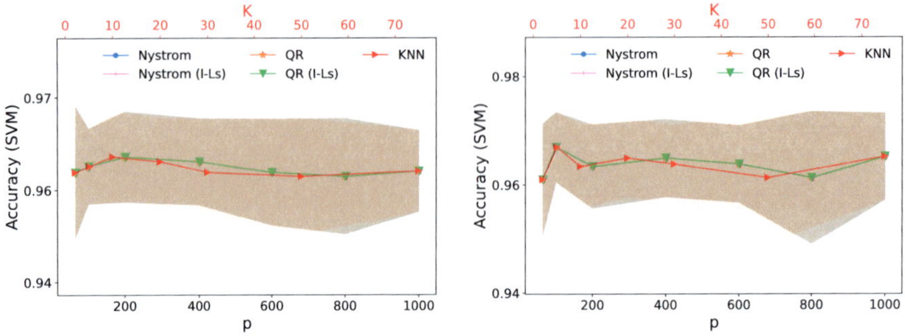

Fig. 20 Accuracy of SVM classification for KNN as a function of K (top axis) and Nyström methods as a function of p, for $\sigma^2 = 1.0$ (left) and $\sigma^2 = 10.3$ (right), on the digits dataset. The results are averaged over 30 random trials and computed means are reported. The shaded region in the plots represents the standard deviation calculated over the random trials. Note that Nyström methods completely overlap and only QR, which lays on top of original Nyström, is visible in the plots

based decompositions, 25 top eigenvectors were used for the spectral clustering, while only five top eigenvectors were used for Nyström methods. It is clear in Fig. 19 that projecting on the eigendecomposition of the Nyström methods produces good results, with around 90% accuracy, and these are much better than what is obtained with KNN. Nevertheless, in this case, major improvements in accuracy are observed for using a larger number of neighbors K in the KNN method.

Supervised Task

We also evaluate the downstream task of SVM classification and report results in Fig. 20. As observed before, the supervised learning improves the classification results, and again, even when the approximation to L_s computed by the Nyström methods has a larger error than KNN (see Fig. 17), the accuracy results are similar and deemed satisfactory in all cases.

3.2 CT Reconstruction

To test and compare the algorithms in different downstream processing tasks, we use a low-dose CT reconstruction problem with real image data of high dimensionality (256×256). In particular, we follow the MAGIC (manifold and graph integrative convolution network) approach [28], which unrolls a gradient descent algorithm into a neural network, using a convolutional neural network (CNN) to preserve pixel-level features and a graph convolutional network (GCN) to extract the nonlocal features from a patch-based manifold space. The graph is constructed by treating every pixel of the CT image as a node and computing the weight using the Eq. (2) measured by the Euclidean distance between two small patches, whose top-left corner corresponds to the respective nodes. Then, the graph Laplacian is used in the GCN component of MAGIC to define the spectral graph convolution [5]. Here, the matrix composed of eigenvectors of the normalized graph Laplacian, i.e., V in Eq. (9), is analogous to the Fourier transform in standard spatial convolution, following the convolution theorem.

In what follows, we use three methods, KNN, Nyström, and QR-based Nyström, to approximate the computation of L_s for the GCN component of MAGIC and evaluate the obtained reconstructions in terms of peak signal-to-noise ratio (PSNR), structural similarity index measure (SSIM), and computational time. In all cases, we build the similarity matrix using a Gaussian similarity, Eq. (2), with $\sigma^2 = 5.7$. Note that given the high dimensionality of the data, we do not even attempt to build a fully connected graph for this case. We do not run Nyström variants that approximate $(I - L_s)$ since we expect similar performance to the one obtained with the direct L_s approximations. We follow the MAGIC work and use the same architecture and training parameters. For a proof of concept, we enact the following simplifications: (i) we use a reduced set of ten training images, (ii) we train for 50 epochs using a batch size of 2, and (iii) we test the trained model on ten test images different from the training set. We compare results for dose levels of 0.01 and 0.1 (see more details about the dose levels in the original work [28]).

Table 4 displays performance results for the reconstructions for the two dose levels or each of the three methods for computing L_s. The mean and standard deviations over the testing set are reported. Note that PSNR results are computed assuming a signal range in [0, 1], not the actual dynamic range. It can be observed that the results are very similar for all three methods, and of course, better results are obtained for measurements using a large dose level. Specific visual results are shown in Figs. 21 and 22 for dose levels of 0.01 and 0.1, respectively. Results for the lower-dose level have more granular artifacts, while results for the high-dose level are smoother (it may be necessary to zoom over the figures to note the difference). Finally, Fig. 23 shows a comparison of computation times on a GPU cluster (one node, eight NVIDIA GeForce RTX 2080 Ti GPUs), obtained for the three methods when approximating the symmetric normalized Laplacian for the coarse stage of the

Table 4 CT reconstruction comparison under two dose levels (0.01 and 0.1) for $K = 5$ (KNN) and $p = 50$ (Nyström methods)

		PSNR [dB]		SSIM	
Dose level	Method	Mean	Std	Mean	Std
0.01	KNN	35.60	0.38	0.9133	0.0066
	Nyström	35.60	0.38	0.9118	0.0063
	QR	36.04	0.39	0.9252	0.0057
0.10	KNN	41.36	0.36	0.9676	0.0033
	Nyström	41.33	0.37	0.9670	0.0033
	QR	41.13	0.38	0.9654	0.0036

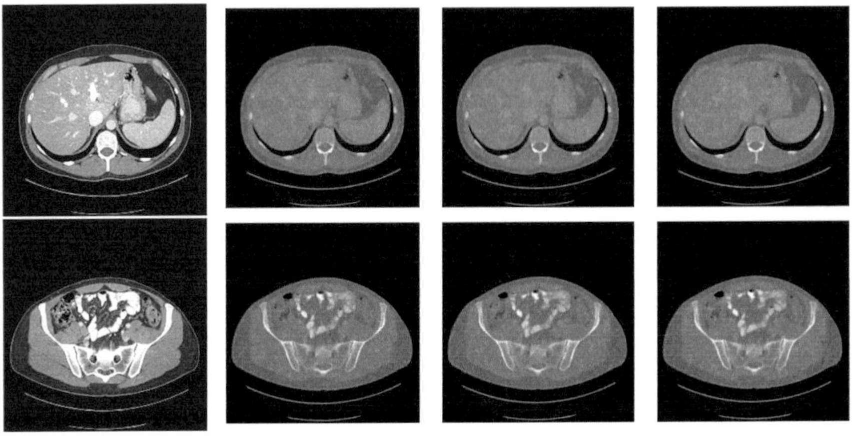

Fig. 21 Visual results of CT reconstruction under 0.01 dose level. From left to right: ground truth, KNN, Nyström, and QR

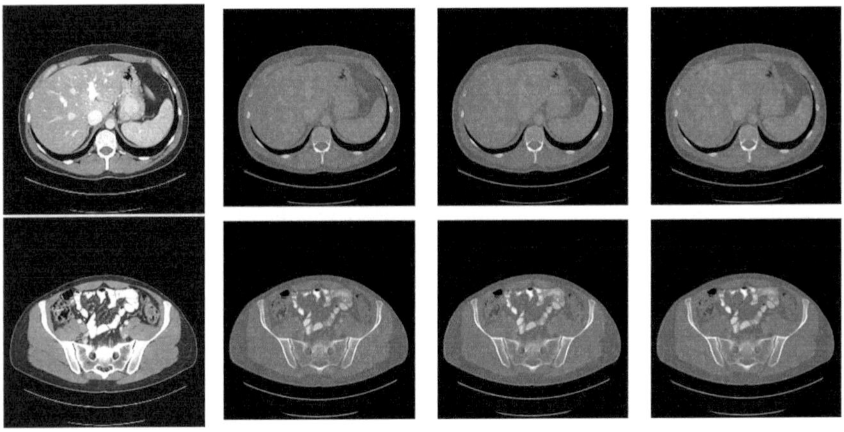

Fig. 22 Visual results of CT reconstruction under 0.1 dose level. From left to right: ground truth, KNN, Nyström, and QR

MAGIC reconstruction, using different numbers of p sampled data patches for the Nyström methods and different numbers of K patch neighbors for the KNN method. It is seen, consistent with results presented in previous sections, that the Nyström methods considerably reduce the computation time without significantly decreasing performance. Also, note that the QR-based Nyström method is slightly faster than the original Nyström method, which aligns with the observation in the synthetic case.

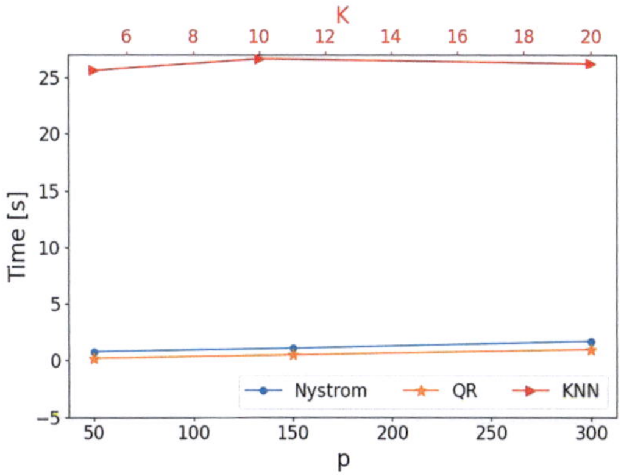

Fig. 23 Computation times for KNN as a function of K (top axis) and Nyström methods as a function of p when approximating the symmetric normalized Laplacian for the coarse stage of the MAGIC reconstruction

4 Conclusions

Through extensive numerical experimentation, including benchmarks as well as high-dimensional real datasets, we confirm the advantages of the Nyström methods for approximating the eigendecomposition of the symmetric Laplacian. Briefly, these methods provide accurate approximations of the eigenvalues and eigenvectors of a fully connected graph. Additionally, significant time savings are achieved by computing approximations based on eigendecompositions using subsets of data samples. The direct computation of eigenvalues and eigenvectors also facilitates the analysis of the graph structure, which is beneficial for downstream tasks such as clustering, classification, or graph-based signal filtering. We also observe that the QR method is slightly faster than the original Nyström method. However, the latter can become unstable or yield nonvalid solutions when a "large" number of data samples or a "large" value of σ^2 (resulting in the weight matrix being low rank) is used. It also seems the case that the Nyström approximations to the fully connected graph become worse when a "smaller" value of σ^2 is used. The problem, however, is that typically there is no a priori way to determine what "small" or "large" means in this context since it is heavily dataset-dependent. Overall, the QR-based method seems like a good alternative for more robust and faster approximations. Moreover, variants that approximate $(I - L_s)$ have much smaller approximation errors to the fully normalized symmetric Laplacian. It is also worth noticing that the relative Frobenius distance E_r can provide a somewhat misleading idea of the quality of the approximations, in particular when comparing the relative errors of KNN and Nyström methods. Although Nyström methods that directly approximate L_s seem to have worse errors compared to KNN and Nyström methods that approximate $(I - L_s)$ have much smaller approximation errors, their performance can be similar in downstream tasks.

Acknowledgments The authors would like to acknowledge the support from the Women in Data Science and Mathematics Research Workshop (WiSDM) hosted by IPAM at UCLA from August 7 to 11, 2023, which initiated the collaboration. C. Garcia-Cardona was funded by the Los Alamos

National Laboratory LDRD Program Director's Initiative (DI) project 20230771DI. Y. Liu is partially supported by NSF 2213436. Y. Lou is partially supported by NSF CAREER 2414705. S. Tang is partially supported by NSF 2111303 and NSF CAREER 2340631.

Competing Interests The authors have no conflicts of interest to declare that are relevant to the content of this chapter.

References

1. Alfke, D., Potts, D., Stoll, M., Volkmer, T.: NFFT meets Krylov methods: fast matrix-vector products for the graph Laplacian of fully connected networks. Front. Appl. Math. Stat. **4**, 61 (2018)
2. Bebendorf, M., Kunis, S.: Recompression techniques for adaptive cross approximation. J. Integral Equ. Appl. **21**(3), 331–357 (2009)
3. Belongie, S., Fowlkes, C., Chung, F., Malik, J.: Spectral partitioning with indefinite kernels using the nyström extension. In: Computer Vision—ECCV 2002: 7th European Conference on Computer Vision Copenhagen, Denmark, May 28–31, 2002 Proceedings, Part III 7, pp. 531–542. Springer, Berlin (2002)
4. Bertozzi, A.L., Flenner, A.: Diffuse interface models on graphs for classification of high dimensional data. Multiscale Model. Simul. **10**(3), 1090–1118 (2012)
5. Bruna, J., Zaremba, W., Szlam, A., LeCun, Y.: Spectral networks and deep locally connected networks on graphs (2014). arXiv preprint arXiv:1312.6203
6. Budd, J., van Gennip, Y., Latz, J.: Classification and image processing with a semi-discrete scheme for fidelity forced Allen–Cahn on graphs. GAMM-Mitteilungen **44**(1), e202100004 (2021)
7. Budd, J.M., van Gennip, Y., Latz, J., Parisotto, S., Schönlieb, C.B.: Joint reconstruction-segmentation on graphs. SIAM J. Imaging Sci. **16**(2), 911–947 (2023)
8. Chen, B., Lou, Y., Bertozzi, A.L., Chanussot, J.: Graph-based active learning for nearly blind hyperspectral unmixing. IEEE Trans. Geosci. Remote Sensing **61**, 1–16 (2023)
9. Chen, Y., Qi, L., Zhang, X.: The fiedler vector of a Laplacian tensor for hypergraph partitioning. SIAM J. Sci. Comput. **39**(6), A2508–A2537 (2017)
10. Cheng, X., Rachh, M., Steinerberger, S.: On the diffusion geometry of graph Laplacians and applications. Appl. Comput. Harmon. Anal. **46**(3), 674–688 (2019)
11. Chung, F.R.: Spectral Graph Theory, vol. 92. American Mathematical Society (1997)
12. Dong, W., Moses, C., Li, K.: Efficient k-nearest neighbor graph construction for generic similarity measures. In: Proceedings of the 20th International Conference on World Wide Web, pp. 577–586 (2011)
13. Dong, X., Thanou, D., Frossard, P., Vandergheynst, P.: Learning Laplacian matrix in smooth graph signal representations. IEEE Trans. Signal Process. **64**(23), 6160–6173 (2016)
14. Fowlkes, C., Belongie, S., Chung, F., Malik, J.: Spectral grouping using the Nyström method. IEEE Trans. Pattern Anal. Mach. Intell. **26**(2), 214–225 (2004)
15. Hart, R., Yu, L., Lou, Y., Chen, F.: Improvements on uncertainty quantification for node classification via distance based regularization. In: Advances in Neural Information Processing Systems, vol. 36 (2024)
16. Hsu, C.W., Chang, C.C., Lin, C.J., et al.: A practical guide to support vector classification (2003)
17. McCollough, C.: TU-FG-207A-04: overview of the low dose ct grand challenge. Med. Phys. **43**(6), 3759–3760 (2016)
18. Merkurjev, E., Sunu, J., Bertozzi, A.L.: Graph MBO method for multiclass segmentation of hyperspectral stand-off detection video. In: IEEE International Conference on Image Processing, pp. 689–693 (2014)

19. Ortega, A., Frossard, P., Kovačević, J., Moura, J.M., Vandergheynst, P.: Graph signal processing: overview, challenges, and applications. Proc. IEEE **106**(5), 808–828 (2018)
20. Ozaki, K., Shimbo, M., Komachi, M., Matsumoto, Y.: Using the mutual k-nearest neighbor graphs for semi-supervised classification on natural language data. In: Proceedings of the Fifteenth Conference on Computational Natural Language Learning, pp. 154–162 (2011)
21. Pedregosa, F., Varoquaux, G., Gramfort, A., Michel, V., Thirion, B., Grisel, O., Blondel, M., Prettenhofer, P., Weiss, R., Dubourg, V., et al.: Scikit-learn: machine learning in python. J. Mach. Learn. Res. **12**(Oct), 2825–2830 (2011)
22. Perraudin, N., Paratte, J., Shuman, D., Martin, L., Kalofolias, V., Vandergheynst, P., Hammond, D.K.: Gspbox: a toolbox for signal processing on graphs (2014). ArXiv e-prints
23. Qin, J., Lee, H., Chi, J.T., Drumetz, L., Chanussot, J., Lou, Y., Bertozzi, A.L.: Blind hyperspectral unmixing based on graph total variation regularization. IEEE Trans. Geosci. Remote Sensing **59**(4), 3338–3351 (2020)
24. Shi, B., Han, L., Yan, H.: Adaptive clustering algorithm based on KNN and density. Pattern Recognit. Lett. **104**, 37–44 (2018)
25. Tauzin, G., Lupo, U., Tunstall, L., Pérez, J.B., Caorsi, M., Medina-Mardones, A., Dassatti, A., Hess, K.: giotto-tda: a topological data analysis toolkit for machine learning and data exploration (2020)
26. Von Luxburg, U.: A tutorial on spectral clustering. Stat. Comput. **17**, 395–416 (2007)
27. Wu, Y., Ianakiev, K., Govindaraju, V.: Improved k-nearest neighbor classification. Pattern Recogn. **35**(10), 2311–2318 (2002)
28. Xia, W., Lu, Z., Huang, Y., Shi, Z., Liu, Y., Chen, H., Chen, Y., Zhou, J., Zhang, Y.: Magic: manifold and graph integrative convolutional network for low-dose ct reconstruction. IEEE Trans. Med. Imaging **40**(12), 3459–3472 (2021)
29. Zhang, B., Srihari, S.N.: Fast k-nearest neighbor classification using cluster-based trees. IEEE Trans. Pattern Anal. Mach. Intell. **26**(4), 525–528 (2004)

Part III
Dimensionality Reduction

Supervised Dimension Reduction via Local Gradient Elongation

Jannatul Ferdous Chhoa, Longxiu Huang, Anna Little, Aimee Maurais, Kirsten D. Morris, Maria D. van der Walt, Geetika Verma, and Rongrong Wang

1 Introduction

This chapter explores a geometric approach for supervised dimension reduction (SDR), where we assume we have features $x_1, \ldots, x_n$ together with observations of a response variable $y_1, \ldots, y_n$, where $y_i \approx f(x_i)$ for some unknown function f. In general, the goals are twofold: (1) obtain a low-dimensional representation of the data using an embedding/process guided by the response variable Y, which

J. F. Chhoa
Department of Mathematics, University of Houston, Houston, TX, USA
e-mail: jchhoa@cougarnet.uh.edu

L. Huang
Department of Computational Mathematics, Science and Engineering, Department of Mathematics, Michigan State University, East Lansing, MI, USA
e-mail: huangl3@msu.edu

A. Little (✉)
Department of Mathematics, University of Utah, Salt Lake City, UT, USA
e-mail: little@math.utah.edu

A. Maurais
Center for Computational Science and Engineering, Massachusetts Institute of Technology, Cambridge, MA, USA
e-mail: maurais@mit.edu

K. D. Morris
Department of Mathematics, University of Nebraska-Lincoln, Lincoln, NE, USA
e-mail: kmorris11@huskers.unl.edu

M. D. van der Walt
Department of Mathematics and Computer Science, Westmont College, Santa Barbara, CA, USA
e-mail: mvanderwalt@westmont.edu

C. Garcia-Cardona, H. Lee (eds.), *Advances in Data Science*, Association for Women in Mathematics Series 37, https://doi.org/10.1007/978-3-031-87804-6_9

leads to more effective exploratory analysis, and (2) perform prediction of unknown labels, which can be significantly improved by appropriate choice of SDR algorithm. In the linear case, SDR is a well-studied problem: therein, the goal is to restrict features to the *active subspace*, i.e., project out any data dimensions that contain no information useful for predicting Y [15, 16, 54], and apply a nonparametric regression for prediction in the active subspace. Seminal works include sliced inverse regression (SIR) [18], which discovers the active subspace by finding the conditional expectation of the predictors, conditioned on the response, as well as [17, 21, 64, 65], which eliminate some of the restrictive assumptions of SIR. Unfortunately, these useful tools fail when the underlying relevant domain is nonlinear. One possible approach in the case of a nonlinear active manifold is to apply nonparametric manifold regression [7, 12]; however, these methods are not targeted to the regime where the data manifold has many coordinates irrelevant to Y, and thus are not truly nonlinear SDR methods. A number of methods have thus been developed, which extend supervised dimension reduction to the nonlinear/manifold case, including neural-network-based approaches [24, 38, 67], methods designed for discrete labels [1, 13, 20, 23, 25, 28, 34, 53, 62, 66], and methods applicable to a continuous response variable [5, 10, 13, 51, 52, 57] as considered in the current chapter; see also [11, 26, 40, 47, 50].

We propose a novel geometric approach to nonlinear SDR that utilizes the local gradients of a (generally continuous) response variable to stretch the data in directions useful for prediction and shrink the data in uninformative directions. More specifically, when the response $y \in \mathbb{R}$ is univariate, we define (locally) the following metric:

$$d_{Y,\tau}(x_i, x_j)^2 = (1-\tau)\|x_i - x_j\|^2 + \tau\|y_i - y_j\|^2 . \tag{1}$$

The parameter $0 \le \tau \le 1$ controls the extent to which the labels y impact the distance. In practice, if the labels are noisy, i.e., if $y_i = f(x_i) + \epsilon_i$, it may be advantageous to use the following formulation:

$$d_{\nabla Y,\tau}(x_i, x_j)^2 = (1-\tau)\|x_i - x_j\|^2 + \frac{\tau}{2}\left(\langle \nabla y_i, x_i - x_j\rangle^2 + \langle \nabla y_j, x_i - x_j\rangle^2\right) \tag{2}$$

where $\nabla y_i \approx \nabla f(x_i)$ is an approximation of the gradient of the response at x_i. Note in the noiseless case, these two definitions are locally essentially identical for smooth C^2 functions, since by Taylor's Theorem:

G. Verma
RMIT University, Melbourne, VIC, Australia

R. Wang
Department of Computational Mathematics, Science & Engineering, Michigan State University, East Lansing, MI, USA

$$
\begin{aligned}
y_i - y_j &= \langle \nabla y_i, x_i - x_j \rangle + O(\|x_i - x_j\|^2) \\
&= \langle \nabla y_j, x_i - x_j \rangle + O(\|x_i - x_j\|^2) \\
\implies (y_i - y_j)^2 &= \langle \nabla y_i, x_i - x_j \rangle^2 + O(\|x_i - x_j\|^3) \\
&= \langle \nabla y_j, x_i - x_j \rangle^2 + O(\|x_i - x_j\|^3) \\
\implies (y_i - y_j)^2 &= \frac{1}{2}\left(\langle \nabla y_i, x_i - x_j \rangle^2 + \langle \nabla y_j, x_i - x_j \rangle^2\right) + O(\|x_i - x_j\|^3) .
\end{aligned}
$$

However, writing the metric as (2) leads to some insight, since it illustrates that the metric is elongating in the direction of the gradient. When τ is small, ∇y does not impact the metric, and one recovers Euclidean distance; when τ is large, local connections are adjusted to shrink distances in the directions of $\nabla y^{\perp}$. When data points are sampled from a Riemannian manifold $(\mathcal{M}, g)$ and $\tau < 1$, this stretching in fact corresponds to a new Riemannian metric tensor $\tilde{g}$ on $\mathcal{M}$ defined by the following modified inner product $\{U, V\}_x$ on the tangent plane $T_x\mathcal{M}$:

$$
\{U, V\}_x := \tau \langle \nabla f(x), U \rangle_x \langle \nabla f(x), V \rangle_x + (1 - \tau)\langle U, V \rangle_x
$$

for $U, V \in T_x\mathcal{M}$, where $\langle U, V \rangle_x$ is the inner product corresponding to the original Riemannian metric $g(x)$.

Equation (2) is thus a local approximation of the geodesic distance under $\tilde{g}$. Utilizing the theory in [6] for anisotropic kernels, the work [5] proposes an iterative nonlinear SDR algorithm for the τ small case. Although it is based on insightful geometric principles, the algorithm is too complex to be practical in real data applications, and the theoretical framework is not applicable when $\tau = 1$; in this case, there is a collapse of geometry since $\tilde{g}$ is no longer full-rank and $\mathcal{M}$ becomes a sub-Riemannian manifold. This chapter explores the utility of the gradient elongated metric (2) for supervised dimension reduction, focusing specifically on the two key tasks of *visualization* and *prediction*.

2 Methodology

We first leverage (2) to develop an algorithm for visualization as described in Sect. 2.3; we then propose an algorithm for prediction of unlabeled data points by combining (2) with Laplacian learning as described in Sect. 2.4.

2.1 Notation and Assumptions

Throughout the chapter, we assume that a set of n feature vectors $X = \{x_1, \ldots, x_n\}$ are sampled from a compact Riemannian manifold $\mathcal{M}$ of intrinsic dimension d embedded in $\mathbb{R}^D$. We let $y_1, \ldots, y_n$ denote the corresponding labels, which can be observed or unobserved, and either discrete or continuous. We let $I_{\text{label}}/I_{\text{label}}^C$ denote the indices of the labeled/unlabeled points, $m = |I_{\text{label}}|$ denote the number of labeled points, and $X_{\text{label}} = \{x_i : i \in I_{\text{label}}\}$ and $y_{\text{label}} = \{y_i : i \in I_{\text{label}}\}$ denote the feature and response values of the labeled points. We let $\text{NN}_k(x, X)$ denote the set of k Euclidean nearest neighbors of x in the set X.

In the noiseless setting, we assume $y_i = f(x_i)$ for some function f and that we have access to (x_i, y_i) for $i \in I_{\text{label}}$. However, we will also evaluate the prediction methodology proposed in Sect. 2.4 in the presence of feature or label noise and thus consider the following two noise models.

Model 1 (Noisy Labels) *We assume*

$$y_i = f(x_i) + \sigma_y \eta_i \,, \tag{3}$$

where the η_i *are independent standard normal random variables,* $\sigma_y > 0$ *is the noise level, and the* x_i *are sampled from* $\mathcal{M}$*. We assume access to* (x_i, y_i) *for* $i \in I_{label}$ *but only to* x_i *for* $i \in I_{label}^C$*.*

Model 2 (Noisy Features) *We assume*

$$y_i = f(m_i) \quad , \quad x_i = m_i + \sigma_x \xi_i \,, \tag{4}$$

where the ξ_i *are independent multivariate normal random vectors with mean zero and covariance* I_D*,* $\sigma_x > 0$ *is the noise level, and the* m_i *are sampled from* $\mathcal{M}$*. We assume access to* (x_i, y_i) *for* $i \in I_{label}$ *but only to* x_i *for* $i \in I_{label}^C$*.*

2.2 Estimation of $\nabla f(x_i)$

To compute $d_{\nabla Y,\tau}(x_i, x_j)$ as introduced in (2), it is essential to estimate ∇y_i and ∇y_j. We utilize a similar least squares fitting procedure as in locally linear regression [48] as described below. Given the labeled dataset (x_i, y_i) for $i \in I_{\text{label}}$ and the specific data feature x, we denote the k Euclidean nearest neighbors (k-NNs) of x within the labeled dataset as $(x_1(x), y_1(x)), \ldots, (x_k(x), y_k(x))$.

Notice that given a smooth function $f : \mathbb{R}^D \to \mathbb{R}$ and $a, b \in \mathbb{R}^D$ sufficiently close to each other, the approximation $f(a) \approx f(b) + \nabla f(b)^\top (a - b)$ holds. For each neighbor point of x, we use this linear approximation to predict the label $y_i(x) = f(x_i(x)) \approx f(b) + \nabla f(b)^\top (x_i(x) - b) = f(b) - \nabla f(b)^\top b + \nabla f(b)^\top x_i(x) \equiv c + \nabla f(b)^\top x_i(x)$, $i = 1, \ldots, k$, where we set $c = f(b) - \nabla f(b)^\top b$. To identify c and $\nabla f(b)$, we use the following least squares fitting

$$\hat{c}, \hat{G} = \arg \min_{c \in \mathbb{R}, G \in \mathbb{R}^D} \sum_{i=1}^{k} \|y_i(x) - c - G^\top x_i(x)\|^2$$

and assign $\nabla f(b) = \hat{G}$ to be the estimated gradient.

The closed-form solution of the above minimization is

$$\hat{G} = \begin{bmatrix} (\bar{x} - x_1(x))^\top \\ \vdots \\ (\bar{x} - x_k(x))^\top \end{bmatrix}^\dagger \begin{bmatrix} \bar{y} - y_1(x) \\ \vdots \\ \bar{y} - y_k(x) \end{bmatrix}.$$

where $\bar{x} = \frac{1}{k}\sum x_i(x)$ and $\bar{y} = \frac{1}{k}\sum y_i(x)$. In our numerical experiments, we eliminate the small singular values from the coefficient matrix $\begin{bmatrix} (\bar{x} - x_1(x))^\top \\ \vdots \\ (\bar{x} - x_k(x))^\top \end{bmatrix}$ prior to calculating the pseudo-inverse to improve the stability of our algorithm.

2.3 Visualization

We create a kNN graph $\mathcal{G} = (X, E, W)$ of local connections based off of the features x_i, in which each point is connected to its Euclidean nearest neighbors, but the corresponding edges are weighted according to the gradient-adjusted metric (2), i.e., the edge weights depend on the response y. More specifically, if x_i, x_j are Euclidean nearest neighbors, we define $W_{ij} = d_{\nabla Y,\tau}(x_i, x_j)$; else $W_{ij} = 0$. We then define a new metric by computing shortest path distances within this graph:

$$\ell_{\nabla Y,\tau}(a, b) := \inf_{(x_0,\ldots,x_s)} \sum_{i=0}^{s-1} d_{\nabla Y,\tau}(x_i, x_{i+1}), \tag{5}$$

where the infimum is taken over all sequences of points $x_0, \ldots, x_s$ in X with $x_0 = a$, $x_s = b$, and consecutive x_i connected in $\mathcal{G}$. Note when $\tau = 1$ and the number of sample points $n \to \infty$, we expect that $\ell_{\nabla Y}$ converges to the following geodesic distance (see, e.g., [4]):

$$\mathcal{L}_{\nabla Y}(a, b) = \inf_\gamma \int_0^1 |\langle \nabla f(\gamma(t)), \gamma'(t)\rangle|\, dt = \inf_\gamma \int_0^1 \left\| \frac{df(\gamma(t))}{dt} \right\| dt, \tag{6}$$

where the infimum is taken over all differentiable curves $\gamma : [0, 1] \to \mathbb{R}^D$ satisfying $\gamma(0) = a$ and $\gamma(1) = b$. As seen in (6), curves that remain in a level set of f will be measured as having distance zero: one can travel along the level sets "for free," but cost is incurred when the value of f changes, i.e., when the path has a component in the direction of ∇f.

Once the (supervised) distances (5) are computed between all points in X, we use these distances within unsupervised dimension reduction algorithm for visualization. Specifically, we employ these distances within both classic multidimensional scaling (CMDS), a linear method introduced in [60] (see [8] for an overview), as well as the t-distributed stochastic neighbor embedding (t-SNE), a nonlinear method developed in [32], to obtain reduced-dimensional visualizations of X.

Related work on data-driven metrics and unsupervised dimension reduction The proposed algorithm can be thought of as a generalization of the classic Isomap [59] algorithm, which approximates manifold geodesic distance $d_{\mathcal{M}}(a, b)$ by classic shortest path distance in a graph of local connections, i.e., by $\inf_{(x_0,\ldots,x_s)} \sum_{i=0}^{s-1} \|x_i - x_{i+1}\|$, where the infimum is taken over paths $(x_0, \ldots, x_s)$ connecting a, b. Recent work has also focused on the analysis of *power weighted* shortest path distances , [22, 27, 33, 42, 61], where the distances are computed as $\inf_{(x_0,\ldots,x_s)} \sum_{i=0}^{s-1} \|x_i - x_{i+1}\|^p$ for some $p \geq 1$. If data points are i.i.d. samples from a probability measure with density ρ on $\mathcal{M}$, then this discrete distance (appropriately normalized) converges to a density-reweighted geodesic distance $\mathcal{L}_\rho^p(a, b) = \inf_\gamma \int \rho(\gamma(t))^{\frac{1-p}{d}} |\gamma'(t)|\, dt$, where the infimum is once again over all differentiable curves on $\mathcal{M}$ connecting a, b. Such metrics stretch the manifold geometry according to the data density, an adjustment that can be highly useful for clustering [43] as well as topological data analysis [22]. This chapter investigates a similar geometric approach, but the manifold is stretched according to a response variable instead of a density function. Alternatively, one could adjust data geometry utilizing a diffusion process [14, 39, 44, 58]. The novelty of our approach is utilizing the metric (6), and although we use CMDS and t-SNE for visualization, this choice is somewhat arbitrary, and the metric could be combined with other embedding algorithms such as metric MDS [9], Laplacian Eigenmaps [3], diffusion maps [14], UMAP [2], etc. In addition, recent studies have explored alternative metrics for graph Laplacian embeddings, including the use of optimal transportation [63] and Wasserstein-based isometric mappings [29]. These approaches offer more flexible and robust geometric frameworks for analyzing high-dimensional data, which align with the objectives of our supervised dimensionality reduction model.

Related work on supervised dimension reduction We employ a supervised version of dimensionality reduction techniques. The majority of SDR algorithms are designed for discrete label information. For example, these algorithms might define dissimilarity according to:

$$\mathrm{dis}(x_i, x_j) = \begin{cases} \sqrt{1 - e^{-\frac{\|x_i - x_j\|^2}{\beta}}} & x_i, x_j \text{ in the same class} \\ \sqrt{e^{-\frac{\|x_i - x_j\|^2}{\beta}}} - \alpha & x_i, x_j \text{ in different classes} \end{cases}$$

for parameters α, β, where β generally depends on the feature distances and α determines the degree of supervision. In contrast, the proposed method of this chapter is very natural in the case of a continuous response, i.e., for problems of prediction and not just classification. Other methods for continuous response variables include [5, 10, 13, 57]. In addition, embedding methods based on random forest proximities have been proposed, which can be applied to either discrete or continuous labels [30, 41, 51, 52]. However since decision trees leverage individual features, resulting methods may not perform as well when the response is a linear combination of features (e.g., $y = x_1 + x_2$) or a more complex nonlinear function of features. Although random forests can partially address this issue by aggregating multiple trees, our method may offer an advantage in this setting.

The two works most relevant to the current chapter are [10] and [13]. The work [10] utilizes the same mathematical framework we are suggesting (Figs. 2c, f, 3c, and f are visualizations of an *active manifold* as described in [10]). However, unlike [10], which defines an active manifold as a submanifold of the original domain by following a gradient flow on a high-dimensional grid, we propose to compute an active manifold via simple embeddings of gradient-based path distances. This new approach allows one to lift some of the restrictive assumptions in [10] such as connected level sets (indeed, this restriction rules out some very interesting examples like evolutionary processes). The work [13] defines an SDR embedding by minimizing

$$C(Q) = \rho \mathrm{KL}(P||Q) + (1-\rho)\mathrm{KL}(O||Q)$$

over embedding coordinates $\{z_i\} \subseteq \mathbb{R}^p$, with $p < D$, where $P, O, Q \in \mathbb{R}^{m\times m}$ are similarity matrices computed over the features $\{x_i\}_{i=1}^m$, labels $\{y_i\}_{i=1}^m$, and embedding coordinates $\{z_i\}_{i=1}^m$ with entries given by

$$p_{j|i} \propto \exp\left(-\|x_i - x_j\|^2/2\sigma_i^2\right), \quad o_{j|i} \propto \exp\left(-\|y_i - y_j\|^2\right),$$
$$q_{ij} \propto (\|z_i - z_j\|^2 + 1)^{-1}, \quad i \neq j$$

and $p_{i|i} = o_{i|i} = q_{ii} = 0$. $\mathrm{KL}(\cdot\|\cdot)$ is the discrete Kullback-Leibler divergence, and $\rho \in [0, 1]$ weighs the contributions of the feature divergence and label divergence to C. The authors refer to this method as supervised t-SNE (St-SNE) as it is an extension of t-SNE (t-SNE can be viewed as St-SNE with $\rho = 1$). As we are proposing, they balance between finding an embedding Q that is representative of the features (first term) with finding an embedding that is representative of the response (second term). However, we demonstrate that this approach will not be capable of simultaneously discarding irrelevant features and preserving the geometry of y—as ρ decreases, all points with similar y-values will be "glued together," and the underlying structure of the response variable will be lost. See Figs. 2 and 3 and the accompanying descriptions for an illustration.

2.4 Prediction

We propose a new method for transductive semi-supervised learning, which combines the gradient-elongated metric (2) with Laplacian regularization, an effective interpolation method used in machine learning for estimating values on a graph structure. Specifically, given n data points consisting of both labeled and unlabeled features, Laplacian regularization predicts the unknown labels by solving:

$$y^{\mathrm{LL}} = \arg\min_{y \in \mathbb{R}^n} \| P_{\mathrm{label}} y - y_{\mathrm{label}} \|_2^2 + \lambda y^T L y, \tag{7}$$

where P_{label} is a projection matrix that selects the labeled data from the full label vector y, y_{label} is the vector of known labels, and $L \in \mathbb{R}^{n \times n}$ is the graph Laplacian matrix defined using pairwise distances of features of all the data points. The method thus extends the given label information in a smooth way across the graph and is particularly useful in areas involving social networks [35], sensor networks [68], or image restoration [56]. The graph Laplacian is used to impose a smoothness constraint on the interpolation process, ensuring that the interpolated values change gradually along the graph's edges. This regularization helps in producing more accurate and reliable interpolations, especially in scenarios where the data points are sparsely distributed across the graph. However, if the knowledge of the graph structure used in graph Laplacian regularized interpolation is not accurate, it may greatly impact the accuracy of the interpolation.

We apply the Laplacian regularization procedure but define $L = L(d_{\nabla Y,\tau})$ using the local gradient metric (2)[1] with partially observed label information incorporated. Specifically, we define a graph of local connections among the features weighted by $W_{ij} = \exp\left(-\frac{d_{\nabla Y,\tau}(x_i,x_j)^2}{\epsilon}\right)$, where gradients are approximated for all features using the training data and the method of Sect. 2.2, and $\epsilon > 0$ is a scale parameter. Even though the gradient approximations are not very accurate when the number of training points is small, we demonstrate the method still provides a significant improvement over local linear regression and standard Laplacian regularization.

For comparative analysis, we evaluate our prediction approach against several existing methods, specifically K-nearest neighbors (KNN) regression [31], local linear regression (LLR) [48], traditional Laplacian learning (TLL) [45], and nonlinear level-set learning (NLL) [67]. We have summarized these methodologies as follows:

1. KNN regression (kNN): Each unlabeled data point is paired with its k nearest neighbors among the labeled points, determined by Euclidean distance. The label

[1] In the numerical experiments, the local gradient metric is calculated using the normalized data $\frac{x_i - \bar{x}}{\sqrt{\frac{1}{m}\sum_{i=1}^m \|x_i - \bar{x}\|^2}}$ and $\frac{y_i - \bar{y}}{\sqrt{\frac{1}{|I_{label}|}\sum_{i=1}^{|I_{label}|} \|y_i - \bar{y}\|^2}}$ where $\bar{x} = \frac{1}{m}\sum_{i=1}^m x_i$, $\bar{y} = \frac{1}{|I_{label}|}\sum_{i=1}^{|I_{label}|} y_i$.

for the unlabeled point is assigned based on the average label of these k nearest labeled neighbors.

2. Local linear regression (LLR): the label of the point x is defined as

$$y(x) = \frac{1}{k} 1_k^\top y_{\text{knn}} + \left(x - \frac{1}{k} X_{\text{knn}} 1_k\right)^\top \left(X_{\text{knn}}^\top - \frac{1}{k} X_{\text{knn}}^\top 1_k^\top 1_k\right)^\dagger (y_{\text{knn}} - 1_k 1_k^\top y_{\text{knn}})$$

where $1_k \in \mathbb{R}^k$ is a vector with all 1s, X_{knn} is the matrix with each column corresponding to the feature information of the k nearest neighbors (kNNs) of x, and y_{knn} denotes the label vector for the kNNs of x.

3. Traditional Laplacian learning (TLL): we approach the task of estimating labels for unlabeled data by solving the optimization problem (7), where $L = D - W$ is a graph Laplacian matrix, with W being the similarity matrix defined on all the data points and D being the associated degree matrix. The construction of the similarity matrix is carried out in the following manner:

   (a) Initialize $W_{ij} = \exp\left(-\frac{\|x_i - x_j\|^2}{\epsilon}\right)$, where $\|x_i - x_j\|$ signifies the Euclidean distance between points x_i and x_j and $\epsilon > 0$ is a scale parameter as before.
   (b) For every row in W, maintain only the k highest values, setting the remainder to zero and ensuring that each row contains at most k non-zero elements.
   (c) Symmetrize W by constructing $\widetilde{W} \in \mathbb{R}^{n \times n}$ according to $\widetilde{W}_{ij} = \max(W_{ij}, W_{ji})$, $i, j = 1 \ldots, n$. Note that other approaches can also be used to symmetrize W, including $\widetilde{W} = W + W^T$ or $\widetilde{W}_{ij} = \min(W_{ij}, W_{ji})$. Finally, set $W = \widetilde{W}$.

4. Supervised Laplacian learning (SLL): This approach employs Laplacian learning method for label estimation of unlabeled data, with the modification that $d_{\nabla Y}(x_i, x_j)$ (2) is substituted for the Euclidean distance in construction of the similarity matrix W.
5. Nonlinear level-set learning (NLL): In the NLL approach of [67], a neural network is trained to identify a nonlinear embedding $g : \mathbb{R}^D \to \mathbb{R}^D$ of input data $x \in \mathbb{R}^D$ such that the first few coordinates $g(x)_{1:p}$, where $p < D$, are highly predictive of $f(x) \in \mathbb{R}$. The remaining coordinates $g(x)_{p+1:D}$ are not predictive of f, and $g(x)_{1:p}$ can be used as a reduced-dimensional embedding of x. The embedding $g : \mathbb{R}^D \to \mathbb{R}^D$ is parametrized as a neural network and trained using a loss that drives the first few coordinates of the embedding, $g(\cdot)_{1:p}$ to capture directions orthogonal to level sets of f, and the last coordinates $g(\cdot)_{p+1:D}$ to capture directions parallel to the level sets of f. Under these conditions, the value of f will change with perturbations in the direction of $g(x)_{1:p}$ but will not change with perturbations in the directions of $g(x)_{p+1:D}$. Evaluation of the loss requires evaluations of f and its gradient ∇f; hence, for our experiments in Sect. 3, we approximate the needed gradients using the method of Sect. 2.2

3 Numerical Results

3.1 Visualization

In this section, our goal is to visualize the geometry or structure of given datasets, as described in Sect. 1. We start by considering two toy datasets. Specifically, we consider a tree with three branches (Small Tree) and a tree with seven branches (Big Tree). The datasets are created by first sampling points m_i along a one-dimensional tree structure, defining the label of m_i to correspond to the geodesic distance from the root of the tree to m_i; noise was then added to produce the noisy data points x_i, which are plotted in Fig. 1 (here we consider noise according to Model 2); the plots are colored by the response variable.

Results for Small Tree are shown in Fig. 2. In the first two rows of Fig. 2, we show the results of applying CMDS and t-SNE (respectively) to visualize in two dimensions the path distance $\ell_{\Delta Y,\tau}$ for different values of τ in (2). For comparison, the third row of Fig. 2 shows a visualization obtained with St-SNE from [13] for different values of ρ. Note how, as supervision increases (increasing τ in our method with CMDS and t-SNE or decreasing ρ in St-SNE), our method is able to denoise the tree while preserving the underlying structure of the data. In contrast, the St-SNE visualization does not remain faithful to the true geometry—the branches of the tree are glued together.

Various visualizations of the Big Tree dataset are shown in Fig. 3. As before, the first two rows show CMDS and t-SNE embeddings of the path distance $\ell_{\Delta Y,\tau}$ for different values of τ; the third row shows a St-SNE visualization for different values of ρ. Note again the visual confirmation that our method is able to faithfully represent the true geometry of the data even as τ increases, as opposed to St-SNE with decreasing ρ. Also note that a three-dimensional CMDS embedding is necessary to accurately display the finer branches at the endpoints of the big branches. Note we also computed the MDS embedding of the random forest-based

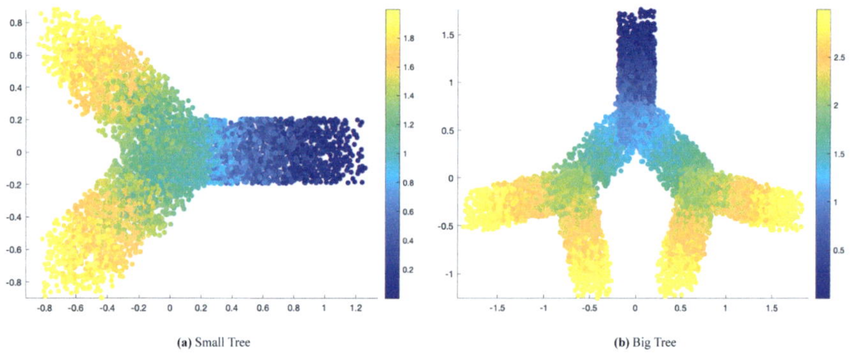

Fig. 1 Noisy tree datasets, colored by the response variable

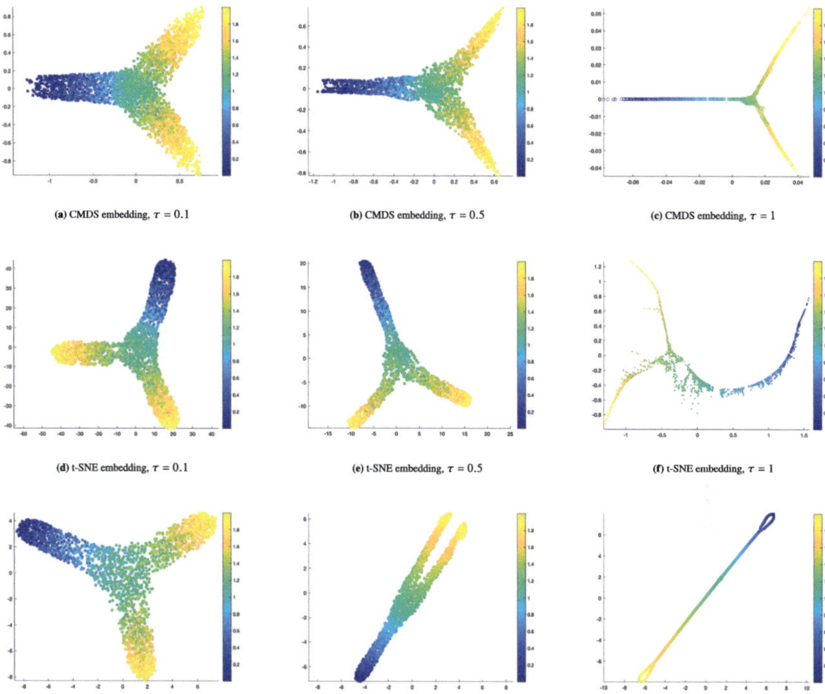

(a) CMDS embedding, $\tau = 0.1$ (b) CMDS embedding, $\tau = 0.5$ (c) CMDS embedding, $\tau = 1$

(d) t-SNE embedding, $\tau = 0.1$ (e) t-SNE embedding, $\tau = 0.5$ (f) t-SNE embedding, $\tau = 1$

(g) St-SNE, $\rho = 1$ (h) St-SNE, $\rho = 0.5$ (i) St-SNE, $\rho = 0$

Fig. 2 Two-dimensional visualization of the path distance $\ell_{\Delta Y,\tau}$ calculated on the small tree dataset by applying a CMDS embedding (top row) and a t-SNE embedding (middle row) for different values of τ. The last row shows a comparison with St-SNE for different values of ρ

proximities (RF-GAP) proposed in [52] to the small and big tree datasets; see Fig. 8 in Appendix "Further Visualization Results". However, the geometry of the tree structure is lost.

In addition, we display visualization results on two real-world datasets: one concerning severe acute respiratory syndrome-coronavirus-2 (SARS-CoV-2) and one concerning the differentiation of embryoid bodies (EB) [44]. In a recent study published in [55], the authors want to quantify the neurological phenotypes induced by the SARS-CoV-2 spike protein in neurons, as measured by in-vitro multi-well micro-electrode arrays. To this end, a visualization of how much different instances and exposures of the SARS-CoV-2 spike protein affect neurons is of great value. Figure 4 displays a visualization of $\ell_{\Delta Y,\tau}$ with $\tau = 0.8$ (top row) as measured on the SARS-CoV-2 dataset referenced in [55], by applying a t-SNE embedding. In each subfigure, blue represents instances of control neurons, while yellow represents exposed neurons, with the exposure ranging from 1 to 100 ng. Note how the separation becomes clearer as the spike protein exposure increases, which supports the hypothesis that the neurons are affected under exposure to the SARS-CoV-2

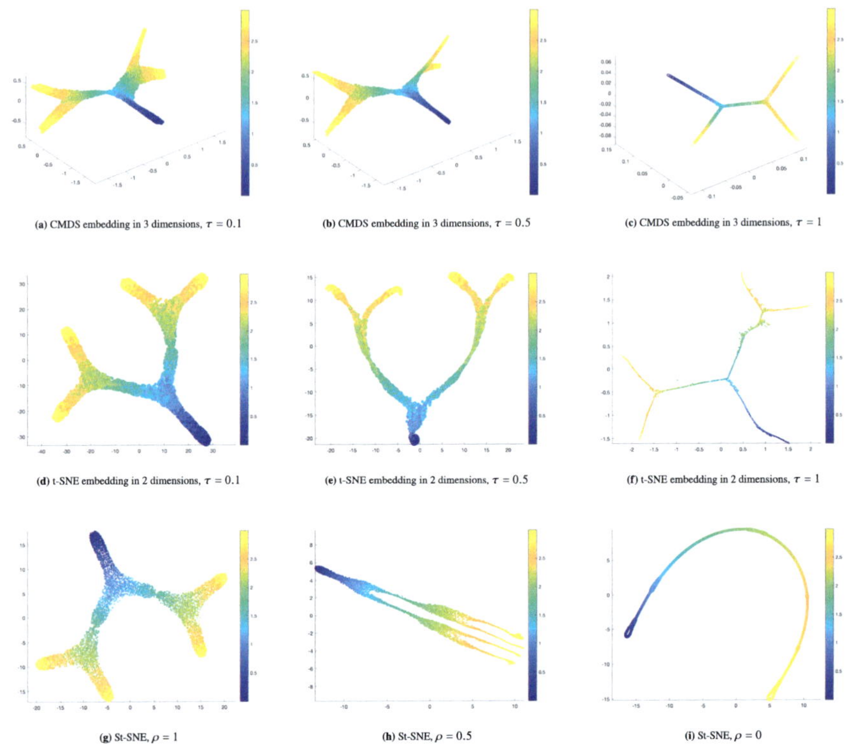

Fig. 3 Visualization of the path distance $\ell_{\Delta Y,\tau}$ calculated on the big tree dataset by applying a CMDS embedding (top row) and a t-SNE embedding (middle row) for different values of τ. The last row shows a comparison with St-SNE for different values of ρ

spike protein. For comparison, we also show in Fig. 4 visualizations obtained with t-SNE (second row), St-SNE with $\rho = 0.75$ (third row), and MDS of RF-GAP [52] (bottom row). The fourth row of Fig. 4 shows visualizations obtained with PCA; our method more clearly depicts the separation as the exposure increases while maintaining the underlying geometry observed in the unsupervised case.

Next we apply our method to the EB dataset, which tracks the development of human stem cells as they differentiate into various embryoid bodies. Measurements are taken every 3 days over a 27-day period, and cells were sequenced with the 10x chromium platform; see [44] for more details. We consider a subset of 35,000 cells and use the day the cell was sequenced (i.e., time) as the response variable; the goal of applying our methodology is to emphasize the features that are changing in time, i.e., the ones relevant to differentiation, and to de-emphasize features not changing in time, which are not relevant for a visualization of differentiation. Figure 5 (top row) shows the results of applying CMDS to visualize $\ell_{\Delta Y,\tau}$ for various values of τ: as τ is increased, the separation of the time periods becomes more clear, as expected, but one also sees a larger spread of values for higher label values, which reflects the

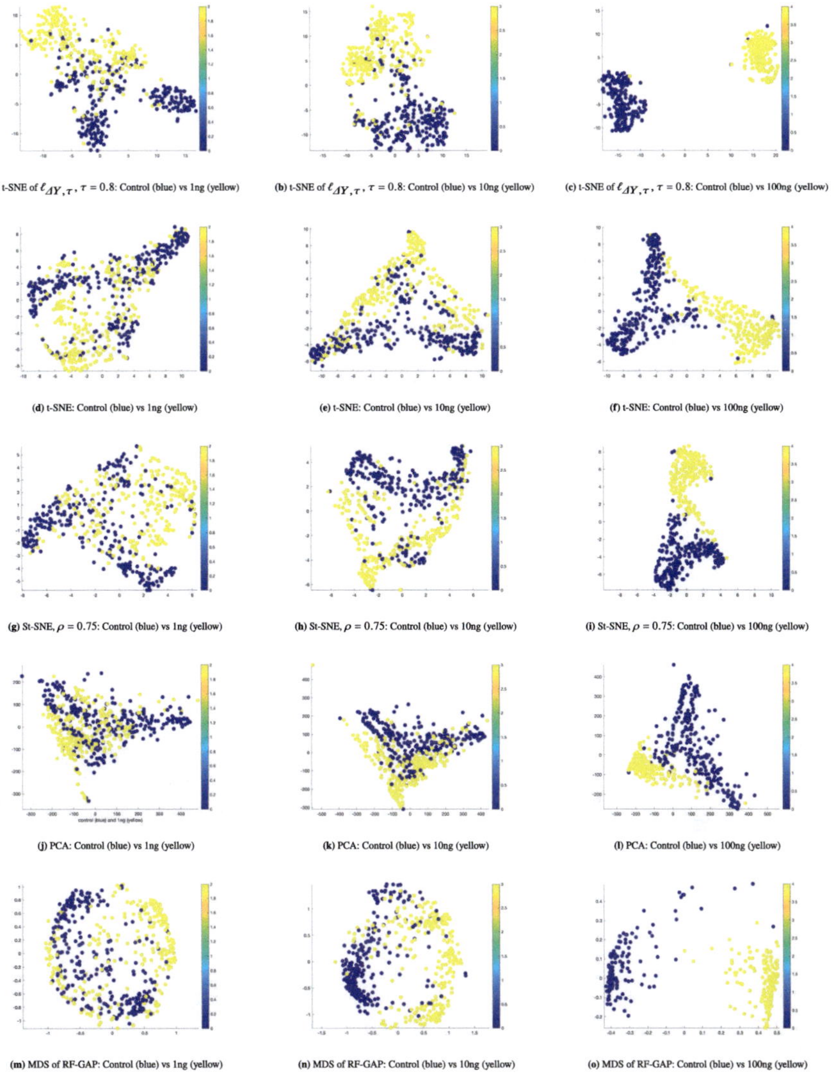

(a) t-SNE of $\ell_{\Delta Y,\tau}$, $\tau = 0.8$: Control (blue) vs 1ng (yellow)
(b) t-SNE of $\ell_{\Delta Y,\tau}$, $\tau = 0.8$: Control (blue) vs 10ng (yellow)
(c) t-SNE of $\ell_{\Delta Y,\tau}$, $\tau = 0.8$: Control (blue) vs 100ng (yellow)
(d) t-SNE: Control (blue) vs 1ng (yellow)
(e) t-SNE: Control (blue) vs 10ng (yellow)
(f) t-SNE: Control (blue) vs 100ng (yellow)
(g) St-SNE, $\rho = 0.75$: Control (blue) vs 1ng (yellow)
(h) St-SNE, $\rho = 0.75$: Control (blue) vs 10ng (yellow)
(i) St-SNE, $\rho = 0.75$: Control (blue) vs 100ng (yellow)
(j) PCA: Control (blue) vs 1ng (yellow)
(k) PCA: Control (blue) vs 10ng (yellow)
(l) PCA: Control (blue) vs 100ng (yellow)
(m) MDS of RF-GAP: Control (blue) vs 1ng (yellow)
(n) MDS of RF-GAP: Control (blue) vs 10ng (yellow)
(o) MDS of RF-GAP: Control (blue) vs 100ng (yellow)

Fig. 4 Visualization of the SARS-CoV-2 dataset. In each case, blue indicates the control, and yellow indicates the SARS-CoV-2-exposed neurons

fact that the stem cells are developing into a variety of embryoid bodies. Although supervised t-SNE (third row) clearly separates classes as ρ is decreased, it fails to reflect this geometry of small-to-large variances for larger values of ρ, and thus the embedding is not as biologically meaningful. Combining our metric $\ell_{\Delta Y,\tau}$ with t-SNE (second row) yields cleaner class separation than our metric with CMDS (top row), but it does not reflect the global geometry as accurately as CMDS. Finally, we also compare with the MDS embedding of RF-GAP proximities [52]; see Fig. 5(i).

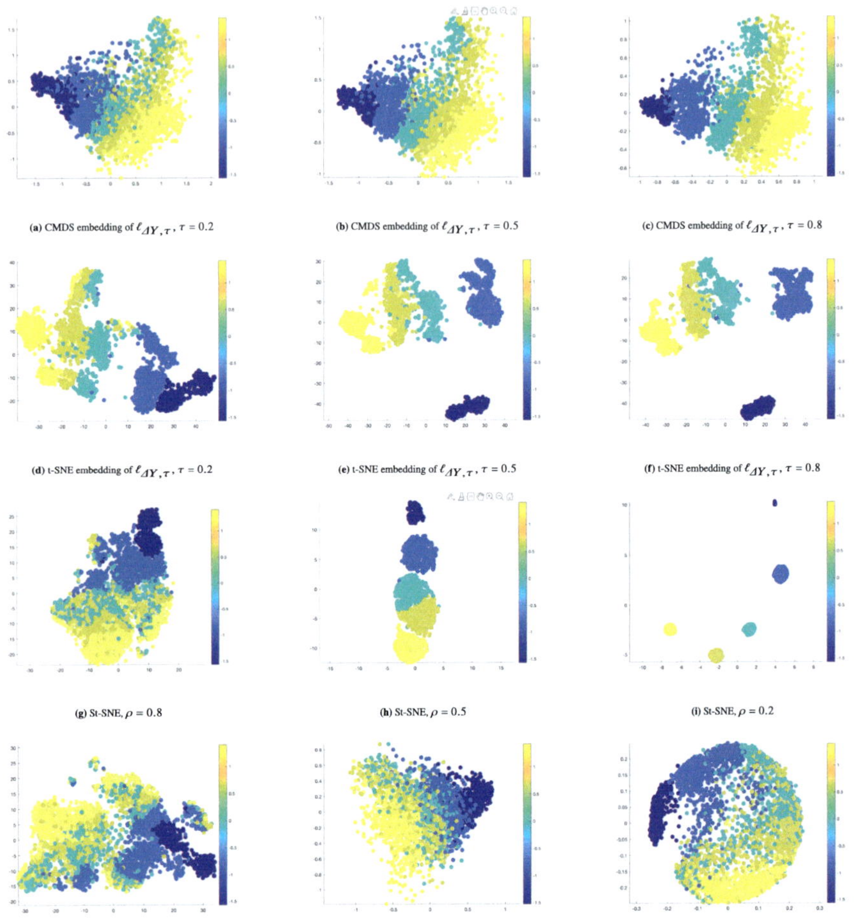

(a) CMDS embedding of $\ell_{\Delta Y,\tau}$, $\tau = 0.2$ (b) CMDS embedding of $\ell_{\Delta Y,\tau}$, $\tau = 0.5$ (c) CMDS embedding of $\ell_{\Delta Y,\tau}$, $\tau = 0.8$

(d) t-SNE embedding of $\ell_{\Delta Y,\tau}$, $\tau = 0.2$ (e) t-SNE embedding of $\ell_{\Delta Y,\tau}$, $\tau = 0.5$ (f) t-SNE embedding of $\ell_{\Delta Y,\tau}$, $\tau = 0.8$

(g) St-SNE, $\rho = 0.8$ (h) St-SNE, $\rho = 0.5$ (i) St-SNE, $\rho = 0.2$

(j) t-SNE (k) PCA (l) MDS embedding of RF-GAP proximities

Fig. 5 Visualization of the EB dataset. Colors indicate the number of days the embryonic bodies have grown, where the progression from blue to yellow corresponds to a progression from 1 day to 5 days

To quantitatively assess the quality of supervised dimension reduction algorithms is a difficult task, but we attempt to do so by calculating the variable importance correlation metric proposed in [51]. This metric computes scores quantifying the importance of each predictor variable for predicting the response and also for predicting the embedding coordinates and then calculates the correlation between the importance scores; the goal is to assess the preservation of the structure of the predictor variables. The correlation metrics of all of the EB embeddings shown in Fig. 5 are given in Table 1. The highest score was obtained by supervised t-SNE with small ρ (0.777); we emphasize however that it is clear from Fig. 5(i) that the

Table 1 Variable importance correlation metric from [51] on EB dataset [44]

CMDS of $\ell_{\Delta Y,\tau}$	$\tau = 0.2$	0.695
	$\tau = 0.5$	0.729
	$\tau = 0.8$	0.733
t-SNE of $\ell_{\Delta Y,\tau}$	$\tau = 0.2$	0.693
	$\tau = 0.5$	0.672
	$\tau = 0.8$	0.651
Supervised t-SNE	$\rho = 0.8$	0.627
	$\rho = 0.5$	0.650
	$\rho = 0.2$	0.777
MDS of RF-gap		0.644
PCA (unsupervised)		0.591
t-SNE (unsupervised)		0.582

embedding oversimplifies the geometry, collapsing all the various embryoid bodies into the small yellow cluster; the second highest score was obtained by our gradient-elongated metric with CMDS for large τ (0.733).

3.2 Prediction

In this section, we investigate the utility of the SLL method proposed in Sect. 2.4 for prediction of unknown labels on both synthetic and real-world datasets.

3.2.1 Synthetic Datasets

We first consider prediction on the following six synthetic datasets. For each dataset, we sample $n = 1000$ points; we assume access to 100 labeled points and measure predictive performance on the remaining points. The datasets are described below:

(i) Small Tree d8: dataset is generated by forming a small tree (three branches) in the first two dimensions and then sampling from a $d = 8$-dimensional tube about this tree. The response variable is $y = 8(x_2 - 1)^3$, i.e., it depends on the tree structure/direction of elongation.

(ii) Big Tree d4: dataset is generated by forming a large tree (seven branches) in the first two dimensions and then sampling from a $d = 4$-dimensional tube about this tree. The response variable is $y = (5x_4)^3$, i.e., it does not depend on the tree structure/direction of elongation.

(iii) Cube: data is sampled from a $d = 5$-dimensional unit cube, and the response variable is defined by $y = 4(x_1 + x_2)^4$.

(iv) Sphere: data is sampled from the $d = 4$-dimensional unit sphere $\mathbb{S}^4$, and the response is defined by $y = \theta^4$, where θ is the angle formed with a fixed polar cap.

(v) Swiss Roll: data is sampled from a $d = 2$-dimensional Swiss Roll, and the response variable is $y = \ell_1^2$, where ℓ_1 is an intrinsic manifold coordinate.
(vi) Annulus: data is sampled from a $d = 4$-dimensional annulus; specifically, all points satisfy $1 \leq \|x\| \leq 3$, and the response variable is $y = r^2 = \|x\|^2$.

3.2.2 Noiseless Experiments on Synthetic Data

In this section, we present the outcomes of our experiments where we compared supervised Laplacian learning (SLL) with prevalent methods such as k-nearest neighbors regression (kNN), local linear regression (LLR), and traditional Laplace learning (TLL) on the synthetic datasets without noise. For each dataset and for each method, we test a wide range of parameters as outlined in Appendix "Local and Laplacian-Based Methods" and in this section report the lowest relative error across all parameter settings for each dataset/method combination.

We also compare our proposed method to the result of using a one-dimensional embedding generated by the nonlinear level set learning (NLL) of [67] as an input to local linear regression (LLR) or k-nearest neighbors regression (kNN). Specifically, because the output $y \in \mathbb{R}$ in each of the synthetic datasets depends only on a one-dimensional function of the input coordinates $x \in \mathbb{R}^D$, we use NLL to learn an embedding $g : \mathbb{R}^D \to \mathbb{R}^D$ and then use $g(x)_1 \in \mathbb{R}$ as a predictor for $y \in \mathbb{R}$ within LLR and kNN. For further experimental details, see Appendix "Nonlinear Level-Set Learning".

For each example and parameter configuration, the dataset was randomly partitioned into a training set of $N_{\text{train}} = 100$ points, used to train the prediction methods, and a test set of $N_{\text{test}} = 900$ points over which the mean relative error of the prediction method was computed. The relative errors and runtimes reported in Table 2 correspond to optimal choices of parameters—that is, the parameters yielding minimal mean relative error—for each method/example and are averages over ten independent trials for each method/example combination, i.e., ten different partitions of the data into training and test sets.

The computational tasks in this study were executed on a MacBook Pro or MacBook Air equipped with an Apple M1 chip, featuring 8 cores split between 4 performance cores and 4 efficiency cores, and 16 GB of RAM. From Table 2, we see that the SLL approach demonstrated minimal relative error among all methods across all datasets. The runtime required for SLL is also quite reasonable, clocking in at <0.1s on a standard laptop computer across all examples—this runtime stands in particular stark contrast to the NLL method, which requires training of a neural network.

On all of the above examples, supervised Laplacian learning (SLL) outperforms all competing methods in the noiseless setting, except on the Swiss Roll where SLL and TLL work equally well (perhaps because the intrinsic dimension is so small) and on the Annulus where SLL and the NLL-based methods work equally well. The minimum relative error among all examples (9.82%) is achieved on the Swiss Roll dataset, which also features the fewest optimal nearest neighbors and the smallest τ

Table 2 Best relative error and runtime performance in label prediction. The **best results** are emphasized in bold

	Relative error						Runtime (sec)					
	kNN	LLR	TLL	NLL-kNN	NLL-LLR	SLL	kNN	LLR	TLL	NLL-kNN	NLL-LLR	SLL
Small tree	0.3417	0.1692	0.5015	0.2406	0.2235	**0.1300**	**0.0013**	0.0567	0.0793	591	1568	0.0780
Big tree	0.7669	0.4125	0.7991	0.07967	0.1006	**0.1160**	**0.0014**	0.0457	0.0551	367	999	0.0774
Cube	0.5010	0.3055	0.6169	0.6005	0.6393	**0.1607**	**0.0013**	0.0458	0.0513	993	993	0.0826
Sphere	0.4013	0.3311	0.5881	0.7199	0.7205	**0.1828**	**0.0013**	0.0443	0.0807	497	497	0.0697
Swiss roll	0.5065	0.4343	0.1038	0.6390	0.5621	**0.0982**	**0.0036**	0.0413	0.0532	653	683	0.0528
Annulus	0.7895	0.7029	0.8139	0.2920	0.2843	**0.2743**	**0.0012**	0.0417	0.0536	407	407	0.0641

Table 3 Relative error in noiseless label prediction for fixed λ, τ, and ϵ (average over ten trials). The **best results** are emphasized in bold

$\lambda = 0.01, \tau = 1,$	Relative error			
$\epsilon = 0.005$	kNN	LLR	TLL	SLL
Small tree	0.3719	0.1781	0.9678	**0.1309**
Big tree	0.7838	0.4138	0.7916	**0.1530**
Cube	0.4488	0.3971	0.9034	**0.1782**
Sphere	0.3471	0.2225	0.8582	**0.1641**
Swiss roll	0.4857	0.3648	0.1186	**0.1071**
Annulus	0.7434	0.6859	0.8773	**0.2723**

value. We posit that this predictive power and parsimony in parameters may be due to the simpler, lower-dimensional structure of the Swiss Roll. The relative errors for the other datasets ranged between 12.42 and 27.43%.

To compare methods across the same parameters, we fix $\lambda = 0.01$, $\tau = 1$, and $\epsilon = 0.005$ and compare relative error accordingly as seen in Table 3. Across all methods we either see the best relative error with the SLL method or have a tie for lowest relative error for the Swiss Roll dataset.

3.2.3 Noisy Experiments on Synthetic Data

In this section, we compare the prediction efficacy of the SLL method against kNN, LLR, and TLL on the synthetic datasets in the presence of additive noise, either on the labels as described by Model 1 or the features as described by Model 2. We do not include a comparison to the NLL-driven prediction methods in the noisy setting, as we saw previously that their performance was no better than SLL (and sometimes quite worse) and their runtimes were roughly four orders of magnitude higher than SLL, owing to the need to train a neural network.

With σ denoting the noise level ($\sigma = \sigma_x$ in the case of noisy features and $\sigma = \sigma_y$ in the case of noisy labels), we incorporate additive noise as given by Models 1 and 2. We do this for the noise on data features or on data labels separately. In each case, we tune the parameters of each method across the collection of parameters as described in Appendix "Local and Laplacian-Based Methods" and then choose the best parameters for each method. However, we do fix $\tau = 1$ and $\lambda = 10^{-2}$ in all experiments, as optimal performance of our method is almost always for τ large, and results were insensitive to choice of λ. We then plot the results of the relative error of each method against the varying σ levels in both cases, as seen in Fig. 6 (noise on features) and Fig. 7 (noise on labels).

For both noise models, as the noise level increases, the relative error increases across all methods. However, our proposed SLL method significantly outperforms all other methods across all noise levels, with the exception of nearly identical performance between our SLL method and traditional Laplace learning on the Swiss Roll as seen in Figs. 6e and 7e. As in the noiseless case, we again posit that this may

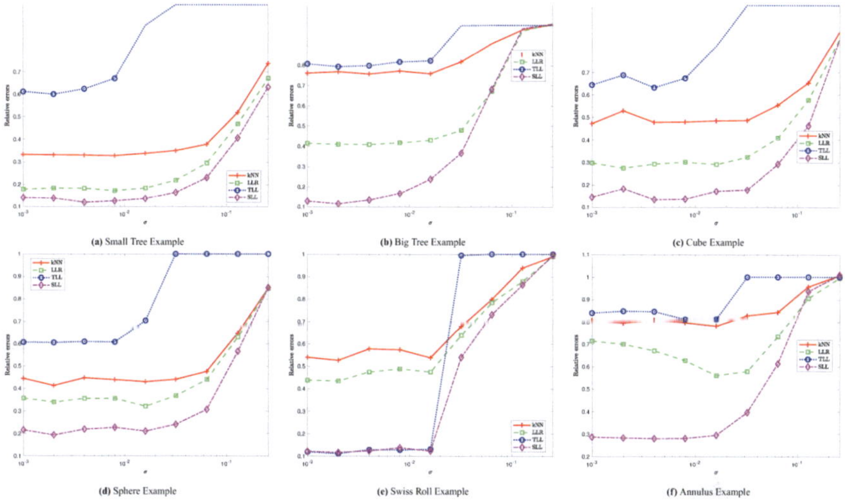

Fig. 6 Relative errors vs. noise levels σ for noise on data features (Model 2)

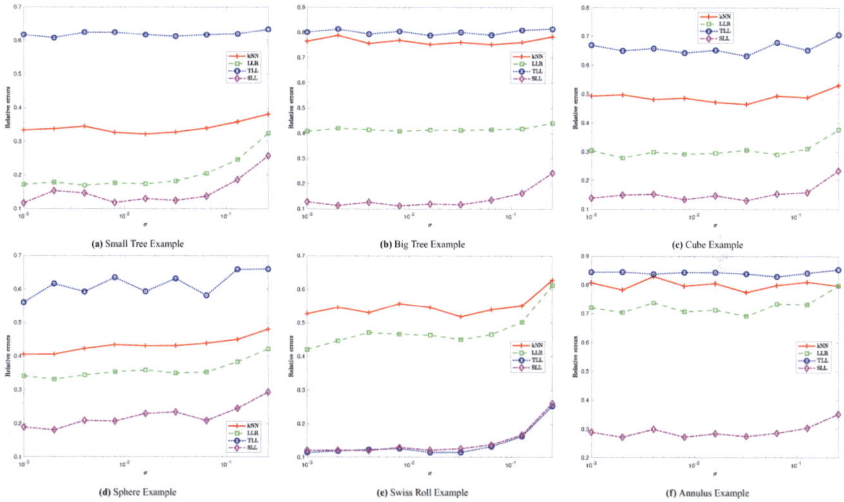

Fig. 7 Relative errors vs. noise level σ for noise on data labels (Model 1)

be due to the low intrinsic dimension of the Swiss Roll. Overall, our SLL method remains highly robust to noise on both the data features and on the labels.

3.2.4 Experiments on Real-World Datasets

Here we consider the problem of predicting house prices and other variables using the following real-world datasets. For each dataset, we assume access to 10% of the

Table 4 The evaluation of relative error and runtime performance in label prediction using various algorithms across real-world datasets for fixed lambda, tau, and epsilon

Lambda $= 0.01$, Tau $= 1$,	Relative error			
Epsilon $= 0.005$	kNN	LLR	TLL	SLL
Abalone	0.7314	0.6984	0.7570	0.6719
Ames housing	0.4699	0.4363	0.8080	0.4002
California housing	0.6046	0.5219	0.6429	0.5056

labels and measure predictive performance on the remaining points. The datasets used in this study are described as follows:

(i) Abalone: The Abalone dataset consists of $4,177$ samples with 8 features and can be accessed from [46]. It is used to predict the age of abalone from physical measurements such as length, diameter, and shell weight. For simplicity, we chose to ignore the categorical "sex" variable in this analysis. This dataset has been a benchmark in regression tasks.
(ii) Ames Housing: This dataset contains $2,930$ samples with 82 selected features. It provides detailed information about residential properties in Ames, Iowa, and is commonly used for regression tasks predicting house prices based on various physical and locational attributes[19]. We accessed this dataset from Kaggle [36], and the features were selected with an absolute correlation greater than 0.3 with the target variable, SalePrice, to ensure the inclusion of only the most relevant predictors.
(iii) California Housing: With $20,433$ samples and 9 features, the California Housing dataset is used to predict house prices based on demographic and geographic data from California [49]. This dataset was accessed from Kaggle [37], and to keep the analysis more straightforward, we excluded the categorical "ocean proximity" variable.

The results across all datasets demonstrate that our method outperformed the other algorithms, although the error was rather high for all methods with so few labels (Table 4). For these datasets, SLL significantly outperformed TLL, but there was only a small improvement over LLR; we conjecture that this is because the response variable is (locally) fairly linear, whereas our synthetic examples were constructed to have strong nonlinearities. Overall, the findings highlight the effectiveness of SLL in leveraging the structural relationships within the data while also reaffirming the competitiveness of traditional methods.

4 Conclusion

This chapter explores a geometric approach to supervised dimension reduction, where local gradient information is used to elongate useful dimensions. By computing and embedding geodesic distances under this local gradient stretching, we obtain supervised visualizations capable of simultaneously denoising the data while preserving global geometric information. By incorporating this metric into a graph Laplacian construction, we obtain a supervised graph Laplacian, which is used for prediction in a Laplacian learning framework. Extensive numerical experiments indicate the utility of this approach when the number of labels is small and the data is noisy. Since shortest path distances can be sensitive to noise, future work will explore whether combining diffusion-based algorithms such as PHATE [44] with our local metric can produce more noise-robust visualizations. Future work will also explore a more rigorous theoretical analysis of the convergence of $d_{\nabla Y,\tau}$ to the continuum limit (6), a theoretical analysis of supervised Laplacian learning, and the application of our prediction methodology on more real-world data.

Acknowledgments Part of this research was performed while the authors were visiting the Institute for Pure and Applied Mathematics (IPAM), which is supported by the National Science Foundation (Grant No. DMS-1925919). AL thanks NSF DMS 2309570, NSF RTG DMS 2136198. AM thanks the NSF GRFP (grant no. 1745302) and ONR MURI N00014-20-1-2595. KM thanks the support of GFSD.

Competing Interests The authors have no conflicts of interest to declare that are relevant to the content of this chapter.

Appendix: Supporting Results

Further Visualization Results

Figure 8 shows the MDS embedding of RF-GAP proximities on the Small Tree and Big Tree datasets.

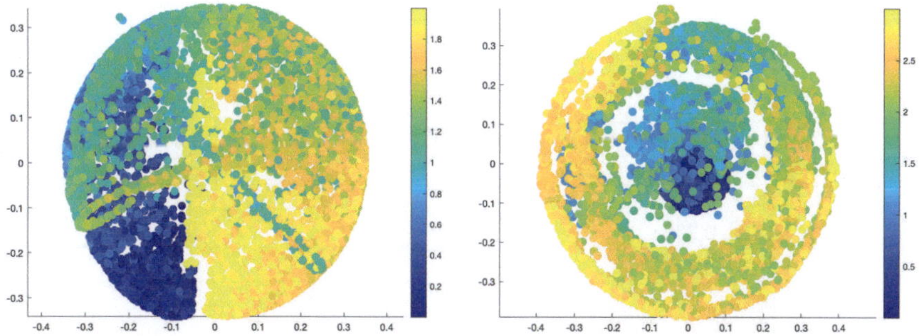

(a) Small Tree (b) Big Tree

Fig. 8 MDS embeddings of RF-GAP proximities on tree datasets

Local and Laplacian-Based Methods

For each synthetic dataset, we tune the parameters of the kNN, LLR, and TLL methods along with our SLL method across a wide range and report the best relative error. The parameter sets are given in Table 5.

Based on this parameter tuning, the optimal parameters returning the lowest relative error for our SLL method are shown in Table 6.

For the noisy experiments, we again tune the parameters of the kNN, LLR, and TLL methods along with our SLL method and report the resulting best relative error. The parameter sets are given in Table 7.

Table 5 Parameters used for the noiseless cases

Parameter	Value
ϵ set	$10^{-3}, 2\times10^{-3}, 3\times10^{-3}, 4\times10^{-3}, 5\times10^{-3}, 6\times10^{-3}, 7\times10^{-3}, 8\times10^{-3}, 9\times10^{-3}, 10^{-2}$
NN set	2, 3, 4, 6, 8, 12, 16, 23, 32, 46, 64, 91
λ set	$10^{-6}, 10^{-4}, 10^{-2}, 1.0, 10^{2}$
τ set	0.1, 0.2, 0.3, 0.4, 0.5, 0.6, 0.7, 0.8, 0.9, 1

Table 6 Optimal parameters for supervised Laplacian learning (SLL)

Data	SLL error	Nearest neighbor	ϵ	λ	τ
Small tree ($d = 8$)	0.1300	23	0.006	0.01	1
Big tree ($d = 4$)	0.1160	91	0.002	0.01	1
Cube ($d = 5$)	0.1607	64	0.003	0.01	1
Sphere ($d = 4$)	0.1828	32	0.006	0.0001	1
Swiss roll ($d = 2$)	0.0982	6	0.008	0.0001	0.1
Annulus ($d = 4$)	0.2743	23	0.006	0.01	1

Table 7 Parameters used for the noisy cases

Parameter	Value
ϵ set	$10^{-3}, 2\times10^{-3}, 3\times10^{-3}, 4\times10^{-3}, 5\times10^{-3}, 6\times10^{-3}, 7\times10^{-3}, 8\times10^{-3}, 9\times10^{-3}, 10^{-2}$
σ_x set	$10^{-3}, 2\times10^{-3}, 4\times10^{-3}, 8\times10^{-3}, 1.6\times10^{-2}, 3.2\times10^{-2}, 6.4\times10^{-2}$, 0.128, 0.256
σ_y set	$10^{-3}, 2\times10^{-3}, 4\times10^{-3}, 8\times10^{-3}, 1.6\times10^{-2}, 3.2\times10^{-2}, 6.4\times10^{-2}$, 0.128, 0.256
NN set	2, 3, 4, 6, 8, 12, 16, 23, 32, 46, 64, 91
λ	10^{-2}
τ	1
Trials	10

Nonlinear Level-Set Learning

For each of the synthetic datasets of Sect. 3.2, we apply the nonlinear level-set learning (NLL) method of [67] in the noiseless setting to obtain a nonlinear embedding $g : \mathbb{R}^D \to \mathbb{R}^D$ such that $g(x)_1 \in \mathbb{R}$ is highly predictive of $y = f(x) \in \mathbb{R}$. As the output $y \in \mathbb{R}$ only depends on a one-dimensional function of the input coordinates $x \in \mathbb{R}^D$ in each example, a one-dimensional embedding of x, which is entirely predictive y, exists in each example. Following [67], we parametrize g as seven-layer reversible neural network with "time-step" $h = 0.25$ and hyperbolic tangent activation. We train the network using the loss in [67] (Equations 9–11), computed over 100 training data points and parametrized with anisotropy weights $\boldsymbol{\omega} = (0, 1, 1, \ldots, 1) \in \mathbb{R}^D$. These anisotropy weights serve to drive the first coordinate of the embedding to be orthogonal to level sets of f, which on an intuitive level corresponds to embedding as much information as possible about how $f(x)$ changes with x into the first coordinate of $g(x)$. We weight the two loss terms in Equation (11) of [67] equally. Training is performed using stochastic gradient descent with learning rate $\alpha = 0.01$ and stopped when the loss drops below 0.0001 or after 20,000 steps are taken, whichever occurs first. After the embedding g has been obtained in this way, we test the efficacy of $g(x)_1$ in predicting $y = f(x)$ via k-nearest neighbors regression (KNN) and local linear regression over the 900 remaining test data points.

As the NLL loss depends on evaluations of ∇f, which for most practical applications we will not have access to, we approximate the gradient of f using the k-neighbors method of Sect. 2.2. We vary the number of neighbors k_{grad} used in gradient computation, as well as the number of neighbors k_{kNN} used in the subsequent prediction via k-nearest neighbors and the number k_{LLR} used in prediction via local linear regression, within the set $\{1, 2, \ldots, 25\}$ and for each example report best kNN and LLR relative error over all combinations $(k_{\text{grad}}, k_{\text{kNN}})$ and $(k_{\text{grad}}, k_{\text{LLR}})$, as averaged over ten independent trials (i.e., ten different partitions of the data into training and test sets). The minimal relative errors for each example, along with the optimal combinations of k_{grad}, k_{kNN}, and k_{LLR} and corresponding training times, can be found in Table 8.

Table 8 Relative error, embedding training times, and optimal parameters for prediction on NLL embeddings via k-nearest neighbors and local linear regression

	k-nearest neighbors regression				Local linear regression			
	Relative error	k_{grad}	k_{kNN}	Training time (s)	Relative error	k_{grad}	k_{LLR}	Training time (s)
Small tree	0.2900	23	2	1256	0.2528	25	6	896
Big tree	0.1519	25	4	568	0.1791	25	11	568
Cube	0.5749	6	4	679	0.6495	6	5	679
Sphere	0.6236	6	8	708	0.6914	6	25	708
Swiss roll	0.6437	4	10	492	0.5645	11	25	488
Annulus	0.3006	17	7	493	0.2971	22	23	503

References

1. Amouzgar, M., Glass, D.R., Baskar, R., Averbukh, I., Kimmey, S.C., Tsai, A.G., Hartmann, F.J., Bendall, S.C.: Supervised dimensionality reduction for exploration of single-cell data by hss-lda. Patterns **3**(8), 100536 (2022)
2. Becht, E., McInnes, L., Healy, J., Dutertre, C.A., Kwok, I.W., Ng, L.G., Ginhoux, F., Newell, E.W.: Dimensionality reduction for visualizing single-cell data using umap. Nat. Biotechnol. **37**(1), 38–44 (2019)
3. Belkin, M., Niyogi, P.: Laplacian eigenmaps for dimensionality reduction and data representation. Neural Comput. **15**(6), 1373–1396 (2003)
4. Bernstein, M., De Silva, V., Langford, J.C., Tenenbaum, J.B.: Graph approximations to geodesics on embedded manifolds. Technical Report, Citeseer (2000)
5. Berry, T., Harlim, J.: Iterated diffusion maps for feature identification. Appl. Comput. Harmonic Analy. **45**(1), 84–119 (2018)
6. Berry, T., Sauer, T.: Local kernels and the geometric structure of data. Appl. Comput. Harmonic Analy. **40**(3), 439–469 (2016)
7. Bickel, P.J., Li, B., et al.: Local polynomial regression on unknown manifolds. In: Complex Datasets and Inverse Problems, pp. 177–186. Institute of Mathematical Statistics, Beachwood (2007)
8. Borg, I., Groenen, P.: Modern multidimensional scaling: theory and applications. J. Educ. Meas. **40**(3), 277–280 (2003)
9. Borg, I., Groenen, P.J.: Modern multidimensional Scaling: Theory and Applications. Springer Science & Business Media, Cham (2007)
10. Bridges, R.A., Gruber, A.D., Felder, C., Verma, M., Hoff, C.: Active manifolds: A non-linear analogue to active subspaces. arXiv preprint arXiv:1904.13386 (2019)
11. Chen, C., Zhang, L., Bu, J., Wang, C., Chen, W.: Constrained Laplacian eigenmap for dimensionality reduction. Neurocomputing **73**(4–6), 951–958 (2010)
12. Cheng, M.Y., Wu, H.T.: Local linear regression on manifolds and its geometric interpretation. J. Am. Stat. Assoc. **108**(504), 1421–1434 (2013)
13. Cheng, Y., Wang, X., Xia, Y.: Supervised t-distributed stochastic neighbor embedding for data visualization and classification. INFORMS J. Comput. **33**(2), 566–585 (2021)
14. Coifman, R.R., Lafon, S.: Diffusion maps. Appl. Comput. Harmonic Analy. **21**(1), 5–30 (2006)
15. Constantine, P.G., Dow, E., Wang, Q.: Active subspace methods in theory and practice: applications to kriging surfaces. SIAM J. Sci. Comput. **36**(4), A1500–A1524 (2014)
16. Constantine, P.G., Emory, M., Larsson, J., Iaccarino, G.: Exploiting active subspaces to quantify uncertainty in the numerical simulation of the hyshot ii scramjet. J. Comput. Phys. **302**, 1–20 (2015)
17. Cook, R.D.: On the interpretation of regression plots. J. Am. Stat. Assoc. **89**(425), 177–189 (1994)
18. Cook, R.D., Weisberg, S.: Sliced inverse regression for dimension reduction: comment. J. Am. Stat. Assoc. **86**(414), 328–332 (1991)
19. De Cock, D.: Ames, Iowa: Alternative to the Boston housing data as an end of semester regression project. J. Stat. Educ. **19**(3) (2011)
20. De Ridder, D., Kouropteva, O., Okun, O., Pietikäinen, M., Duin, R.P.: Supervised locally linear embedding. In: International Conference on Artificial Neural Networks, pp. 333–341. Springer (2003)
21. Dennis Cook, R.: Save: a method for dimension reduction and graphics in regression. Commun. Stat.-Theory Methods **29**(9–10), 2109–2121 (2000)
22. Fernández, X., Borghini, E., Mindlin, G., Groisman, P.: Intrinsic persistent homology via density-based metric learning. J. Mach. Learn. Res. **24**(75), 1–42 (2023)
23. Geng, X., Zhan, D.C., Zhou, Z.H.: Supervised nonlinear dimensionality reduction for visualization and classification. IEEE Trans. Syst. Man Cybern. Part B (Cybern.) **35**(6), 1098–1107 (2005)

24. Ghosh, T., Kirby, M.: Supervised dimensionality reduction and visualization using centroid-encoder. J. Mach. Learn. Res. **23**(1), 901–934 (2022)
25. Goldberger, J., Hinton, G.E., Roweis, S., Salakhutdinov, R.R.: Neighbourhood components analysis. In: Advances in Neural Information Processing Systems, 17 (2004)
26. Grey, Z., Constantine, P.: Characterizing subspaces of engineering shapes using differential geometry. In: 2018 AIAA Non-Deterministic Approaches Conference, p. 1176 (2018)
27. Groisman, P., Jonckheere, M., Sapienza, F.: Nonhomogeneous Euclidean first-passage percolation and distance learning. Bernoulli **28**(1), 255–276 (2022)
28. Hajderanj, L., Weheliye, I., Chen, D.: A new supervised t-SNE with dissimilarity measure for effective data visualization and classification. In: Proceedings of the 8th International Conference on Software and Information Engineering, pp. 232–236 (2019)
29. Hamm, K., Henscheid, N., Kang, S.: Wassmap: Wasserstein isometric mapping for image manifold learning. SIAM J. Math. Data Sci. **5**(2), 475–501 (2023)
30. Hastie, T., Tibshirani, R., Friedman, J., Hastie, T., Tibshirani, R., Friedman, J.: Random forests. In: The Elements of Statistical Learning: Data Mining, Inference, and Prediction, pp. 587–604. Springer, New York (2009)
31. Hastie, T., Tibshirani, R., Friedman, J.H., Friedman, J.H.: The Elements of Statistical Learning: Data Mining, Inference, and Prediction, vol. 2. Springer, Berlin (2009)
32. Hinton, G., Van Der Maaten, L.: Visualizing data using t-SNE journal of machine learning research. J. Mach. Learn. Res. **9**, 2579–2605 (2008)
33. Hwang, S., Damelin, S., Hero, A.: Shortest path through random points. Ann. Appl. Probabil. **26**(5), 2791–2823 (2016)
34. Jiang, Q., Jia, M.: Supervised Laplacian eigenmaps for machinery fault classification. In: 2009 WRI World Congress on Computer Science and Information Engineering, vol. 7, pp. 116–120. IEEE (2009)
35. Jiang, B., Lin, D.: Graph Laplacian regularized graph convolutional networks for semi-supervised learning. arXiv preprint arXiv:1809.09839 (2018)
36. Kaggle: Ames housing dataset. https://www.kaggle.com/datasets/prevek18/ames-housing-dataset. Accessed: 2024-09-18
37. Kaggle: California housing prices. https://www.kaggle.com/datasets/camnugent/california-housing-prices. Accessed: 2024-09-20
38. Le, L., Patterson, A., White, M.: Supervised autoencoders: Improving generalization performance with unsupervised regularizers. In: Advances in Neural Information Processing Systems, 31 (2018)
39. Lederman, R.R., Talmon, R.: Learning the geometry of common latent variables using alternating-diffusion. Appl. Comput. Harmonic Analy. **44**(3), 509–536 (2018)
40. Lee, K.Y., Li, B., Chiaromonte, F., et al.: A general theory for nonlinear sufficient dimension reduction: formulation and estimation. Ann. Stat. **41**(1), 221–249 (2013)
41. Liaw, A.: Classification and regression by randomforest. R News (2002)
42. Little, A., McKenzie, D., Murphy, J.M.: Balancing geometry and density: path distances on high-dimensional data. SIAM J. Math. Data Sci. **4**(1), 72–99 (2022)
43. Manousidaki, A., Little, A., Xie, Y.: Clustering and visualization of single-cell RNA-seq data using path metrics. PLOS Comput. Biol. **20**(5), e1012014 (2024)
44. Moon, K.R., Van Dijk, D., Wang, Z., Gigante, S., Burkhardt, D.B., Chen, W.S., Yim, K., Elzen, A.v.d., Hirn, M.J., Coifman, R.R., et al.: Visualizing structure and transitions in high-dimensional biological data. Nat. Biotechnol. **37**(12), 1482–1492 (2019)
45. Nadler, B., Srebro, N., Zhou, X.: Semi-supervised learning with the graph Laplacian: The limit of infinite unlabelled data. Adv. Neural Inf. Process. Syst. **22**, 1330–1338 (2009)
46. Nash, W., Sellers, T., Talbot, S., Cawthorn, A., Ford, W.: Abalone. UCI Machine Learning Repository (1994). https://doi.org/10.24432/C55C7W
47. Nilsson, J., Sha, F., Jordan, M.I.: Regression on manifolds using kernel dimension reduction. In: Proceedings of the 24th International Conference on Machine Learning, pp. 697–704. ACM (2007)

48. Nottingham, Q.J., Cook, D.F.: Local linear regression for estimating time series data. Comput. Stat. Data Analy. **37**(2), 209–217 (2001)
49. Pace, R.K., Barry, R.: Sparse spatial autoregressions. Stat. Probabil. Lett. **33**(3), 291–297 (1997)
50. Raducanu, B., Dornaika, F.: A supervised non-linear dimensionality reduction approach for manifold learning. Pattern Recogn. **45**(6), 2432–2444 (2012)
51. Rhodes, J.S., Cutler, A., Wolf, G., Moon, K.R.: Random forest-based diffusion information geometry for supervised visualization and data exploration. In: 2021 IEEE Statistical Signal Processing Workshop (SSP), pp. 331–335. IEEE (2021)
52. Rhodes, J.S., Cutler, A., Moon, K.R.: Geometry-and accuracy-preserving random forest proximities. IEEE Trans. Pattern Analy. Mach. Intell. **45**(9), 10947–10959 (2023)
53. Ribeiro, B., Vieira, A., Carvalho das Neves, J.: Supervised isomap with dissimilarity measures in embedding learning. In: Progress in Pattern Recognition, Image Analysis and Applications: 13th Iberoamerican Congress on Pattern Recognition, CIARP 2008, Havana, Cuba, September 9–12, 2008. Proceedings 13, pp. 389–396. Springer (2008)
54. Russi, T.M.: Uncertainty quantification with experimental data and complex system models. University of California, Berkeley (2010)
55. Salvador, M., Tseng, N., Park, C., Williams, G., Vethan, A., Thomas, G., Baker, J., Hemry, J., Hammond, E., Freeburg, P., et al.: SARS-CoV-2 spike protein reduces burst activities in neurons measured by micro-electrode arrays. Ann. Med. Surgery **85**, 10–1097 (2023)
56. Sha, L., Schonfeld, D., Wang, J.: Graph Laplacian regularization with sparse coding for image restoration and representation. IEEE Trans. Circ. Syst. Video Technol. **30**(7), 2000–2014 (2020). https://doi.org/10.1109/TCSVT.2019.2913411
57. Szlam, A.D., Maggioni, M., Coifman, R.R.: Regularization on graphs with function-adapted diffusion processes. J. Mach. Learn. Res. **9**(8), 1711–1739 (2008)
58. Talmon, R., Cohen, I., Gannot, S., Coifman, R.R.: Diffusion maps for signal processing: a deeper look at manifold-learning techniques based on kernels and graphs. IEEE Signal Process. Mag. **30**(4), 75–86 (2013)
59. Tenenbaum, J.B., De Silva, V., Langford, J.C.: A global geometric framework for nonlinear dimensionality reduction. Science **290**(5500), 2319–2323 (2000)
60. Torgerson, W.S.: Multidimensional scaling: I. theory and method. Psychometrika **17**(4), 401–419 (1952)
61. Trillos, N.G., Little, A., McKenzie, D., Murphy, J.M.: Fermat distances: Metric approximation, spectral convergence, and clustering algorithms. J. Mach. Learn. Res. **25**(176), 1–65 (2024)
62. Vlachos, M., Domeniconi, C., Gunopulos, D., Kollios, G., Koudas, N.: Non-linear dimensionality reduction techniques for classification and visualization. In: Proceedings of the Eighth ACM SIGKDD International Conference on Knowledge Discovery and Data Mining, pp. 645–651 (2002)
63. Wang, W., Ozolek, J.A., Slepčev, D., Lee, A.B., Chen, C., Rohde, G.K.: An optimal transportation approach for nuclear structure-based pathology. IEEE Trans. Med. Imag. **30**(3), 621–631 (2010)
64. Xia, Y.: A constructive approach to the estimation of dimension reduction directions. Ann. Stat. **35**, 2654–2690 (2007)
65. Xia, Y.: A multiple-index model and dimension reduction. J. Am. Stat. Assoc. **103**(484), 1631–1640 (2008)
66. Zhang, S.Q.: Enhanced supervised locally linear embedding. Pattern Recogn. Lett. **30**(13), 1208–1218 (2009)
67. Zhang, G., Zhang, J., Hinkle, J.: Learning nonlinear level sets for dimensionality reduction in function approximation. In: Advances in Neural Information Processing Systems, 32 (2019)
68. Zhu, X., Rabbat, M.: Graph spectral compressed sensing for sensor networks. In: 2012 IEEE International Conference on Acoustics, Speech and Signal Processing (ICASSP), pp. 2865–2868 (2012). https://doi.org/10.1109/ICASSP.2012.6288515

Reducing NLP Model Embeddings for Deployment in Embedded Systems

Karolyn Babalola, Arnaja Mitra, and Jing Qin

1 Introduction

The latest advancements in natural language processing (NLP) have revolutionized the way technology represents and replicates human communication. In particular, transformer-based models have marked a major turning point in NLP by enabling large-scale statistical representations of semantics and syntactic rules [36]. Transformer-based technologies underlie many state-of-the-art large language models, such as ChatGPT [6] and its variants, LLaMa [33], and BERT [10]. They significantly outperform previous NLP approaches by networking several layers of encoding networks that comprise hundreds of millions to billions of parameters. The growth of NLP technology has undoubtedly relied heavily on the increasing availability of high-performance computing resources. To overcome the associated computational bottlenecks, our hypothesis focuses on finding a balance between model efficiency and performance, with the goal of developing smaller, faster models that maintain high levels of accuracy.

The compute-intensive structure of state-of-the-art NLP models has restricted its application in resource-limited environments. Embedded systems, such as FPGAs [5], network-restricted applications, the Internet of Things (IoT), and edge devices, often impose restrictions that make deploying large language models virtually

K. Babalola
Chief Technology Office, AI IMT, Booz Allen Hamilton, Mclean, VA, USA

A. Mitra
Department of Mathematics, University of Maryland, College Park, MD, USA
e-mail: amitra13@umd.edu

J. Qin (✉)
Department of Mathematics, University of Kentucky, Lexington, KY, USA
e-mail: jing.qin@uky.edu

C. Garcia-Cardona, H. Lee (eds.), *Advances in Data Science*, Association for Women in Mathematics Series 37, https://doi.org/10.1007/978-3-031-87804-6_10

impossible without significant reductions in model size. To that end, various methods, such as distillation and pruning, exist to achieve this reduction.

Distillation is a size-reduction method that involves transferring knowledge from a "teacher" model to a smaller "student" model, which is a copy of the teacher model with several layers removed. Distillation occurs by training the student on the loss over the soft target probabilities of the teacher. DistilBERT [28] and TinyBERT [16] are well-known examples of distilled BERT models. While the original base BERT model comprises 110M parameters and uses 431 MB of memory, DistilBERT contains 66M parameters requiring approximately 259 MB of memory, and TinyBERT has a remarkable 14.5M parameters, which would require approximately 60 MB of memory. Despite their reduced size, distilled models retain comparable performance as the base BERT model on bench-marking tasks. Consequently, several pre-trained versions of distilled models have been made available for fine-tuning on more specific NLP tasks such as intent classification, sentiment classification, and named-entity recognition [30].

Pruning is another technique for reducing the size of a model by removing model weights and their respective synapses. In practice, one takes a pre-trained model, such as BERT or RoBERTa [18], and selectively masks or ablates the weights within the attention heads of each encoding layer. This can be achieved using greedy methods, such as magnitude weight pruning [13], or those mentioned in [20], and/or entropy methods as explored in [29].

Distilled and pruned models have shown considerable robustness in their overall performance but still exhibit some degradation compared to their original models. Because previous studies focus on maximizing benchmarking performance of reduced models, it is unclear about how to effectively balance model reduction and fine-tuning techniques for specific NLP tasks in terms of size and performance trade-offs. Recent investigations have begun to explore the performance trade-offs of iterative model reduction using distillation and pruning [29]. In addition, effective dimensionality reduction techniques have been shown for fixed word embeddings in text analysis applications [26, 27, 31]. Building on this promise, we hypothesize that because categorical features of the corpus are preserved in the pre-trained BERT models' token embedding, dimension reduction methods that preserve a significant percentage of the variance could help maintain performance as model size decreases. In this work, we aim to explore the trade-offs of model reduction by applying dimensionality-reduction techniques to the embedding layer of BERT models.

1.1 NLP Tasks in Embedded Systems

The process of documenting NLP task performance trade-offs for size-reduced transformer-based models was initiated in [29]. In this case, three tasks, intent classification, sentiment classification, and named-entity recognition, were chosen based on a simple robot arm apparatus that the authors chose. However, the chosen tasks embody a range of use cases that have relevant applications in compute-constrained environments.

Intent classification (IC) is a natural language understanding (NLU) task that involves classifying a user's intention from either a predefined set of utterances or a summary of text input. In [29], they were using intent-classification to control the output of a robot arm based on a user's input. They tested the accuracy of the chosen models using a well-defined public IC dataset called *HurIC* [35]. HurIC is a pre-defined corpus of utterances used to define human-robot interaction in house service robots. Interestingly enough, the HurIC dataset has a well-defined classification; thus, models of many sizes produce relatively high performances.

The sentiment classification NLP task is defined as interpreting a predefined set user's utterances into a set of sentiment labels. In [29], the authors use this to classify user emotion in combination with intent to determine how to program the actions of a robot. They use the dataset called GoEmotions[1] to test their models' performance in different environments [7]. This dataset has been tested in several other use cases and provides a useful test bed to expand beyond the use case in [29]. Furthermore, GoEmotion presents a relatively challenging task that produces a range of performance outcomes for different models.

Finally, in [29], named-entity recognition (NER) is used to enable the robot arm use case to target entities in its environment on which to direct user instructions. This is a widely used NLP task with many applications. In this use case, they tested entity recognition using two different datasets, i.e., CoNLL-2003 [32] and WNUT17 [8]. WNUT17 is a dataset comprised of user-generated social-media data that was developed for the task of identifying previously unseen and unusual entities in ongoing discussions. The authors chose the WNUT17 due to its similarity to user utterance data and its shared task of recognizing unknown entities; however, the overall task of WNUT17 is quite difficult and tends to produce low performance for many models. CoNLL-2003 is a dataset of annotated entities generated from Reuters news stories published between August 1996 and August 1997. CoNLL has similarly high-performance outcomes as the HuRIC, but it provides a useful benchmarking task for NER.

While [29] documents the performance trade-offs of the three aforementioned NLP tasks, we focus our study testing a range of dimensionality-reduction techniques on NER and particularly the CoNLL-2003 dataset, since its relatively high performance lends itself well to demonstrating degradation.

[1] https://huggingface.co/datasets/go_emotions

1.2 Extended Exploration of Model Size Reduction and Performance

In [29], the authors tested the performance of two large BERT models, i.e., BERT and RoBERTa, and two distilled models, namely, DistilBERT and TinyBERT; all models were fine-tuned on the aforementioned NLP tasks [29]. These models' performance stats were contrasted with their own custom-pruned models. The custom models were pruned by measuring the entropy of the attention heads within a given each layer of the model and masking the heads with the lowest entropy. If the masked heads produced an F1 score above a predefined threshold, it would be removed, and the masking and removal process would iterate until the minimum F1 threshold was reached. This process resulted in models ranging in sizes from 75.9M parameters and 303.5 MB to as small as 34.1M parameters and 136.4 MB.

This investigation replicates the methodological approach demonstrated in [29] to reduce BERT models to fit certain resource and performance constraints. However, rather than pruning BERT models, this study investigates whether one could achieve comparable performance by manipulating the token embedding layer and, thus, models' overall *hidden* size. The initial reasoning for this approach is a predicated fact that reducing the embedding size by any factor is advantageous because it automatically reduces the model's total parameter space by the same factor, e.g., reducing the embedding layer from 768 by a factor of 3–256 would reduce the DistilBERT parameter space to 22M. This chapter discusses the results and challenges faced while exploring this idea.

The remainder of the chapter is organized as follows. In Sect. 2, we introduce the token embedding layer in BERT type of models, briefly present the four dimension reduction methods that we employ, and our proposed embedding dimension-reduced NLP pipeline. A variety of numerical experiments on named-entity recognition test with DistilBERT are conducted and described in Sect. 3 to discuss the impacts of reduced dimension, batch size, and learning rate on recognition accuracy and training runtime. Finally, we conclude the chapter and outline future work in Sect. 4.

2 Proposed Methodology

In this work, we apply various dimension reduction methods to reduce the model token embedding size, so as to produce new embedding vectors that maximize the variance of its components in a smaller vector space. In particular, we chose principal components analysis (PCA), truncated singular value decomposition (TSVD), agglomerative clustering (AC), and uniform manifold approximation and projection (UMAP).

2.1 Embeddings in BERT

Tensor representations of chunks of text, referred to as *tokens*, form foundation of quantitative NLP. Such numerical representations range from large, sparse categorical representations of words in a corpus, such as the one-hot encoding [22], to corpus frequency representations such as TF-IDF [1]. Selecting an appropriate numerical representation of words and tokens has often been the first step in optimizing the performance of a task-specific NLP model. For instance, historically topic models, such as LDA [4], often relied on generating a bag-of-words representation that infers a “global” word distribution and presumes exchangeability [2].

More recent topic models have replaced sparse representation of words with embeddings. Generated by neural network models, embeddings are vectors that encode categorical features of words or tokens into a multidimensional space [3]. Algorithms such as Word2Vec [21] and GloVe [25] have been used to train on corpuses in specific domains and are a means of sharing pre-trained embeddings to perform vector-based search tasks or to precede a downstream classifier or clustering model. Even topic models received more recent updates, using embedding representations [11, 17].

BERT transformer models leverage token embeddings; however, they are a fixed part of its first layer (cf. Fig. 1). The token embedding is summed with a position embedding and a token type embedding in the first layer. The position embedding encodes the position of each word in a fixed input sequence, typically 512. The token type embedding encodes to which sequence a token belongs; for instance, in sequence-to-sequence training, the token type would label the inputs as either sequence “A” or sequence “B,” whether the sequence is the input or the output respectively. Unlike word2vec and GloVe, the token embedding in BERT models is initialized as Wordpiece embeddings [37] and trained with the entire model to generate a contextual language representation in output of the final (hidden) layer. The models are then further fine-tuned for specific tasks, such as classification and question/answering. Therefore, pre-trained BERT models have their own unique token embeddings. Nonetheless, the token embedding alone still effectively represents the categorical features of the corpus; this can be demonstrated

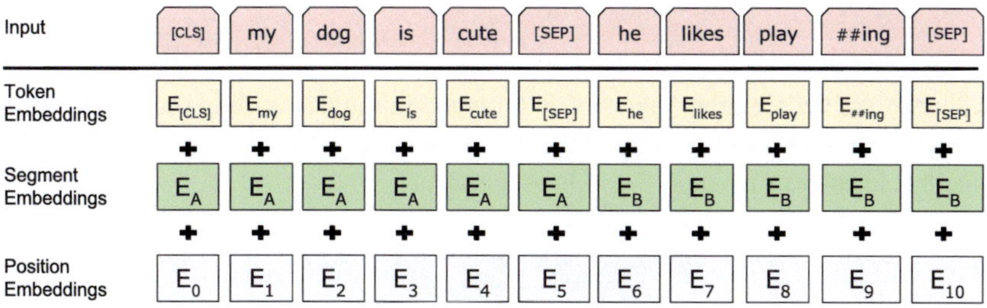

Fig. 1 BERT input representation highlighting the token embeddings [9]

by contrasting the cosine similarity of embeddings corresponding to words in the corpus that are similar in meaning to those that are not [9].

Previous studies have shown effective dimensionality reduction for fixed word embeddings with text analysis applications [26, 27, 31]. Extending this idea, we assume that since categorical features of the corpus are preserved in the pre-trained BERT models' token embedding, it follows that dimension reduction methods that preserve a significant percentage of the variance could help maintain performance as the model size decreases.

2.2 Dimension Reduction Methods

In this section, we briefly introduce the dimension reduction methods employed in this study. For a given token embedding matrix $X \in \mathbb{R}^{m \times n}$, where m is the number of tokens and n is the dimension of token features, we intend to reduce its dimensionality. The default dimension $n = 768$ is quite large, which not only leads to the increased storage requirement but also slows down computation or retrieval speeds. There are many dimensionality reduction approaches that have been developed to convert high-dimensional datasets into their low-dimensional representations while retaining the most important data features.

PCA was first proposed in [24] as an analogue of the principal axis theorem in mechanics and was later independently developed and named by Harold Hotelling [14]. As one of the most popular dimensionality reduction methods, PCA uses a linear transformation to project the high-dimensional embedding matrix into its low-dimensional representations with the size $m \times k$, where $k < n$ such that the variance of the low-dimensional representations is maximized.

Another method that we apply is truncated singular value decomposition (TSVD), which comes naturally with the singular value decomposition of a matrix and has been widely used for low-rank matrix approximation [12]. If an embedding matrix X is assumed to be low rank with a relatively small rank $k < \min\{m, n\}$, then we can use the following way to approximate X:

$$\widehat{X} = \sum_{i=1}^{k} \sigma_i \mathbf{u}_i^T \mathbf{v}_i,$$

where σ_i are the singular values of X arranged in the descending order and $\mathbf{u}_i$ and $\mathbf{v}_i$ are the left and right singular vectors. In fact, we can show that $\widehat{X}$ achieves the smallest Frobenius norm in the following sense:

$$\widehat{X} = \operatorname*{argmin}_{Y \in \mathbb{R}^{m \times n}:\, \mathrm{rank}(Y) \leq k} \|X - Y\|_F .$$

As one important hierarchical clustering method, agglomerative clustering (AC) [15] is a "bottom-up" approach, where each data point starts as its cluster and pairs of clusters are merged as one moves up the hierarchy. The iterative algorithm terminates until all points belong to a single cluster or a stopping criterion is satisfied. In this method, the similarity between two clusters is typically measured by the distance between the closest members of the clusters, which is called the *single linkage*; the farthest members, which are called the *complete linkage*; and the average distance between all pairs of members, called the *average linkage*, or other metrics. In addition, agglomerative clustering yields a *dendrogram*, i.e., a tree structure indicating the merges that can generate the final clustering result. In this work, we apply agglomerative clustering to reduce the feature numbers in the token embedding matrix while preserving the most distinguishable features.

More recently, uniform manifold approximation and projection (UMAP) has been proposed as a powerful nonlinear dimensionality reduction technique [19]. By visualization, it resembles the T-distributed stochastic neighbor embedding (t-SNE) [34], but it assumes that the data is uniformly distributed on a locally connected Riemannian manifold with the Riemannian metric being locally constant or approximately locally constant. This assumption may not be valid in NLP, which prevents the training on the reduced dataset from achieving high accuracy (see Sect. 3 for more justifications). Note that unlike PCA, UMAP can preserve nonlinear overall variance of the dataset, but it typically projects the data onto 2D or 3D spaces for visualization. In our experiments, UMAP is much slower than the other counterparts for training, so we exclude it from comparison.

2.3 Embedding Reduced NLP Method

In this work, we extract the pre-trained token embedding from a fine-tuned DistilBERT model and retrain a reconfigured DistilBERT model with token embedding reduced using the dimensionality reduction methods described above. We explore whether the information retained in the token embedding layer is significant enough to help maintain some portion of the performance of the original model size. In lieu of our initial goal, to train and test reduced models on the three NLP tasks stated in Sect. 1.1, we opted to focus on the NER task while adding a demonstration of the effect of hyperparameter tuning on the performance.

3 Numerical Experiments

In this section, we illustrate the influence of dimension reduction methods applied to the token embedding matrix of DistilBERT on the overall performance of an NER task, including the accuracy and training runtime. Although hyperparameter optimization may seem less critical for models trained infrequently and deployed in

Table 1 Overview of the CoNLL-2003 dataset

	English data			German data		
Data type	Articles	Sentences	Tokens	Articles	Sentences	Tokens
Training set	946	14,987	203,621	553	12,705	206,931
Development set	216	3,466	51,362	201	3,068	51,444
Test set	231	3,684	46,435	155	3,160	51,943

embedded systems, it is still essential for achieving optimal performance. A well-tuned model improves prediction accuracy and generalizability, directly enhancing application reliability. While training may require more resources, optimizing for efficient inference is crucial, as this phase runs continuously in embedded environments. Even small improvements in tuning can significantly impact runtime, power efficiency, and accuracy, which is particularly important for applications where low latency and power consumption are critical. Motivated by these facts, our investigation here aims to future-proof the DistilBERT model for potentially more resource-constrained or larger-scale applications where runtime and hardware efficiency become more pronounced.

Our main question arises: Can reducing the size of a model by manipulating the embedding size using dimensionality reduction methods produce performance results comparable to that of the original model? In addition, we observe whether we can map reduction methods to expected performance trade-offs. To answer this, we train the model DistilBERT to perform one of the NLP tasks mentioned above, i.e., named-entity recognition (NER), on one benchmark dataset.

All the numerical experiments are implemented on Python 3 in a desktop computer with Intel CPU i9-9960X RAM 64 GB and GPU Dual Nvidia Quadro RTX5000 with Windows 10 Pro.

Throughout the section, we focus on the CoNLL-2003 dataset [32], which comprises eight files covering two languages: English and German. These files are annotated with four types of named entities: persons, locations, organizations, and miscellaneous entities that do not belong to the previous three groups. For each language, there exists a training file, a development file, a test file, and a large file with unannotated data, offering a standardized framework for training and evaluating NER models. The English data was sourced from news articles within the Reuters corpus, spanning stories from August 1996 to August 1997. Likewise, the German data was extracted from the August 1992 issues of the German newspaper Frankfurter Rundschau. Refer to Table 1 for the distribution of various categories and subsets in the CoNLL-2003 dataset.

We have fine-tuned the DistilBERT model using the aforementioned dataset collected from Hugging Face[2] to perform the NLP task NER while maintaining the following hyperparameter setup: batch size among $\{8, 16, 32, 64\}$, number of epochs as 7, learning rate among $\{10^{-6}, 10^{-5}, 10^{-4}\}$, and weight decay rate

[2] https://huggingface.co/

between 0.01 and 0.3. To evaluate the performance, we use the standard metrics for classification: precision, recall, F1 score, and accuracy [23]. Let FP, FN, TP, and TN denote the respective number of false positives, false negatives, true positives, and true negatives. Then the four metrics are defined as follows:

- *Precision* is the ratio of relevant instances among the retrieved instances, given by $Pr = TP/(TP + FP)$.
- *Recall* is the ratio of relevant instances that were retrieved, given by $Re = TP/(TP + FN)$.
- *F1 Score* (or *F-measure*) is the harmonic mean of precision and recall, given by $F_1 = 2PrRe/(Pr + Re)$.
- *Accuracy* is the ratio of correct predictions to the total number of predictions, defined as $acc = (TP + TN)/(TP + FN + FP + TN)$.

3.1 Experiment 1: Vary Reduced Dimension

In our initial experiment, we set the batch size to 8 and the learning rate to 10^{-4}. We first ran the DistilBERT without data dimension reduction as a baseline in Table 2. We then evaluated each dimension reduction method using various dimensions: 128, 256, 512, and 768. Table 3 shows the overall precision, recall, F1 score, accuracy, and training runtime for PCA. Similar results for TSVD, AC, and UMAP are presented in Tables 4, 5, and 6, respectively.

Tables 2, 3, 4, 5 and 6 show that data compression may cause very modest drops in accuracy despite very substantial drops in F1 score. The F1 score, which emphasizes both false positives (incorrectly predicted entities) and false negatives (missed entities), is particularly sensitive to small errors in named-entity recognition (NER) tasks. As data dimensionality decreases, imbalances in entity types, e.g.,

Table 2 Evaluation results without dimension reduction with batch size 8 and learning rate 10^{-4}

Dim	Overall precision	Overall recall	Overall F1	Overall accuracy	Train runtime (s)
128	0.516295	0.515718	0.516006	0.882060	2101.14
256	0.572505	0.612149	0.591664	0.903490	2738.20
512	0.633252	0.664280	0.648395	0.918455	4375.02
768	0.920866	0.928180	0.924508	0.981270	6615.40

Table 3 Evaluation results from PCA with batch size 8 and learning rate 10^{-4}

Dim	Overall precision	Overall recall	Overall F1	Overall accuracy	Train runtime (s)
128	0.526883	0.525115	0.525997	0.884395	2012.77
256	0.579988	0.623783	0.601089	0.905063	2627.11
512	0.634869	0.656673	0.645587	0.918646	4107.24
768	0.721791	0.739233	0.730408	0.941570	6290.71

Table 4 Evaluation results from TSVD with batch size 8 and learning rate 10^{-4}

Dim	Overall precision	Overall recall	Overall F1	Overall accuracy	Train runtime (s)
128	0.521172	0.527352	0.524244	0.883156	2000.89
256	0.581499	0.626580	0.603199	0.905031	2590.51
512	0.631465	0.651527	0.641339	0.918360	4135.47
768	0.851968	0.852444	0.852206	0.964573	6260.57

Table 5 Evaluation results from AC with batch size 8 and learning rate 10^{-4}

Dim	Overall precision	Overall recall	Overall F1	Overall accuracy	Train runtime (s)
128	0.526924	0.525450	0.526186	0.883156	2007.38
256	0.576366	0.613827	0.594507	0.904157	2588.36
512	0.634869	0.656673	0.645587	0.918646	4162.72
768	0.721791	0.739233	0.730408	0.941570	6357.39

Table 6 Evaluation results from UMAP with batch size 8 and learning rate 10^{-4}

Dim	Overall precision	Overall recall	Overall F1	Overall accuracy	Train runtime (s)
128	0.482526	0.312004	0.378966	0.860740	2221.97
256	0.481790	0.375881	0.422296	0.874323	2957.02
512	0.517582	0.439646	0.475442	0.887620	4749.07
768	0	0	0	0.789108	6948.93

some entities appearing more frequently than others, become more pronounced. Accuracy remains relatively stable as the model correctly classifies the majority of entities, particularly the common ones, even if rare entities are occasionally misclassified or missed. However, the F1 score drops more substantially due to the greater impact on recall and precision for rare entities, which in turn lowers the harmonic mean.

As the dimension increases, longer running times are required for the training step, but the accuracy scores improve. Among the four methods, UMAP exhibits the poorest performance in terms of accuracy, also requiring slightly more training time. When the original hidden dimension of 768 is used, UMAP produces zero precision, recall, and F1 score. This is likely due to the fact that the manifold assumption of UMAP may not be satisfied for the token embedding matrix.

On the other hand, PCA, TSVD, and AC perform similarly in terms of accuracy and runtime. When the reduced dimension size is 256, these three methods can achieve an accuracy of above 90%, but with much faster training times.

3.2 *Experiment 2: Vary Batch Size*

In our second experiment, we kept the learning rate fixed at 10^{-4} and the reduced dimension at 256. Tables 7 and 8 present the overall F1 and accuracy for each

Table 7 Overall F1 for all the methods with learning rate 10^{-4} and dimension 256

Batch size	PCA	TSVD	AC	UMAP
8	0.601089	0.524244	0.526186	0.378966
16	0.575301	0.573156	0.57065	0.418090
32	0.554822	0.552738	0.556845	0.350840
64	0.500000	0.494786	0.503551	0.27224

Table 8 Overall accuracy for all the methods with learning rate 10^{-4} and dimension 256

Batch size	PCA	TSVD	AC	UMAP
8	0.905063	0.905031	0.904157	0.874323
16	0.898550	0.898994	0.899392	0.871130
32	0.893720	0.893974	0.894758	0.856133
64	0.878866	0.874593	0.877611	0.836847

Table 9 Train runtime (s) for all the methods with learning rate 10^{-4} and dimension 256

Batch size	PCA	TSVD	AC	UMAP
8	2627.11	2590.51	2588.36	2957.02
16	3616.34	3631.47	3830.19	3671.07
32	3206.81	2603.91	2576.84	3717.51
64	2257.94	2251.65	2567.62	2315.40

Table 10 Overall F1 for all the methods with batch size 8 and dimension 256

Learning rate	PCA	TSVD	AC	UMAP
10^{-6}	0.106690	0.109293	0.095072	0
10^{-5}	0.392026	0.396161	0.389151	0.103087
10^{-4}	0.601089	0.603199	0.594507	0.422296

method with varying batch sizes in $\{8, 16, 32, 64\}$. The corresponding training runtimes are shown in Table 9. It can be observed that as the batch size increases, the accuracy of each method generally decreases. In addition, the training time tends to be longest when the batch size is 32.

3.3 *Experiment 3: Vary Learning Rate*

In this experiment, we kept the batch size fixed at 8 and the reduced dimension at 256 and varied the learning rate among $10^{-6}, 10^{-5}, 10^{-4}$. Tables 10 and 11 show the overall F1 and accuracy for all the methods, while Table 12 displays the corresponding training runtimes. One can observe that as the learning rate increases from 10^{-6} to 10^{-4}, the accuracy consistently improves for each method with less training runtime. Therefore, a learning rate of 10^{-4} appears to be the optimal choice for these methods. Furthermore, UMAP still performs the poorest in terms of accuracy compared to all the other methods. It is important to note that since the learning rate changes dynamically during training, testing the model with a fixed or static learning rate may not provide illuminating insights into its final performance. As a result, the learning rate may be the least informative metric at this stage.

Table 11 Overall accuracy for all the methods with batch size 8 and dimension 256

Learning rate	PCA	TSVD	AC	UMAP
10^{-6}	0.798275	0.800086	0.796797	0.789108
10^{-5}	0.854417	0.854656	0.852686	0.802103
10^{-4}	0.905063	0.905031	0.904157	0.874323

Table 12 Train runtime (s) for all the methods with batch size 8 and dimension 256

Learning rate	PCA	TSVD	AC	UMAP
10^{-6}	6290.81	6329.54	5486.01	5011.44
10^{-5}	2922.44	2930.25	3890.68	4596.04
10^{-4}	2627.11	2590.51	2588.36	2957.02

4 Conclusions and Future Work

This investigation presents one of many potential approaches to developing prescriptive methods for reducing the size of large language models and enabling their deployment to resource-constrained systems. Reducing token embedding dimensions seems to be a relatively simple and direct approach to manipulating the size of a BERT model, as reducing the embedding by any factor, n, reduces the entire model size by n. Furthermore, the token embeddings provide a quantitative representation of the underlying corpus, as evidenced by the proximity of similar word vectors. However, there are many caveats to reducing only the token embeddings in the first layer of a BERT model.

First, to change the embedding size of a BERT model in Hugging Face (PyTorch[3]), one has to change the embedding size in the model class—`config`. This results in a random re-initialization of all model weights for the pre-trained model. This re-initialization is necessary because the embedding dimension, also known as the hidden dimension, is a fundamental component that supports the entire structure of the model. The embedding/hidden size is the largest component of the weight matrices within each attention head. In simple terms, each attention head essentially maps the importance of each word in a sequence to every other word, with the words represented by their embeddings. When the embedding size is changed, the model weights must be re-initialized, leading to the loss of most of the benefits of pre-training. Apart from the dimension-transformed embedding matrix, and any residual influence from previous weights, the model has to be retrained from scratch. We used a fine-tuning training method, and therefore the reduced overall performance as measured by accuracy and F1 score, was somewhat limited. To improve the results, one could tweak the hyperparameters to extend training, but this raises the question of the overall cost-benefit trade-off unless a more exhaustive and expensive full model training protocol is followed.

Among the selected dimensionality-reduction methods, PCA, TSVD, and AC exhibit similar performance; however, UMAP significantly underperforms. This

[3] https://github.com/pytorch/pytorch

may be due to several factors, such as the possible failure of UMAP's manifold assumptions to hold for this particular dataset [19]. Nonetheless, since UMAP generally performs well on diverse types of data, a more likely explanation could be the stochastic nature of UMAP and the need for a more thorough exploration of the seed used in generating the UMAP projection. Including a seed-tuning step might have improved its performance. In addition, the non-Euclidean nature of UMAP projections, while effective for visualizing clusters, may have overly distorted the feature space, hindering the performance of BERT-based NER.

In the future, we would explore implementing the dimension reduction technique on the token embedding and then propagating the transformation through each layer of the model. This would require a transparent means of determining how each component of the prior embedding was transformed to generate the new, smaller embedding. We would, then, estimate the complexity of generating the new model as applying numerical transformation to all embedding-length components in each layer of the model. This approach assumes that a significant amount of the model entropy is contained in the token embedding.

Furthermore, we could also explore randomly removing weights from all subsequent layers to fit the new hidden dimension and then fine-tune the model. This approach may help reduce the subsequent retraining time observed with a full model re-initialization.

Acknowledgments The authors would like to thank the Women in Data Science and Mathematics Research Workshop (WiSDM) hosted by UCLA in 2023 for the support of this collaboration and also UCLA IPAM for sponsoring access to an HPC cluster. The research of Qin is supported by the NSF grant DMS-1941197.

References

1. Aizawa, A.: An information-theoretic perspective of tf-idf measures. Inf. Process. Manag. **39**, 45–65 (2003). https://api.semanticscholar.org/CorpusID:45793141
2. Aldous, D.J.: Exchangeability and Related Topics. École d'Été de Probabilités de Saint-Flour XIII — 1983, pp. 1–198. Springer, Berlin (1985)
3. Bengio, Y., Ducharme, R., Vincent, P., Janvin, C.: A neural probabilistic language model. J. Mach. Learn. Res. **3**, 1137–1155 (2003)
4. Blei, D.M., Ng, A.Y., Jordan, M.I.: Latent Dirichlet allocation. J. Mach. Learn. Res. **3**, 993–1022 (2003)
5. Brown, S.D., Francis, R.J., Rose, J., Vranesic, Z.G.: Field-Programmable Gate Arrays, vol. 180. Springer Science & Business Media, Cham (2012)
6. Brown, T., Mann, B., Ryder, N., Subbiah, M., Kaplan, J.D., Dhariwal, P., Neelakantan, A., Shyam, P., Sastry, G., Askell, A., et al.: Language models are few-shot learners. Adv. Neural Inf. Process. Syst. **33**, 1877–1901 (2020)
7. Demszky, D., Movshovitz-Attias, D., Ko, J., Cowen, A., Nemade, G., Ravi, S.: Goemotions: A dataset of fine-grained emotions. In: 58th Annual Meeting of the Association for Computational Linguistics (ACL) (2020)

8. Derczynski, L., Nichols, E., van Erp, M., Limsopatham, N.: Results of the WNUT2017 shared task on novel and emerging entity recognition. In: Proceedings of the 3rd Workshop on Noisy User-Generated Text (2017)
9. Devlin, J., Chang, M., Lee, K., Toutanova, K.: BERT: pre-training of deep bidirectional transformers for language understanding. arXiv: 1810.04805 (2018). http://arxiv.org/abs/1810.04805
10. Devlin, J., Chang, M.W., Lee, K., Toutanova, K.: Bert: Pre-training of deep bidirectional transformers for language understanding. arXiv preprint arXiv:1810.04805 (2018)
11. Dieng, A.B., Ruiz, F.J.R., Blei, D.M.: Topic modeling in embedding spaces. Trans. Assoc. Comput. Linguist. **8**, 439–453 (2020). https://doi.org/10.1162/tacl_a_00325
12. Halko, N., Martinsson, P.G., Tropp, J.A.: Finding structure with randomness: probabilistic algorithms for constructing approximate matrix decompositions. SIAM Rev. **53**(2), 217–288 (2011)
13. Han, S., Pool, J., Tran, J., Dally, W.: Learning both weights and connections for efficient neural network. In: Advances in Neural Information Processing Systems, 28 (2015)
14. Hotelling, H.: Analysis of a complex of statistical variables into principal components. J. Educ. Psychol. **24**(6), 417 (1933)
15. Jain, A.K., Murty, M.N., Flynn, P.J.: Data clustering: a review. ACM Comput. Surveys **31**(3), 264–323 (1999)
16. Jiao, X., Yin, Y., Shang, L., Jiang, X., Chen, X., Li, L., Wang, F., Liu, Q.: Tinybert: Distilling bert for natural language understanding. arXiv preprint arXiv:1909.10351 (2019)
17. Limwattana, S., Prom-on, S.: Topic modeling enhancement using word embeddings. In: 2021 18th International Joint Conference on Computer Science and Software Engineering (JCSSE), pp. 1–5 (2021)
18. Liu, Y., Ott, M., Goyal, N., Du, J., Joshi, M., Chen, D., Levy, O., Lewis, M., Zettlemoyer, L., Stoyanov, V.: Roberta: A robustly optimized bert pretraining approach. arXiv preprint arXiv:1907.11692 (2019)
19. McInnes, L., Healy, J., Melville, J.: Umap: Uniform manifold approximation and projection for dimension reduction. arXiv preprint arXiv:1802.03426 (2018)
20. Michel, P., Levy, O., Neubig, G.: Are sixteen heads really better than one? In: Advances in Neural Information Processing Systems, 32 (2019). https://proceedings.neurips.cc/paper_files/paper/2019/file/2c601ad9d2ff9bc8b282670cdd54f69f-Paper.pdf
21. Mikolov, T., Chen, K., Corrado, G.S., Dean, J.: Efficient estimation of word representations in vector space. In: International Conference on Learning Representations (2013). https://api.semanticscholar.org/CorpusID:5959482
22. Naseem, U., Razzak, I., Khan, S.K., Prasad, M.: A comprehensive survey on word representation models: from classical to state-of-the-art word representation language models. arXiv: 2010.15036 (2020). https://arxiv.org/abs/2010.15036
23. Olson, D.L., Delen, D.: Advanced Data Mining Techniques. Springer Science & Business Media, Cham (2008)
24. Pearson, K.: Liii. on lines and planes of closest fit to systems of points in space. London, Edinburgh Dublin Philosoph. Mag. J. Sci. **2**(11), 559–572 (1901)
25. Pennington, J., Socher, R., Manning, C.D.: Glove: Global vectors for word representation. In: Empirical Methods in Natural Language Processing (EMNLP), pp. 1532–1543 (2014). http://www.aclweb.org/anthology/D14-1162
26. Raunak, V.: Simple and effective dimensionality reduction for word embeddings. arXiv preprint arXiv:1708.03629 (2017)
27. Raunak, V., Gupta, V., Metze, F.: Effective dimensionality reduction for word embeddings. In: Proceedings of the 4th Workshop on Representation Learning for NLP (RepL4NLP-2019), pp. 235–243 (2019)
28. Sanh, V., Debut, L., Chaumond, J., Wolf, T.: Distilbert, a distilled version of bert: smaller, faster, cheaper and lighter. arXiv preprint arXiv:1910.01108 (2019)

29. Sarkar, S., Babar, M.F., Hassan, M.M., Hasan, M., Santu, S.K.K.: Exploring Challenges of Deploying BERT-based NLP Models in Resource-Constrained Embedded Devices. arXiv preprint arXiv:2304.11520 (2023)
30. Silalahi, S., Ahmad, T., Studiawan, H.: Named entity recognition for drone forensic using bert and distilbert. In: 2022 International Conference on Data Science and Its Applications (ICoDSA), pp. 53–58. IEEE (2022)
31. Singh, K.N., Devi, S.D., Devi, H.M., Mahanta, A.K.: A novel approach for dimension reduction using word embedding: an enhanced text classification approach. Int. J. Inf. Manag. Data Insights **2**(1), 100061 (2022)
32. Tjong Kim Sang, E.F., De Meulder, F.: Introduction to the CoNLL-2003 shared task: Language-independent named entity recognition. In: Proceedings of the Seventh Conference on Natural Language Learning at HLT-NAACL 2003 (2003)
33. Touvron, H., Lavril, T., Izacard, G., Martinet, X., Lachaux, M.A., Lacroix, T., Rozière, B., Goyal, N., Hambro, E., Azhar, F., et al.: Llama: Open and efficient foundation language models. arXiv preprint arXiv:2302.13971 (2023)
34. Van der Maaten, L., Hinton, G.: Visualizing data using t-SNE. J. Mach. Learn. Res. **9**(11), 2579–2605 (2008)
35. Vanzo, A., Croce, D., Bastianelli, E., Basili, R., Nardi, D.: Grounded language interpretation of robotic commands through structured learning. Artif. Intell. **278**, 103181 (2020). https://doi.org/10.1016/j.artint.2019.103181
36. Vaswani, A., Shazeer, N., Parmar, N., Uszkoreit, J., Jones, L., Gomez, A.N., Kaiser, L., Polosukhin, I.: Attention is all you need. arXiv preprint arXiv:1706.03762 (2023)
37. Wu, Y., Schuster, M., Chen, Z., Le, Q.V., Norouzi, M., Macherey, W., Krikun, M., Cao, Y., Gao, Q., Macherey, K., Klingner, J., Shah, A., Johnson, M., Liu, X., Kaiser, L., Gouws, S., Kato, Y., Kudo, T., Kazawa, H., Stevens, K., Kurian, G., Patil, N., Wang, W., Young, C., Smith, J.R., Riesa, J., Rudnick, A., Vinyals, O., Corrado, G.S., Hughes, M., Dean, J.: Google's neural machine translation system: Bridging the gap between human and machine translation. arXiv:1609.08144 (2016). https://arxiv.org/pdf/1609.08144

Part IV
Data Analysis and Machine Learning

Automated Extraction of Roadside Slope from Aerial LiDAR Data in Rural North Carolina

Saurya Acharya, Matthew Satusky, and Ashok Krishnamurthy

1 Introduction

Most single-vehicle crashes involve roadway departure, when a vehicle crosses the edge-line of a roadway. According to the National Highway Traffic Safety Administration, roadside and shoulder crashes comprised 50.2 and 43.5% of fatal and injurious crashes, respectively, involving a single vehicle in 2023 [1]. Key elements contributing to roadside crashes are fixed obstacles, such as utility poles or fences, and changes in terrain, like ditches and embankments. Big data and artificial intelligence–based safety research has led to increased focus on identification of objects in the vicinity of the roadway [2], but less attention has been paid to terrain geometry, despite evidence that terrain slope has been shown to factor heavily in crash fatality models [3–5].

While urban landscapes tend to be more human-engineered and therefore have less severe grade changes, rural areas have highly variable morphological features surrounding roadways. The rural landscape of the United States comprises a significant portion of the nation's road infrastructure, accounting for 68% of road miles, totaling over 6 million miles as of 2020. In 2021, 40% of motor vehicle traffic fatalities in the United States occurred in rural areas, resulting in a rate 1.5 times higher than urban areas per 100 million miles driven [6].

S. Acharya
Department of Computer Science, University of North Carolina at Chapel Hill, Chapel Hill, NC, USA
e-mail: saurya@email.unc.edu

M. Satusky · A. Krishnamurthy (✉)
Renaissance Computing Institute, University of North Carolina at Chapel Hill, Chapel Hill, NC, USA
e-mail: satusky@renci.org; ashok@renci.org

C. Garcia-Cardona, H. Lee (eds.), *Advances in Data Science*, Association for Women in Mathematics Series 37, https://doi.org/10.1007/978-3-031-87804-6_11

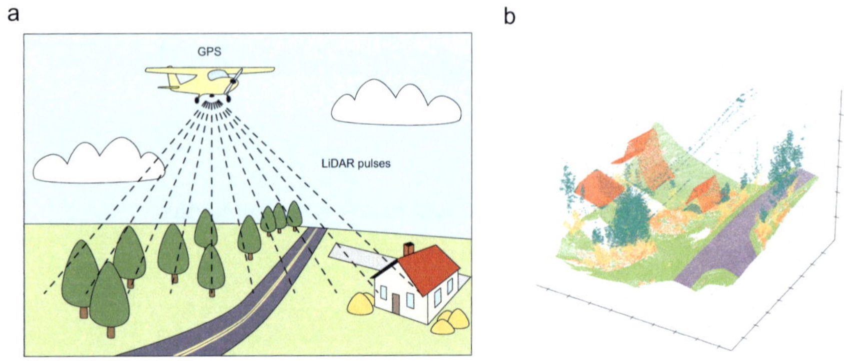

Fig. 1 Aerial LiDAR data form a top-down elevation map of the scanned area. (**a**) An aircraft is flown over an area, sending LiDAR pulses to the terrain below. The return time and intensity are recorded along with GPS coordinates. (**b**) A 3D point cloud is generated using the GPS coordinates (x, y) and the elevation calculated from return times (z). The points can be classified using return intensity and other methods. The data in the example point cloud shown are classified as ground (light green), vegetation (low = pink, medium = orange, high = dark green), road (purple), and buildings (red)

In North Carolina, a substantial proportion of crashes involving vulnerable road users occur in agricultural settings, such as farms, woods, and pastures. In 2023, 34% of these incidents resulted in serious injuries and fatalities, more than industrial, commercial, residential, and institutional areas [7]. Previous attempts to assess roadside hazards in rural areas identified the side slope, or slope of roadside terrain, as a significant component in crash prediction. However, these same studies identified a lack of side slope data for two-lane rural roads in many Department of Transportation (DOT) databases [4, 8, 9].

Because physical surveys are cost- and time-prohibitive, light detection and ranging (LiDAR) is a compelling alternative for assessing roadside terrain grade. LiDAR is a remote sensing method where return time and intensity of light pulses are used to create a surface map of the surroundings. Roadway LiDAR data are typically collected by Mobile Laser Scanning (MLS), where the sensors are attached to a road vehicle or aerial scanning by a sensor-equipped aircraft, generating a topographical map [8]. The resulting point clouds also contain additional information such as the intensity, terrain type, and GPS time (Fig. 1).

Due to its widespread availability, LiDAR technology has been integrated into numerous applications from environmental monitoring to urban planning. For instance, the North Carolina Emergency Risk Management Office, in collaboration with the North Carolina DOT and other stakeholders, has spearheaded initiatives to acquire statewide aerial LiDAR data. This concerted effort has resulted in the comprehensive coverage of the entire state, enabling detailed analyses and informed decision-making processes [10]. These kinds of initiatives are not only in North

Carolina; other states have similarly recognized the value of LiDAR data and undertaken similar projects with support from federal initiatives. According to the US Geological Survey, there are efforts to create and maintain consistent elevation databases in all 50 states and Puerto Rico [11]. Although federal initiatives may exhibit lower accuracy and density compared to state-level endeavors, they nonetheless contribute significantly to the widespread availability of LiDAR datasets on a national scale.

The objective of this preliminary study is to identify stretches of roadway with potentially hazardous side slopes, particularly on secondary roads in rural areas. We propose creating an open-source pipeline for processing aerial LiDAR data. The benefits of this approach are a reduction of time and cost, since many states have already undertaken high-quality scans across large regions. This study aims to leverage existing topographical survey data from the North Carolina DOT. Additionally, we aim to utilize open-source Python libraries to limit licensing and training constraints imposed by commercial geospatial data analysis packages.

2 Prior Work

Balado et al. [12] developed a deep learning model, PointNet, to segment ditches, embankments, and guardrails from MLS point clouds. The model correctly identified 92% of road points, but displayed reduced accuracy on embankment and ditch points, with 88.3% and 65.4%, respectively. The variability in accuracy was attributed to class imbalances in the dataset, as well as foliage interference with embankment points.

Shams et al. [13] tested the effectiveness of airborne and mobile terrestrial LiDAR scanning systems in measuring roadway cross slopes (the slope from midline to edgeline). Mobile LiDAR data required 3 months of collection from five different vendors, while a single vendor provided the aerial LiDAR data by performing 15 flight line passes. As a result, the study found that both aerial and mobile terrestrial LiDAR scanners have cross slope accuracies comparable to conventional manual surveying methods. However, data collection was a costly process in comparison to the widely available datasets used in this method.

Rua et al. [14] used LiDAR data and Monte Carlo simulations to identify areas susceptible to rock slides. To verify their results, human experts were employed to manually measure the cross slopes using ArcGIS, which was a time- and cost-intensive process. In the end, they were able to identify 95% of the slopes found by the experts, and the disagreements were borderline cases.

Jayaler and Zhou [4] gathered 5 years of Illinois runoff road crash data and were able to define a reliability index to measure roadside safety on two-lane roads. They utilized a roadside hazard rating (RHR) system from Zegeer et al. [15] to identify a correlation between the calculated reliability indexes of side slopes and crash severity.

3 Methodology

In this preliminary work, we attempt to establish a minimum LiDAR spatial resolution to reliably collect side slope grades as a continuous feature along a given stretch of road. Furthermore, we aim to directly segment the roadway from the LiDAR data rather than using precalculated polylines. Our method involves converting a block of LiDAR data into a pixelated image, segmenting the roadway, identifying the directional and perpendicular vectors of each road segment, and finally collecting LiDAR data in areas immediately adjacent to each roadway segment. Our analysis is entirely automated using Python, with the exception of selecting test scene boundaries.

3.1 Development Scenes

Aerial LiDAR data for Buncombe County, North Carolina, (North Carolina DOT Division 13) were provided by the North Carolina DOT. The data had a nominal pulse spacing of 8 points per square meter (ppsm) with a 95% non-vegetated vertical accuracy of 0.64 ft. The coordinate points were pre-classified by terrain or object type, including vegetation/stratum, buildings, and roads. For methods development, four example scenes were selected, each covering a 409.6×409.6 ft area and containing roughly 1 million LiDAR points. Scenes were selected for variable road geometries.

3.2 Defining Road Segments and Slope Collection Areas

Each LiDAR point contains world coordinates (x, y), elevation (z), and an associated class, as outlined in Table 1. To reduce the amount and dimensionality of data, the three-dimensional LiDAR points were rasterized into a 512×512 pixel image, with each pixel covering a 0.8×0.8 ft area. Pixels containing any road points were classified as road; otherwise, they were assigned the classification value that occurred most within their (x, y) coordinates. Pixels containing no LiDAR points were assigned values using nearest-neighbor interpolation from surrounding pixels. Road pixels were isolated as a binary image, and road boundary pixels were identified using Sobel edge detection. The boundary was split into two connected components to separate opposite sides of the road.

Figure 2a and b illustrate our approach for a single road segment. One road edge is designated the reference edge (r), from which pixels are sampled at

Table 1 LiDAR classifications

Classification code	Description
1	Default
2	Ground
3	Low Veg/Strata
4	Medium Veg/Strata
5	High Veg/Strata
6	Buildings (Automated)
7	Low points
9	Water (Hydro cleaned area)
10	Breakline proximity
11	Withheld (high points)
13	Roads
14	Bridges
17	Overlap default
18	Overlap ground
25	Overlap water

a user-specified interval ("segment length" or "length"). A segment is defined by two consecutive sampled pixels (r_{start} and r_{end}), a directional vector (d), and perpendicular vectors crossing the road (p) and away from the road (p'). Bresenham's line algorithm [16] is used to identify the pixels o_{start} and o_{end}, where p intersects the opposing edge of the road (o) when originating from r_{start} and r_{end}, respectively (Fig. 2a). The mean coordinates of the start and end pairs define the segment centerline.

The rectangular region adjacent to the reference edge is created by the vertices r_{start} and r_{end}, and the coordinates are sampled a user-specified distance ("width") along p' from each reference pixel (Fig. 2b). The process is repeated for the opposing side of the road, using o_{start} and o_{end} as the origin coordinates and p as the directional vector. The region boundaries are then converted to the world coordinate reference frame from pixel coordinates.

For shoulder slope calculations, the raw LiDAR data are filtered to include only points labeled "ground" within the bounding boxes of the adjacent regions. The remaining data are rotated to an orthographic elevation projection viewed along the road segment centerline (Fig. 3a). Plotting the rotated x- and z-coordinates results in a two-dimensional cross section of the elevation data adjacent to the segment, where x is the distance from the centerline and y is the elevation (Fig. 3b). Linear regression was used to determine the slope of the elevation data on each side of the road independently, and the sign of the slope to the left of the road is reversed to normalize the direction of elevation change relative to the road (i.e., an increase or decrease in elevation is positive or negative, respectively). All steps are repeated for each identified segment.

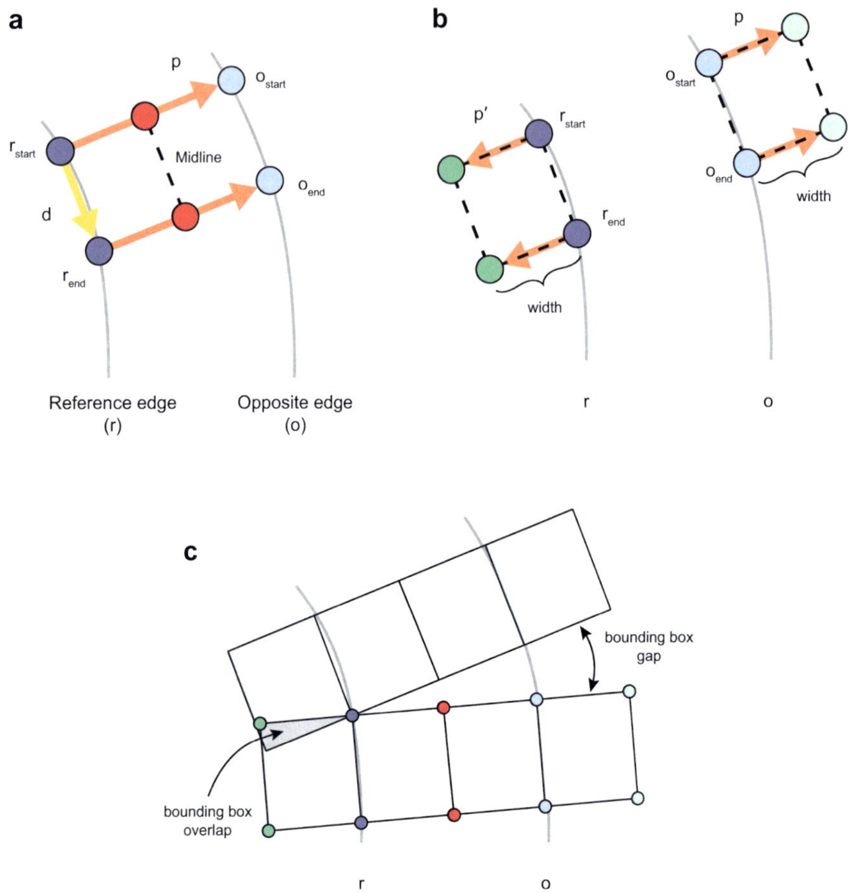

Fig. 2 A schematic representation of the data sampling method. (**a**) A segment is defined by reference pixels (dark blue circles), pixels on the opposite edge (light blue circles), centerline coordinates (red circles), a directional vector (yellow arrow), and perpendicular vectors. (**b**) Rectangular regions adjacent to the segment are created by moving outward from the segment edge pixels (blue circles) to points a given width away (green circles). (**c**) Data within the adjacent regions (gray boxes) are used for slope calculation, and the process is repeated for each segment. Processing consecutive segments of a curved road results in overlaps on the inside and gaps on the outside of the curve

4 Results

To determine the most effective size for each sample region, we used the shorter edge (containing the least pixels) of the road as the reference edge and systematically tested adjacent region sizes with all length and width dimension combinations between 3 and 30 pixels (inclusive), resulting in 784 test conditions. Because no ground-truth data were available to directly compare the accuracy of the derived

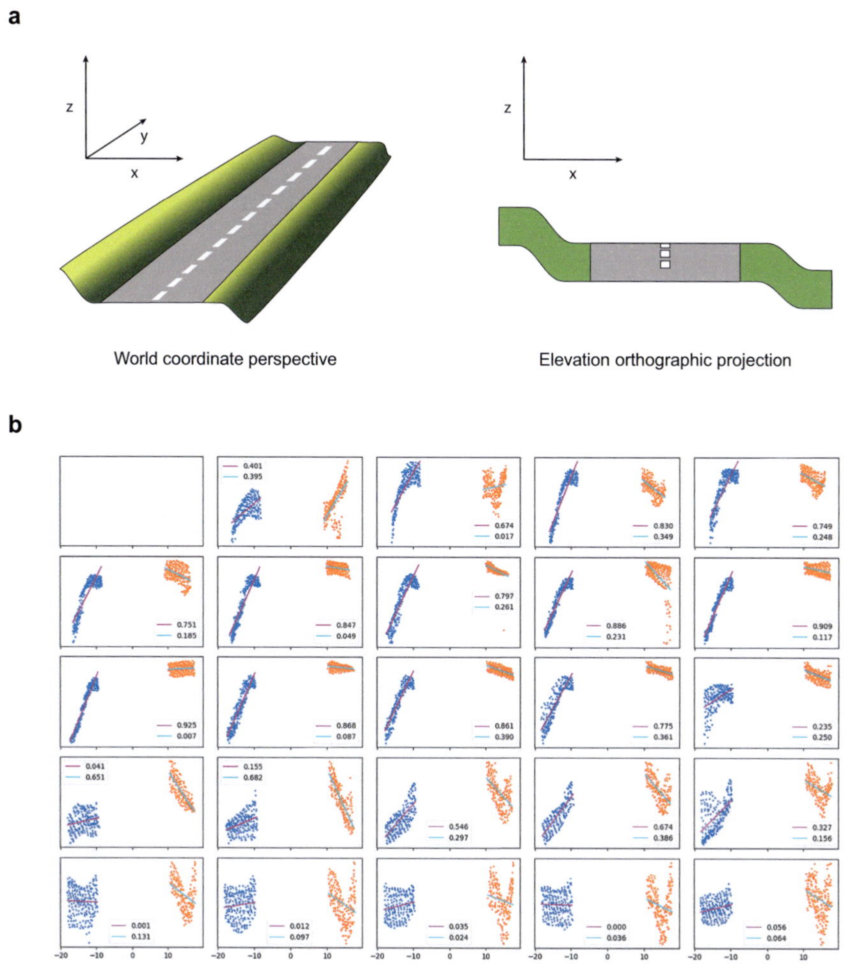

Fig. 3 Fitting shoulder slopes to a length of road. (**a**) Data points adjacent to a road segment are rotated from three dimensions in the world perspective (left) to two dimensions using an orthographic elevation projection (right), where the viewing vector is the centerline of the segment at ground level. The resulting x-axis is the distance from the center of the road (ft), and the y-axis is the original z-axis (elevation in ft). (**b**) A visualization of example outputs for consecutive segments. Data points are colored based on the side of the road compared to the segment vector (blue = left, orange = right). The whitespace in the middle results from filtering road points. Linear slope fits are shown for each side (red = left, green = right), and the R-squared values are displayed in the legend

slopes, the R-squared for the fit to the rotated data points for each shoulder segment was used as a proxy metric. The logic in using this metric was that data collected within sub-optimally large regions would reflect larger-scale terrain shifts, while insufficiently large areas would collect only a few highly varied points. Either of these scenarios would be reflected in the goodness-of-fit parameter.

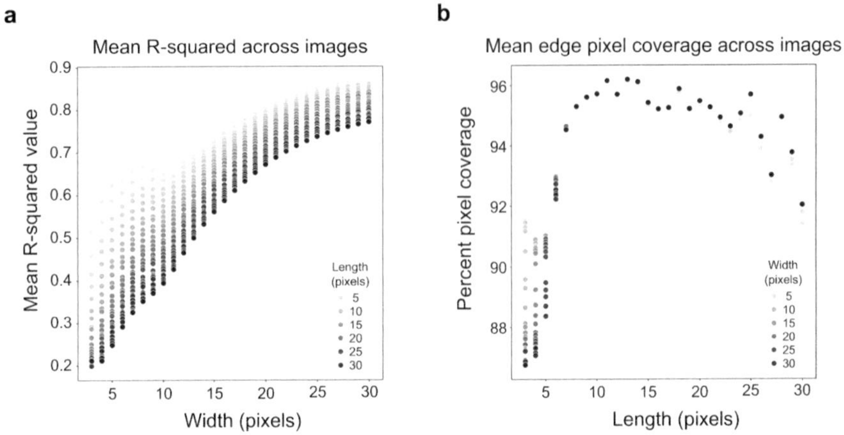

Fig. 4 Comparison of performance of length and width combinations. All combinations of segment lengths and region widths from 3 to 30 pixels (784 total) were applied to all four test scenes. (**a**) Mean R-squared values across all segments plotted as a function of width and colored by length (lighter = shorter). (**b**) Mean percentage of pixels in both edges for which slopes were calculated as a function of segment length and colored by width (lighter = more narrow)

4.1 Fitting Shoulder Slopes

The mean R-squared values across the four test scenes is shown in Fig. 4a. There is a general positive trend in R-squared as the region width increases, suggesting wider areas may smooth over smaller features like ditches. On the contrary, increasing the segment length shows a consistent decrease in R-squared, suggesting the segment length parameter is more sensitive to terrain features. This effect is more pronounced at widths smaller than 10 pixels (8 ft), with a clearly defined peak at 8 pixels in length and 3 pixels in width (6.4 ft and 2.4–3.4 ft, respectively; R-squared = 0.683).

4.2 Shoulder Coverage

One potential drawback of our approach is coverage loss due to gaps between adjacent rectangular regions on the outside edge of curves (Fig. 2c) and areas with sparse ground data points, either from roadside foliage or unclassified data. For each segment, the contiguous pixels between the start and end coordinate on each edge were assigned the associated slope value with that shoulder region, if one was successfully calculated. Total coverage was defined as the percentage of pixels along each road edge with an assigned slope.

The effect of region size on the edge pixel coverage is shown in Fig. 4b. Longer segments displayed more coverage than shorter ones, plateauing around 96% from 10 to 20 pixels before gradually falling off. The width only affected coverage of the shortest (<7 pixels) and longest (>20 pixels) segments. Specifically, smaller widths performed better at lower lengths, while larger widths performed better at longer lengths.

The amount of coverage loss associated with gaps can be determined by the dependence of the loss on the reference edge. We base coverage on the edge pixels between the start and end of a segment. Because we directly sample the pixels on the reference edge, the only data loss on that edge should occur due to sparse data (e.g., from foliage), whereas gaps between bounding boxes will only occur on the opposite edge, as those start/end coordinates are calculated from the reference points. This relationship remains true regardless of which side is designated the reference (Fig. 5a). To demonstrate this effect, we repeated the analysis with the reference and opposing edges reversed (now the longer and shorter edges, respectively) and observed the coverage of each side of the road independently. The average coverage of either road edge is dramatically higher for shorter segment lengths when that side is the reference edge (Fig. 5b,c). This result is likely due to shorter segment lengths requiring more bounding boxes and consequently more opportunity to generate gaps. In both cases, the coverage converges around a 10 pixel segment length. The opposite edge coverage (Fig. 5b, red; c, blue) looks nearly identical in both cases, but the mean coverage of the reference edge in each scenario shows that the longer edge has less coverage across all segment lengths (Fig. 5d). Because there should be no gaps on the reference edge, this difference must be due to other factors like roadside foliage.

4.3 Qualitative Performance Analysis

Given the lack of available ground-truth slope data, we are unable to quantify the accuracy of our tool. However, we can qualitatively examine each of the four test scenes to get a sense of how well the tool is performing. We selected a region size of 3 pixels (2.4–3.4 ft) long and 8 pixels (6.4 ft) wide because it yielded the peak R-squared value for smaller areas (Fig. 4a) and had approximately 98% and 94% coverage of the reference and opposing edges, respectively (Fig. 5a). Slopes were binned based on three vertical-to-horizontal-grade ratios previously used to identify road hazards [4, 15]. Specifically, high safety risk is classified as 1V:2H, moderate as 1V:3H, and low as 1V:4H. Figure 6 shows the outputs of the pipeline for the four scenes with the binned slopes overlaid on the road edges.

Three immediate trends are apparent between scenes. First, the overhead images (top left in each panel) indicate that across all images, there are more trees directly adjacent to the longer side of the road, providing a likely explanation for the difference observed between the reference edge coverage in our opposing tests (Fig. 5d). Second, the distributions of slopes (top right in each panel) show that the vast majority of shoulders are flat. Intuitively, most roadways do not have extreme terrain slopes next to them, and this observation serves as a sanity check. Finally, negative slopes appear more frequently than positive, although these scenes were

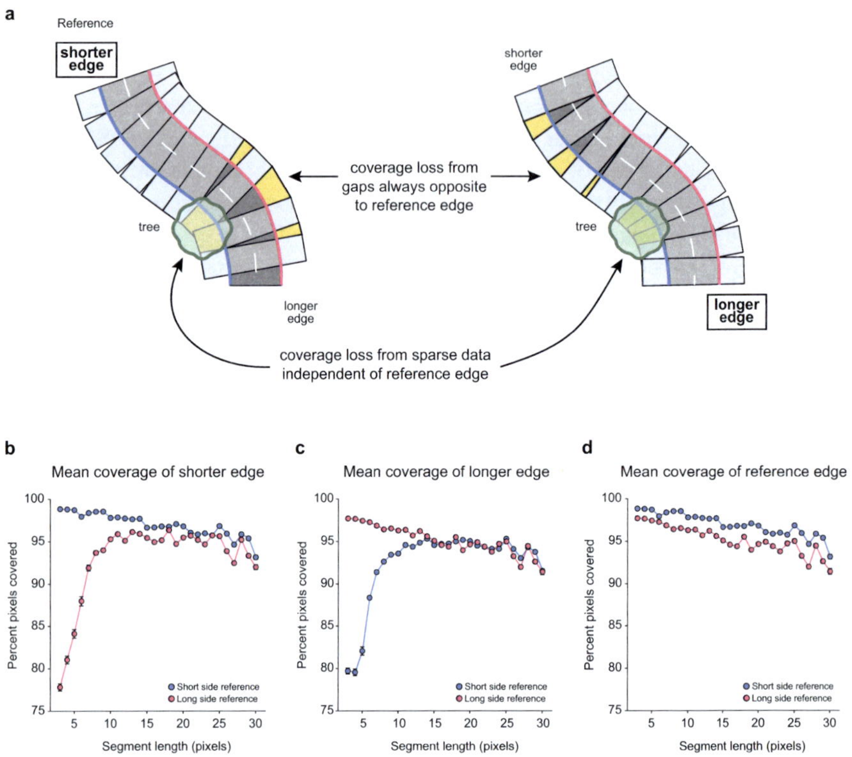

Fig. 5 Coverage dependence on reference edge selection. (**a**) Gaps between bounding boxes only occur on the opposite side of the road. Any data loss on the reference edge is due to sparse data. (**b**) The mean fraction of pixels covered on the shorter side of the road across segment lengths when it is designated the reference (blue) or opposite (red) edge. (**c**) The mean fraction of pixels covered on the longer side of the road across segment lengths when it is designated the reference (red) or opposite (blue) edge. (**d**) The mean fraction of pixels covered on the reference edge across segment lengths when referencing the shorter (blue) or longer (red) edge. All error bars are standard error from the mean

selected for differences in road geometries without underlying knowledge of the surrounding terrain, and this pattern could be due to coincidence.

Looking at the scenes individually, the tool correctly identifies regions of higher slope. Scene 1 has almost entirely flat shoulders, with the exception of one region of negative slope that has a darker region near its center (Fig. 6a, top left). When looking at the terrain overlay (Fig. 6a, bottom), we can see a wide area of low elevation that approaches the road in this dark region. Similar features are detected in Scenes 2 and 3 (Fig. 6b,c, bottom). Prominent sections of positive slope identified in Scenes 2 and 3 also align to features in the terrain as well (in the case Scene 2, the red section corresponds to an uphill driveway entrance according to the street view, which is not shown). Scene 4 is a unique case because it is a private driveway, but interestingly, there is a region of dark red near the house at the bottom of the

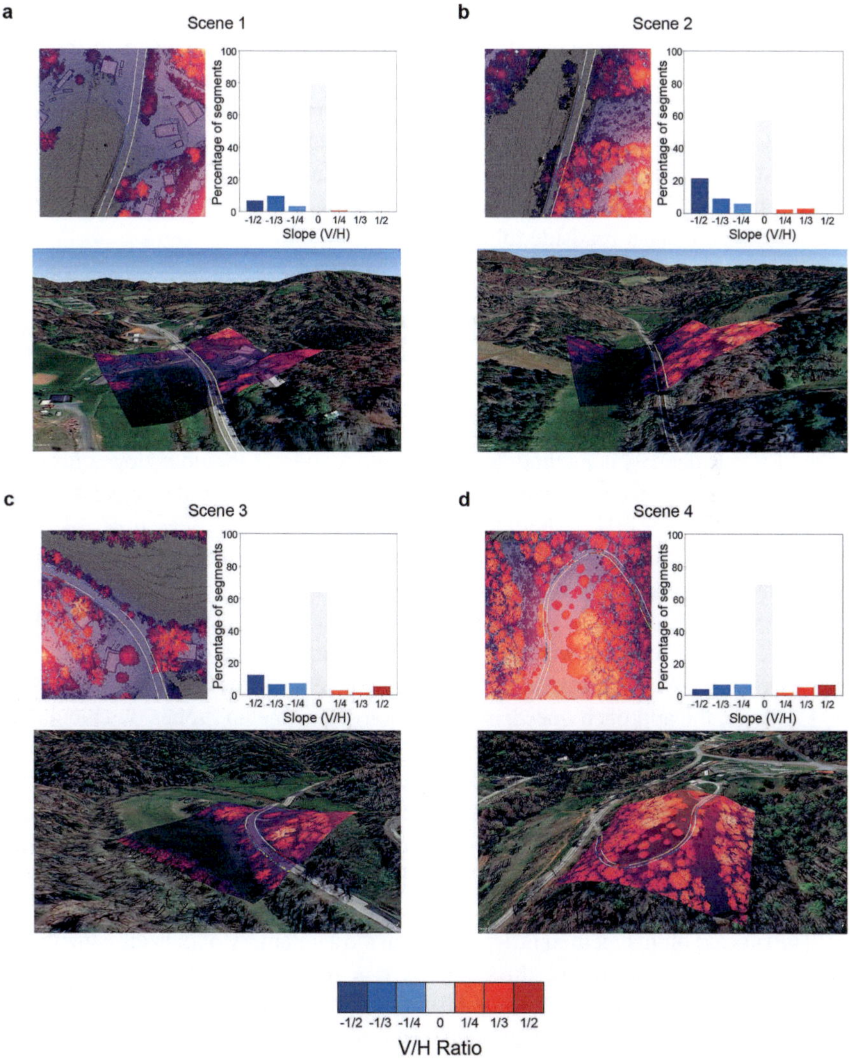

Fig. 6 Results of shoulder slope calculations for (**a**) Scene 1, (**b**) Scene 2, (**c**) Scene 3, (**d**) Scene 4. Each panel consists of a scatterplot of LiDAR elevation data with slopes overlaid on the road edges (top left), the percentage of edge pixels belonging to each vertical to horizontal ratio category (top right), and the scatter plot overlaid on the terrain map from Google Earth (Google, Mountain View, CA) (bottom). Brighter scatter points indicate higher elevation. Negative slopes are depicted in blue, and positive slopes are red, with darker hues indicating more extreme slopes. Gray pixels indicate areas with no slope, and yellow areas are where the slope could not be calculated

overhead image (Fig. 6d, top left, rectangular feature). That section appears to have a dark mark in the terrain map (Fig. 6d, bottom) where the driveway is cut into the hillside (the terrain image is rotated for easier viewing).

There are aspects where the method can be improved. All images contain a number of yellow regions where the slope could not be calculated, many in areas not obscured by foliage. Furthermore, there are regions where slope is being identified but have oscillating or spotty coverage. This observation is particularly strong in Scene 3 (Fig. 6c), where the negative slope around the outside of the curve has mostly strong slopes interspersed with lower magnitude or missing data, but it is also present in several identified regions in Scene 4 (Fig. 6d). In both images, however, this effect occurs near roadside foliage.

5 Conclusion

In this work, we proposed a method for obtaining roadway side slope grade from aerial LiDAR data. This methodology was used to extract shoulder slopes for rural roads with variable length, curvature, and orientation. Initial results suggest we can find terrain grade at 94% shoulder coverage with a segment length of 2.4–3.4 ft and a sampling area 6.4 ft from the road edge. Unfortunately, we cannot draw too many conclusions about the accuracy of the calculated grades or whether our optimal window sizes truly yield the best results due to the lack of validation data. We are currently working with the NCDOT and outside sources to create a dataset using existing physical survey and MLS data to determine the accuracy.

However, the approach does have limitations. This method relies on calculating the midline of the road, so in its preliminary state, roads with complex shapes such as intersections or roundabouts will present issues. Furthermore, because the data are collected aerially, the density of foliage in the immediate proximity to the road can limit the ground data available for fitting. Combining aerial LiDAR with MLS scans could potentially overcome foliage issues, since MLS would scan laterally beneath any tree canopy.

The success of this method is dependent on the availability of data classifications, since the pipeline currently does not have a means of segmenting roadways from unlabeled data, although some off-the-shelf solutions exist. Additionally, the minimum segment dimensions are dependent on the density of ALS sampling. In the geographical region used for this study, the nominal pulse spacing was 8 ppsm, but other regions in the state have sampling densities between 2 and 30 ppsm. More dense sampling would allow for smaller segment dimensions and potentially higher resolution along the road or more confident slope fitting. However, reducing the point density would require larger segments, leading to variability when analyzing multiple regions. To mediate this variability, the ALS data could be rasterized, thereby normalizing segment dimensions.

There are immediate ways in which the pipeline can be improved. Currently, the length of the segment is determined by indexing the list of edge coordinates rather than using a distance metric. We would like to define the centerline as a mathematical spline to achieve more precise control. Using a single edge as a reference has the drawback of leaving gaps on the other side of the road. To counter this artifact, we may attempt to use both sides of the road separately. To further reduce gaps, the data may be smoothed by using overlapping road segments. This approach could also potentially reduce issues with foliage covering the ground adjacent to the road. Alternatively, we could employ non-rectangular shoulder areas.

Future goals of this project are to extend the application of this method to scenes with more complex road geometries (like intersections) and cover as much of the state as possible. This expansion will allow for a broader understanding of its effectiveness across different road environments and conditions. Overall, this preliminary work displays the potential to identify areas of concern for the NCDOT without collecting additional data and using open-source technology.

Acknowledgments This work was performed for an Honors Carolina student thesis in the University of North Carolina at Chapel Hill (UNC) College of Arts and Sciences. The authors would like to thank the North Carolina Department of Transportation for providing the LiDAR data, as well as the faculty and staff in the UNC Department of Computer Science and the Renaissance Computing Institute for their support.

Competing Interests The authors have no conflicts of interest to declare that are relevant to the content of this chapter.

References

1. National Center for Statistics and Analysis: Rural/urban comparison of motor vehicle traffic fatalities: 2021 data (Traffic Safety Facts. Report No. DOT HS 813 488). National Highway Traffic Safety Administration (2023)
2. Jia, X., Tong, Y., Qiao, H., Li, M., Tong, J., Liang, B.: Fast and accurate object detector for autonomous driving based on improved YOLOv5. Sci. Rep. **13**(1), 9711 (2023). https://doi.org/10.1038/s41598-023-36868-w
3. Roque, C., Moura, F., Lourenço Cardoso, J.: Detecting unforgiving roadside contributors through the severity analysis of ran-off-road crashes. Accident Analy. Prevent. **80**, 262–273 (2015). https://doi.org/10.1016/j.aap.2015.02.012
4. Jalayer, M., Zhou, H.: Evaluating the safety risk of roadside features for rural two-lane roads using reliability analysis. Accident Analy. Prevent. **93**, 101–112 (2016). https://doi.org/10.1016/j.aap.2016.04.021
5. Jalayer, M., Zhou, H., Das, S.: Exploratory analysis of run-off-road crash patterns. In: Alavi, A., Buttlar, W.G. (eds.) Data Analytics for Smart Cities, 1st edn., pp. 183–200. Auerbach Publications, Boca Raton: CRC Press, Taylor & Francis Group [2019] (2018). https://doi.org/10.1201/9780429434983-8
6. Bureau of Transportation Statistics, US Department of Transportation: Rural Transportation Statistics (2022). https://www.bts.gov/rural. Cited 2 Aug 2024
7. Mayhew, B., Troy, S., Murphy, B., Oliver, C.: North Carolina Department of Transportation (NCDOT) 2023 VRUSA Strategy Worksheet. North Carolina Department of Transportation (2023). https://spatial.vhb.com/ncdotshsp/. Cited 2 Aug 2024
8. Jalayer, M., Zhou, H., Gong, J., Hu, S., Grinter, M.: A comprehensive assessment of highway inventory data collection methods. J. Transport. Res. Forum **53**, 73–92 (2014). https://doi.org/10.5399/osu/jtrf.53.2.4219
9. Jalayer, M., Gong, J., Zhou, H., Grinter, M.: Evaluation of remote sensing technologies for collecting roadside feature data to support highway safety manual implementation. J. Transport. Safety Security **7**, 345–357 (2015).

10. North Carolina Department of Transportation: Existing Geospatial Data. (n.d.). https://connect.ncdot.gov/resources/Photogrammetry/Pages/Existing-Geospatial-Data.aspx. Cited 2 Aug 2024
11. United States Geologic Survey, 3D Elevation Program: USGS 3D Elevation Program State Factsheets. (n.d.). https://www.usgs.gov/3d-elevation-program/state-factsheets. Cited 2 Aug 2024
12. Balado, J., Martínez-Sánchez, J., Arias, P., Novo, A.: Road environment semantic segmentation with deep learning from MLS point cloud data. Sensors **19**(16), 3466 (2019). https://doi.org/10.3390/s19163466
13. Shams, A., Sarasua, W.A., Russell, B.T., Davis, W.J., Post, C., Rastiveis, H., Famili, A., Cassule, L.: Extracting highway cross slopes from airborne and mobile LiDAR point clouds. Transport. Res. Record J. Transport. Res. Board **2677**(2), 372–384 (2023). https://doi.org/10.1177/03611981221106482
14. Rúa, E., Núñez-Seoane, A., Arias, P., Martínez-Sánchez, J.: Automatic detection to inventory road slopes using open LiDAR point clouds. Int. J. Appl. Earth Observat. Geoinf. **118**, 103225 (2023). https://doi.org/10.1016/j.jag.2023.103225
15. Zegeer, C., Reinfurt, D., Hummer, J., Herf, L., Hunter, W.: Safety effects of cross-section design for two-lane roads. Transport. Res. Record J. Transport. Res. Board **1195**, 20–32 (1988)
16. Bresenham, J.E.: Algorithm for computer control of a digital plotter. IBM Syst. J. **4**(1), 25–30 (1965). https://doi.org/10.1147/sj.41.0025

A Non-parametric Optimal Design Algorithm for Population Pharmacokinetics

Markus Hovd, Alona Kryshchenko, Michael N. Neely, Julian Otalvaro, Alan Schumitzky, and Walter M. Yamada

1 Introduction

Pharmacokinetic modeling and simulation have become a cornerstone in both drug development and therapeutic drug monitoring. The ability to integrate pre-clinical and clinical data, along with covariates, allows for accurate inference of both drug exposure (pharmacokinetics, PK) and response (pharmacodynamics, PD). These statistics are integral to drug therapy optimization at both the individual and population levels. Two different statistical approaches are common: parametric and non-parametric [6]. While parametric approaches assume that the probability distribution of model parameter values follows predefined distributions such as the normal and log-normal [1, 3, 7, 19, 20], non-parametric approaches are free of this assumption. Rather, the joint parameter value probability distribution consists of discrete support points, each point comprising a vector of values for every parameter and an associated probability based on the likelihood of those parameter values. If

M. Hovd
Oslo University Hospital, University of Oslo, Oslo, Norway
e-mail: markus.hovd@farmasi.uio.no

A. Kryshchenko (✉)
California State University Channel Islands, Camarillo, CA, USA
e-mail: alona.kryshchenko@csuci.edu

M. N. Neely
Children's Hospital Los Angeles, University of Southern California, Los Angeles, CA, USA
e-mail: mneely@chla.usc.edu

J. Otalvaro · W. M. Yamada
Children's Hospital Los Angeles, Los Angeles, CA, USA
e-mail: jotalvaro@chla.usc.edu; wyamada@chla.usc.edu

A. Schumitzky
University of Southern California, Los Angeles, CA, USA

C. Garcia-Cardona, H. Lee (eds.), *Advances in Data Science*, Association for Women in Mathematics Series 37, https://doi.org/10.1007/978-3-031-87804-6_12

desired, the shape of the distribution can be inferred by calculations on the optimized support points and their corresponding probabilities, e.g., covariance, mean or median, or an operation on the points, for example, kernel density estimation. Non-parametric approaches allow for more accurate a priori detection of sub-populations and outliers [5, 17].

The non-parametric adaptive grid (NPAG) algorithm is a well-established non-parametric estimation method widely used in pharmacokinetics and pharmacodynamics (PK/PD) [5]. NPAG is a " throw-and-catch" algorithm. It begins with a viable solution (i.e., the likelihood of the current set of support points is greater than 0), and assuming a better solution can be found on the Euclidean grid surrounding each support point of that viable solution, it casts out new, potentially better support points along each dimension of the grid. Successive cycles will find the local optimum around the viable solution. Confidence is gained as the grid is "adaptive" in both discretization length and position in space. However, due to the nature of the adaptive grid, NPAG is computationally expensive and therefore slow to converge.

With increasingly complex PK-PD models, large in the number of parameters, subjects, or both, algorithm speed becomes critical. This has motivated the development of the current non-parametric estimation technique that can maintain the accuracy of NPAG while significantly improving time to convergence. Addressing the convergence speed issue in non-parametric estimation is crucial for streamlining the PK/PD modeling and analysis workflow, enabling faster and more cost-effective drug development processes. The proposed algorithm in this chapter tackles this challenge by introducing innovative computational methods and optimization strategies to enhance the efficiency of non-parametric parameter estimation.

2 Methods

2.1 *Design of the Non-parametric Optimal Design Estimation Algorithm*

Pharmacokinetic observations can be statistically described using a mixing distribution model, where the probability of random variable arguments (the PK population model) in the PK compartmental model is governed by a mixing distribution.

The task of estimating this mixing distribution from a set of PK observations can be defined as follows. Let $Y_1, \ldots, Y_N$ represent a sequence of independent but not necessarily identically distributed random vectors, constructed from one or more observations from each of N subjects in the population. Additionally, let $\theta_1, \ldots, \theta_N$ denote a sequence of independent and identically distributed random vectors representing unknown parameter values for N subjects. These θ values belong to a compact subset Θ of Euclidean space with a common but unknown distribution F, representing the parameter space of the population model.

The objective is to maximize the likelihood function $L(F)$ with respect to all probability distributions F on Θ. Each θ_i is not observed, but it is assumed that the conditional densities $p(Y_i|\theta_i)$ are known for $i = 1, \ldots, N$. The mixing distribution of Y_i with respect to F is then given by $p(Y_i|F) = \int p(Y_i|\theta_i)dF(\theta_i)$.

Let F^{ML} be the distribution that maximizes $L(F)$. It serves as a consistent estimator of the true mixing distribution. Because of independence of the $\{Y_i\}$, the likelihood function can be written as

$$L(F) = p(Y_1, \ldots, Y_N|F) = \prod_{i=1}^{N} \int p(Y_i|\theta_i)dF(\theta_i) \tag{1}$$

It is important to note that $L(F)$ is a convex function of F. Further, it is shown in [15], under simple hypotheses, that the global maximizer F^{ML} of $L(F)$ is a discrete distribution with at most N support points, where N is the number of subjects in the population and a support point is a vector of model parameter values with nonzero probability.

The problem of finding F^{ML} has been addressed in [22] by the NPAG algorithm that uses the primal-dual interior point method to find optimal weights and an Adaptive Grid algorithm to find optimal support points. It was also addressed in Lesperance and Kalbfleisch [13] by the combination of the Semi-Infinite Programming algorithm to find optimal weights; in the Improved Supervised Descent Method (ISDM) algorithm using the D-function to find optimal support points; and in [21] by the combination of Quadratic Programming algorithm to find optimal weights and ISDM algorithm to find optimal support points.

The algorithm described here is an alternative to the NPAG algorithm and is different from it in the step of finding optimal support points. It utilizes the primal-dual interior-point method for convex programming to find optimal weights of the F^{ML} and introduces the optimization of the directional derivative of the likelihood function to address the search for optimal support points of the F^{ML}. This algorithm was proposed by Dr. Robert Leary in the PAGE conference poster [12].

The design of the NPOD algorithm is summarized in the steps below.

2.1.1 Design Principles and Theoretical Foundation

Traditionally, non-parametric maximum likelihood methods rely on iterative approaches such as the expectation-maximization algorithm, which entails optimizing conditional expectation. However, this process can be quite time-intensive, particularly for problems with high dimensions. To address this, we've developed an enhanced iterative non-parametric optimal design (NPOD) algorithm that streamlines certain optimization stages using directional derivatives, significantly boosting its speed compared to the original version detailed by [17, 22].

2.1.2 Algorithm Implementation

Initialization
The first step of any non-parametric algorithm is the initialization of the n-dimensional parameter space. Importantly, the parameter space must be bounded, as an infinitely large parameter space is both computationally and physiologically impossible. In most cases, the sample space represents an uninformed prior. However, it may also be initialized with the joint distribution obtained from previous searches or other algorithms. In the present implementation of NPOD, we are using a modified version of the Sobol pseudo-random sequence generator based on the work by Burley et al. [2] with an improved hash by Kuo et al. [8, 9].

Likelihood Calculation
The next step is the calculation of the likelihood or the objective function. This is the most computationally expensive step in the algorithm, as it is calculated by solving the differential equation representing the pharmacokinetic model for each subject for each point in the initial grid.

Optimization
Following the calculation of the likelihood, the weight of each support point is recalculated in order to maximize the sum of the likelihood function across all subjects. This is achieved through the use of a primal-dual interior point algorithm [22].

Rank Revealing Function
Another important property of the joint parameter distribution in the non-parametric approach is that the maximum number of support points can at most be equal to the number of subjects. Non-optimal solutions can have more support points than the number of subjects. In this step, we use QR decomposition of the $\Psi = P(Y_i|\theta_k)_{N \times K}$ matrix and remove all the support points that are not in the orthonormal basis for the column space of Ψ matrix. We do this at each cycle to guarantee optimizations never expand uncontrollably.

Support Point Adjustment
It is at this step that the NPOD and NPAG algorithms diverge; while NPAG employs an adaptive grid to suggest new support points in the search space, NPOD employs a directional derivative of the log-likelihood function using Nelder-Mead algorithm [10].

The directional derivatives of the log-likelihood of F in the direction of the atomic density function centered at each support point are denoted as $D_{\delta_\xi}\ell(F)$. The idea originates in a text by Fedorov [4], which covers D-optimal design theory. Another connection to Fedorov's D-optimal design theory and maximum likelihood estimators is provided by Mallet [16]. That paper provides an alternative to Lindsay's approach. In fact, according to Schumitzky [18], Lindsay and Mallet worked jointly to develop the theory that reduced the space of distributions to the space of only discrete distributions with K support points, denoted $\mathcal{F}_K$ (where K is no more than the number of subjects N).

Let F be any distribution on Ω, the space of parameters for ξ. Then define the directional derivative D-Function as

$$D(\xi, F) = \left(\sum_{i=1}^{N} \frac{P(Y_i \mid \xi)}{P(Y_i \mid F)} \right) - N \tag{2}$$

where ξ is a parameter and N is the population size. Lindsay [14] showed that $F^* = F^{ML}$ if and only if

$$\max_{\xi \in \Omega} D(\xi, F^*) = 0. \tag{3}$$

Additionally, in the same paper, Lindsay showed when

$$\max_{\xi \in \Omega} D(\xi, F^*) \neq 0, \tag{4}$$

it is still true that

$$L(F^{ML}) - L(F^*) \leq \max_{\xi \in \Omega} D(\xi, F^*) \tag{5}$$

for $F^*, F^{ML} \in \mathcal{F}_K$.

In NPOD, the updated set of support points is found as follows: for $k = 1, \ldots, K$ where K is the current grid size and $F^{(n)}$ is the current distribution:

$$\theta_k^{(n+1)} = argmax^t_{\xi \in \Omega}(D(\xi, F^{(n)})), \tag{6}$$

$$D(\xi, F^{(n)}) = \left(\sum_{i=1}^{N} \frac{P(Y_i \mid \xi)}{P(Y_i \mid F^{(n)})} \right) - N, \tag{7}$$

$$P(Y_i \mid F^{(n)}) = \sum_{l=1}^{K} (w_l^{(n)} P(Y_i \mid \theta_l^{(n)})) \tag{8}$$

where $argmax^t$ only takes t steps in the Nelder-Mead optimization process [10]. This adjustment plays a pivotal role in enhancing the efficiency of NPOD compared to NPAG, particularly in achieving convergence to local maxima.

The parameter t is regarded as one of the hyperparameters that can be fine-tuned to optimize the performance of the algorithm. Typically, we set t to be less than 5, based on empirical observations and computational experiments. This choice balances the trade-off between computational cost and optimization effectiveness.

By limiting the number of steps in the Nelder-Mead optimization, we can focus the algorithm's search on promising regions of the parameter space while avoiding excessive computational overhead. This targeted approach enables NPOD to converge more efficiently toward local maxima, making it a valuable tool for solving optimization problems in diverse domains.

Once $\theta_k^{(n+1)}$ is determined through the optimization process, a validation step ensues to ensure its integrity within the algorithm. Specifically, it undergoes scrutiny to confirm two critical aspects: firstly, that it constitutes a distinct point and, secondly, that it remains within the predefined boundary conditions.

This validation mechanism serves as a safeguard against redundancy and boundary violations, both of which could potentially compromise the accuracy and reliability of the optimization process. By confirming the uniqueness and adherence to boundary constraints of $\theta_k^{(n+1)}$, the algorithm maintains the integrity of its parameter space exploration, facilitating robust and effective optimization outcomes.

Convergence
The previous steps, excluding initialization, are iteratively repeated until no further improvement can be found, indicating convergence to an optimal solution. Improvement is evaluated by change in the likelihood, for which we consider a change less than 10^{-4} to indicate convergence.

2.1.3 Computational Considerations and Optimizations

Initial Search Space
NPOD is initialized with a sufficiently compact set of support points within the search space. We report results for varying density Sobol sequence initializations in the Results section.

Hyperparameters
The NPOD algorithm, relying on the D-optimization function, is tuned by the number of iterations of the Nelder-Mead algorithm t. In our examples, a value of 5 was used for t, empirically chosen based on experience.

2.2 Software Implementation

Recently, significant efforts have been placed in creating a new framework for pharmacometric algorithm development. While the original NPAG algorithm was written in Fortran, both the NPAG and NPOD algorithm have been rewritten in Rust, a memory-safe and computationally efficient programming language. While the framework itself will be presented in a future work, both algorithms are available to use in the development branch of the Pmetrics code repository [11]. All computations were performed on a MacBook Pro (Apple) equipped with an M3 Max processor with 128 GB of RAM.

2.3 Comparative Analysis of NPOD with NPAG

The natural choice of a comparative algorithm for non-parametric pharmacokinetic modeling is NPAG. To compare the algorithms, we use two datasets, one synthetic in which the real parameter distribution is known and another using real pharmacokinetic data from subjects in which the underlying distribution is not known. We will refer to these as datasets A and B, respectively.

The model used to fit dataset A is shown in Eq. (9), where A is the amount of drug in the central compartment, K_e is the elimination rate constant with a bimodal distribution, and V_d is the apparent volume of distribution with unimodal distribution. The model includes an intravenous infusion, modeled as R_{inf}, the rate of infusion.

$$\frac{dA}{dt} = -K_e \cdot A + R_{inf}, \quad C = \frac{A}{V_d} + \epsilon \tag{9}$$

Dataset A consisted of simulated data with known parameter distribution, and with known measurement noise in the observations $\epsilon \sim \mathcal{N}(0, 0.05 * C)$, previously discussed by Neely et al. [17]. It includes a total of 51 simulated subjects, all of whom received an intravenous infusion of 500 units over a duration of 30 minutes. Each subject was sampled 10 times over 24 hours, at 0.5, 1, 2, 3, 4, 6, 8, 12, 18, and 24 hours from the start of the infusion.

The model used to fit dataset B is shown in Eq. (10), where A_1 represents the absorptive compartment and A_2 represents the amount of drug in the central compartment, from which K_e is the elimination rate constant and V_d is the apparent volume of distribution. The model includes an individual lag-term on the input dose D, modeled as a delayed unit Dirac delta function δ.

$$\frac{dA_1}{dt} = -K_a \cdot A_1 + D * \delta(t - t_{lag}), \quad \frac{dA_2}{dt} = K_a \cdot A_1 - K_e \cdot A_2,$$
$$C = \frac{A_2}{V_d} + \epsilon \tag{10}$$

Dataset B was originally provided as one of the example datasets available in the Pmetrics package for R [17]. It includes data from 20 patients, all of whom received 600 units six times every 24 hours. A total of 139 samples were obtained across all subjects, all following the second-to-last dose.

Any observation has an associated uncertainty, which must be accounted for during parameter estimation. We model uncertainty as ϵ, which is normally distributed with mean zero and standard deviation ω defined by Eq. (12) or (13). First, an error polynomial model is used to estimate the uncertainty (σ) in each measurement (y). This is given in Eq. (11).

$$\sigma = C_0 + C_1 \cdot y + C_2 \cdot y^2 + C_3 \cdot y^3 \tag{11}$$

Additional noise is modelled through either an additive (λ, Eq. (12)), or proportional (γ, Eq. (13)) error model. Each observation is then weighted by the reciprocal of the squared uncertainty, i.e., $1/\omega^2$.

$$\omega = \sqrt{\sigma^2 + \lambda^2} \tag{12}$$

$$\omega = \sigma \cdot \gamma \tag{13}$$

For the simulated dataset A, an additive error model was used with an initial value of $\lambda = 0$ and a flat uncertainty of 5%, i.e., $C_1 = 0.05$, and $C_0 = C_2 = C_3 = 0$. For the real-world dataset B, a proportional error model was used with an initial value of $\gamma = 5$, and $C_1 = 0.02$, $C_2 = 0.05$, $C_3 = -0.002$, and $C_4 = 0$.

Both algorithms were compared on each dataset with various densities of the initial parameter search space, with otherwise equal conditions. Multiples of 51 were used for the number of initial support points, i.e., $K = 51 \cdot 2^x$, where x ranged from 0 to 11, producing initial densities ranging from 51 to 104,448.

3 Results

For dataset A, the convergence rates and the location of support points at convergence for NPAG and NPOD, with an initial count of 104,448 support points, are illustrated in Figs. 1 and 2. The weighted means for K_e (NPAG = 0.187, NPOD = 0.187) and V_d (NPAG = 103.7, NPOD = 103.8) between the two algorithms were almost identical. Furthermore, for all the different initial grid densities evaluated, ranging from 51 to 104,448, NPOD was able to achieve convergence at a much faster rate compared to NPAG, requiring almost one twentieth the number of cycles for a high number of initial points (Table 1). However, the difference in overall computation time is negligible.

Furthermore, the shape of the objective function across cycles is markedly different between NPOD and NPAG, shown in Fig. 1. It is immediately apparent that NPOD has a much steeper convergence.

For dataset B, which, to reiterate, consists of real-world data, NPOD was able to obtain a solution as likely or more compared with NPAG with a lower number of cycles. However, for this dataset, the overall computation time was lower, with up to fivefold difference, as seen in Table 2.

The number of cycles required for convergence is again visualized for both algorithms in Fig. 3.

Table 1 For dataset A, the comparison of number of cycles required for convergence, the value of the objective function obtained, and time taken for various sizes of the initial parameter search space. Abbreviations: LL, the twice negative logarithm of the likelihood, also known as the objective function value

	NPAG				NPOD			
№	Cycles	LL	Support points	Time	Cycles	LL	Support points	Time
51	196	−646.45	48	3.87 s	21	−646.85	47	1.56 s
102	164	−646.38	49	3.40 s	15	−646.85	48	1.47 s
204	132	−646.52	48	3.00 s	13	−646.80	47	1.47 s
408	136	−646.39	48	3.27 s	11	−646.78	48	1.43 s
816	139	−646.43	46	3.31 s	11	−646.81	48	1.56 s
1632	112	−646.45	48	3.45 s	11	−646.83	49	1.75 s
3264	99	−646.59	48	3.83 s	7	−646.77	49	2.19 s
6528	97	−646.48	50	5.50 s	6	−646.77	49	3.33 s
13,056	93	−646.37	48	7.99 s	7	−646.84	49	5.99 s
26,112	85	−646.52	48	13.97 s	6	−646.84	48	10.80 s
52,224	98	−646.49	50	25.39 s	5	−646.83	49	25.01 s
104,448	99	−646.57	48	51.78 s	6	−646.83	48	50.02 s

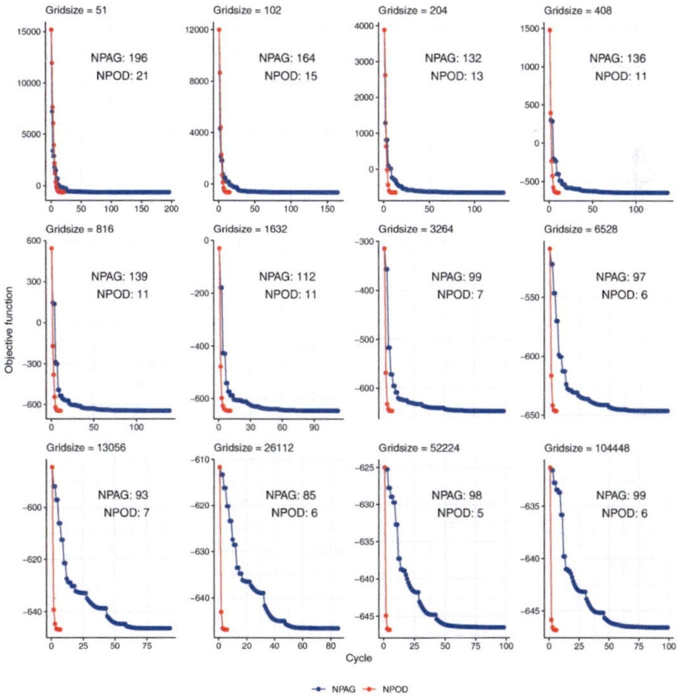

Fig. 1 Comparison of objective function between NPAG and NPOD for different numbers of initial grid points in the parameter search space. Grid sizes are chosen as multiples of the number of subjects ($n = 51$)

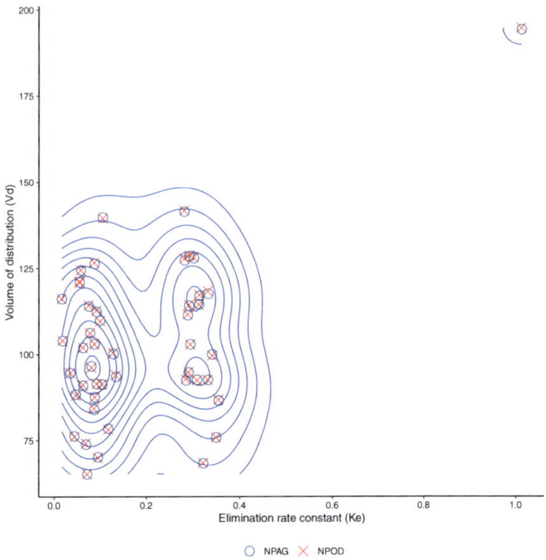

Fig. 2 Kernel density estimate for the joint parameter distribution. The support points estimated by NPAG are shown as blue circles, and those by NPOD is shown as red crosses. The bimodal distribution of K_e is readily apparent, with the univariate distribution of V_d and the inclusion of an extreme outlier

Table 2 For dataset B, the comparison of number of cycles required for convergence, the value of the objective function obtained, and time taken for various sizes of the initial parameter search space. Abbreviations: LL, the twice negative logarithm of the likelihood, also known as the objective function value

	NPAG				NPOD			
№	Cycles	LL	Support points	Time	Cycles	Support LL	points	Time
51	1091	−337.93	19	68.43 s	189	−331.98	19	47.74 s
102	2166	−337.97	20	95.11 s	248	−336.91	17	36.87 s
204	1234	−336.56	20	75.95 s	119	−342.64	17	32.75 s
408	1313	−336.54	19	71.25 s	68	−346.98	18	21.31 s
816	2218	−343.91	20	116.91 s	74	−345.31	17	21.72 s
1632	3034	−343.91	20	110.95 s	69	−345.38	17	22.62 s
3264	1415	−343.92	20	65.83 s	192	−335.75	18	36.29 s
6528	1913	−343.84	20	73.75 s	75	−336.66	18	25.46 s
13,056	1401	−337.91	20	77.60 s	85	−336.09	17	36.40 s
26,112	2169	−336.54	20	122.14 s	82	−334.35	17	56.66 s
52,224	1209	−336.53	20	135.22 s	65	−335.45	18	89.48 s
104,448	2014	−336.53	20	202.92 s	75	−335.38	18	147.74 s

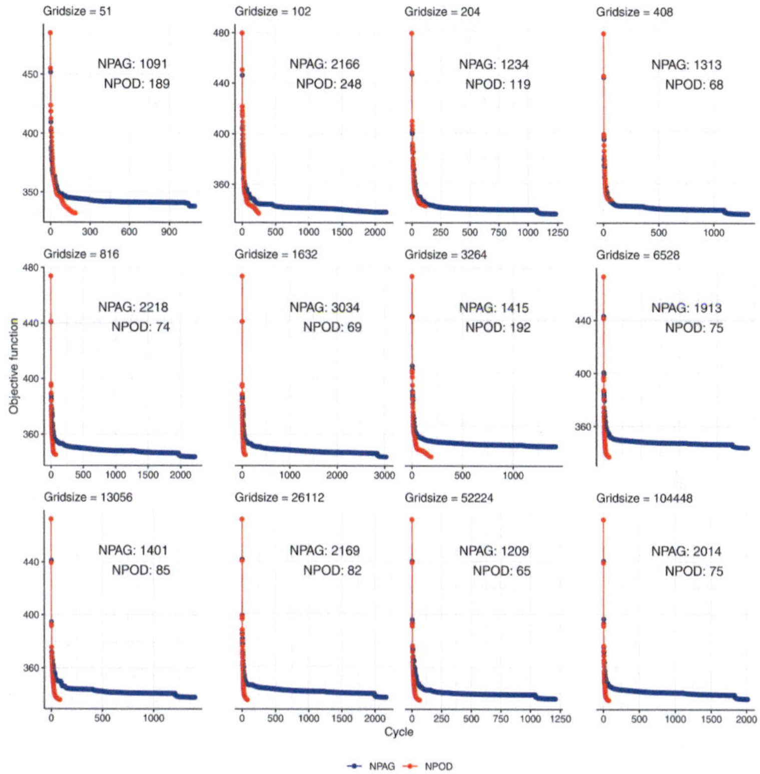

Fig. 3 For dataset B, the comparison of objective function between NPAG and NPOD for different number of initial grid points in the parameter search space. For simplicity, the same grid sizes in the first example was used

4 Discussion

We have developed and demonstrated an algorithm for non-parametric parameter estimation with application to population pharmacokinetics. The algorithm relies on directional derivatives, which constitutes a new approach to parameter estimation in pharmacometrics.

Furthermore, we compared the NPOD algorithm to the current gold-standard non-parametric algorithm, NPAG, on two datasets; one simulated without any noise in the observations and another using real-world data. The two algorithms have some important differences, which is elucidated by the results in Table 1. Most importantly, NPOD was able to determine a solution that was as likely as that of NPAG. The time savings with NPOD is due to the markedly reduced number of cycles, with a difference of up to 20-fold for a simple model on a simulated dataset (dataset A) and more than twice that for a more complex model with real-world data (dataset B). However, the optimization step of NPOD, which is guided by the

D-function value, is more expensive than the adaptive grid in NPAG and as such requires more time in each cycle. Because we expect NPOD to always converge in fewer cycles than NPAG, we also expect that time to convergence will be at worst similar and at best shorter than NPAG, although this remains to be empirically demonstrated as we gain experience with NPOD.

The estimated joint parameter distribution from both NPOD and NPAG was very similar, as seen in Fig. 2, where the support points are close to perfectly overlapping. The simulated dataset included an extreme outlier, whose parameter values deviated greatly from the population weighted mean (K_e, 1.0 vs 0.187; V_d 200.0 vs 103.8). This is especially impressive considering that the single outlier constitutes only 1/51, or approximately 2% of the dataset. This quality is one of the many strengths of NPAG, which is also found for NPOD. The detection of outliers is an important aspect in population pharmacokinetics and one of the chief advantages of non-parametric approaches, compared to parametric [6].

We evaluated the performance of the two algorithms over various densities of the initial search space. While this density does not appear to significantly affect the final objective function value for this simulated dataset, it does affect the number of cycle required to reach convergence. The "throw-and-catch" nature of NPAG ignores the gradient around the current local solution, which NPOD is sensitive to. Importantly, this gradient includes observation noise. NPAG is relatively insensitive to local gradient perturbations resulting from observation errors as it merely compares two potential and spatially distinct solutions at each cycle. However, NPAG is *completely naive* of the intervening space. In either case, both algorithms were capable of obtaining the most likely solution even from a very sparse initial parameter space, equal to the number of subjects.

For Dataset A, NPOD results in a lower log-likelihood (LL) value, indicating potentially more accurate results. A natural questions arises: "Why use NPAG at all?" Some distinctions in how the two algorithms operate might explain their differences. NPAG cycles are generally faster because they add points in a more straightforward manner, without necessarily checking if those points have been added before or if they are crucial to improving the solution. This can lead to quicker iterations but may include less relevant points. In contrast, NPOD evaluates the likelihood surface more thoroughly, proposing new points that are specifically aimed at maximizing the objective function or minimizing the negative likelihood. As a result, every point added by NPOD is highly relevant to refining the outcome. We are planning to investigate the conditions under which one algorithm would consistently be preferable over the other in a future work.

The following procedure is proposed in Yamada et al. [22] for evaluating the global optimality of the final NPAG distribution and estimating its proximity to the optimum using the directional derivative $D(\Theta, F)$ defined above in (2) solely during the last NPAG iteration. As shown in Lindsay [14] and mentioned above in Methods section, if $F^* = F^{ML}$, i.e., NPAG converged to a global maximum of a likelihood function, then $\max_{\xi \in \Omega} D(\xi, F^*) = 0$. We propose the same evaluation steps for the final NPOD distribution and recommend calculating $\max_{\xi \in \Omega} D(\xi, F)$ only at the end of the algorithm using deterministic or stochastic optimization methods.

Future Work

In our experience thus far, NPOD converges at a faster rate when compared with NPAG for a larger population with a given model and prior. In the era of increasingly larger datasets, NPOD may therefore prove advantageous. However, it is not a given that NPOD will always outperform NPAG; therefore, future work will include additional comparisons between NPAG and NPOD to contrast and clarify specific scenarios when one algorithm should be preferred.

We have also observed that the performance of the NPOD algorithm depends on the initial grid or prior. The closer it is to the true solution, the faster the convergence of NPOD. We are planning to explore the ways to improve initial grid point in the future.

At present, formal analyses of convergence guarantees and algorithmic complexity for both NPAG and NPOD are limited. NPOD, being a gradient-based approach, benefits from some theoretical convergence guarantees, particularly when the likelihood surface is smooth and well behaved. Under these conditions, it can converge more efficiently to the maximum likelihood solution when compared with the adaptive grid in NPAG. However, the complexity of the models and datasets, such as in the case of noisy or high-dimensional data, introduces challenges in predicting convergence behavior. For NPAG, the lack of formal convergence guarantees is more pronounced, as its heuristic nature can result in variability in performance depending on the problem being solved. This makes it difficult to derive generalizable results regarding its runtime or convergence. While NPOD has shown promising improvements in speed and accuracy, a deeper theoretical analysis, especially with respect to different types of datasets and models, is something we aim to explore in future studies.

5 Conclusion

We have developed and demonstrated a new algorithm, NPOD, for non-parametric parameter estimation with application to population pharmacokinetics. The algorithm was able to estimate the population joint parameter distribution as accurate as NPAG but requires far fewer cycles to reach convergence. An application of directional derivates represents an important step forward in both the development and application of non-parametric approaches in pharmacometrics.

6 Competing Interests

Michael N. Neely, Julian Otalvaro and Walter M. Yamada were partially supported by NIH-NIAID R01AI173238. The authors have no conflicts of interest to declare that are relevant to the content of this chapter.

Appendix

Algorithm 1 Non-parametric Optimal Design (NPOD) algorithm. Input: $(\boldsymbol{Y}, \boldsymbol{\phi}^0, \boldsymbol{a}, \boldsymbol{b}, \boldsymbol{t}, \Delta_D, \Delta_F, \Delta_e, \Delta_\lambda)$, $\boldsymbol{a}$, and $\boldsymbol{b}$ are the lists of lower and upper bounds, respectively, of Θ, $\boldsymbol{t}$ is the number of Nelder-Mead iterations; Δ_D is the minimum distance allowable between points in the estimated F^{ML}. Output: $(\boldsymbol{\phi}, \boldsymbol{\lambda}, l(\boldsymbol{\lambda}, \boldsymbol{\phi}))$

procedure NPOD($\boldsymbol{Y}, \boldsymbol{\phi}^0, \boldsymbol{a}, \boldsymbol{b}, \Delta_D$) ▷ Estimate F^{ML} given $\boldsymbol{Y}$
  Initialization: $\boldsymbol{\phi} = \boldsymbol{\phi}^0$, $LogLike = -10^{30}$, $\Delta_F = 10^{-2}$, $\Delta_L = 10^{-4}$, $\Delta_\lambda = 10^{-3}$, $n = 0$
  while True **do**
    Calculate $\boldsymbol{\Psi}(\boldsymbol{\phi})$ ▷ $N \times K$ matrix $\{p(Y_i|\phi_k)\}$
    $[\hat{\boldsymbol{\lambda}}(\boldsymbol{\phi}), l(\hat{\boldsymbol{\lambda}}(\boldsymbol{\phi})] \longleftarrow$ PDIP($\boldsymbol{\Psi}(\boldsymbol{\phi})$) ▷ for PDIP see [22]
    $\boldsymbol{\phi} \longleftarrow$ CONDENSE($\boldsymbol{\phi}, \hat{\boldsymbol{\lambda}}(\boldsymbol{\phi}), \Delta_\lambda$) ▷ Alg. 2
    $[\boldsymbol{\phi}, \boldsymbol{\Psi}(\boldsymbol{\phi})] \longleftarrow$ REDUCE($\boldsymbol{\Psi}(\boldsymbol{\phi}), \boldsymbol{\phi}$) ▷ Alg. 3
    $[\hat{\boldsymbol{\lambda}}(\boldsymbol{\phi}), l(\hat{\boldsymbol{\lambda}}(\boldsymbol{\phi})] \longleftarrow$ PDIP($\boldsymbol{\Psi}(\boldsymbol{\phi})$) ▷ PDIP returns G^n ([22])
    $NewLogLike = l(\hat{\boldsymbol{\lambda}}(\boldsymbol{\phi}), \boldsymbol{\phi})$
    if $\mid LogLike - NewLogLike \mid > \Delta_F$ **then**
      return $[\boldsymbol{\phi}, \boldsymbol{\lambda}, NewLogLike]$
    end if
    if $(n >$ MAXCYCLES) **then**
      return $[\boldsymbol{\phi}, \boldsymbol{\lambda}, NewLogLike]$
    end if
    $\boldsymbol{\phi} \longleftarrow$ Dopt($\boldsymbol{\phi}, \boldsymbol{\lambda}, \boldsymbol{\Psi}(\boldsymbol{\phi}), \boldsymbol{a}, \boldsymbol{b}, \boldsymbol{t}, \boldsymbol{\Delta}_D$) ▷ Alg. 4
    $n \longleftarrow n + 1$
    $LogLike \leftarrow NewLogLike$
  end while
end procedure

Algorithm 2 Condense algorithm. Input: $(\boldsymbol{\phi}, \boldsymbol{\lambda}, \Delta_\lambda)$, Output: $\boldsymbol{\phi}^c$ Note: $\boldsymbol{\phi}^c$ is considered a subset of $\boldsymbol{\phi}$

function CONDENSE($\boldsymbol{\phi}, \boldsymbol{\lambda}, \Delta_\lambda$)
  ind = **find** ($\boldsymbol{\lambda} > (\max \boldsymbol{\lambda})\Delta_\lambda$) ▷ Inequality and max are performed component-wise
  $\boldsymbol{\phi}^c = \boldsymbol{\phi}(:, \textbf{ind})$
  return $\boldsymbol{\phi}^c$
end function

Algorithm 3 Reduce algorithm. Input: $(\boldsymbol{\Psi}(\boldsymbol{\phi}), \boldsymbol{\phi})$, Output: $\boldsymbol{\phi}, \boldsymbol{\Psi}(\boldsymbol{\phi})$ Note: both $\boldsymbol{\Psi}(\boldsymbol{\phi})$ and $\boldsymbol{\phi}$ are subsets of the ones used as input

function REDUCE($(\boldsymbol{\Psi}(\boldsymbol{\phi}), \boldsymbol{\phi})$)
  $n\psi(\phi) = norm(\Psi(\phi))$
  $(r, perm) = QR(n\psi(\phi))$
  $keep = []$
  for $i..ncol$
    $ratio = r[i,i]/norm(r[:,i])$
    if $|ratio| > 1e-8$ **push**($perm[i]$) to **keep**
  end for
  $\phi = \phi[keep, :]$
  $\psi = \psi[:, keep]$
  return (ψ, ϕ)
end function

Algorithm 4 Dopt algorithm Input: $(\boldsymbol{\phi}, \boldsymbol{\lambda}, \boldsymbol{\Psi}(\boldsymbol{\phi}), \boldsymbol{a}, \boldsymbol{b}, \boldsymbol{t}, \boldsymbol{\Delta}_D)$, Output: $\boldsymbol{\phi}$

function DOPT($\boldsymbol{\phi}, \boldsymbol{\lambda}, \boldsymbol{\Psi}(\boldsymbol{\phi}), \boldsymbol{a}, \boldsymbol{b}, \boldsymbol{t}$)
  for $k = 1 : K$ **do**
    $\boldsymbol{\phi}_k = argmax^t_{\boldsymbol{\xi} \in \boldsymbol{\Omega}}(D(\boldsymbol{\xi}, \boldsymbol{\lambda}, \boldsymbol{\Psi}))$ ▷ see formula for D in Eq. (6)
    for $ink = 1$: dimension($\boldsymbol{\phi}_k$) **do**
      $\text{new\_dist} = \sum \frac{|\boldsymbol{\phi}_k - \boldsymbol{\phi}(:,ink)|}{b-a}$
      $\text{dist} = \min(\text{dist}, \text{new\_dist})$
    end for
    $\text{up} = \text{sign}(\min(\boldsymbol{\phi}_k - a'))$
    $\text{down} = \text{sign}(\min(b' - \boldsymbol{\phi}_k))$
    if $(\text{dist} > \boldsymbol{\Delta}_D) \wedge (\text{up} > -1) \wedge (\text{down} > -1)$ **then**
      $\boldsymbol{\phi} = [\boldsymbol{\phi}, \boldsymbol{\phi}_k]$
    end if
  end for
  return $\boldsymbol{\phi}$
end function

References

1. Allard, Q., Djerada, Z., Pouplard, C., Repessé, Y., Desprez, D., Galinat, H., Frotscher, B., Berger, C., Harroche, A., Ryman, A., Flaujac, C., Chamouni, P., Guillet, B., Volot, F., Szymezak, J., Nguyen, P., Cazaubon, Y.: Real life population pharmacokinetics modelling of eight factors viii in patients with severe haemophilia a: Is it always relevant to switch to an extended half-life? Pharmaceutics **12**(4), 380 (2020). https://doi.org/10.3390/pharmaceutics12040380
2. Burley, B.: Practical hash-based owen scrambling. J. Comput. Graph. Techniq. **9**(4), 29 (2020)
3. D'Argenio, D.Z., Schumitzky, A., Wang, X.: Population pharmacokinetic mixture models via maximum a posteriori estimation. Comput. Stat. Data Analy. **53**, 3907–3915 (2009)
4. Fedorov, V.: Theory of Optimal Experiments. Academic, New York (1972)
5. Goutelle, S., Woillard, J., Neely, M., Yamada, W., Bourguignon, L.: Nonparametric methods in population pharmacokinetics. J. Clin. Pharmacol. **62**(2), 142–157 (2020–10). https://doi.org/10.1002/jcph.1650. https://www.ncbi.nlm.nih.gov/pubmed/33103785
6. Goutelle, S., Woillard, J., Buclin, T., Bourguignon, L., Yamada, W., Csajka, C., Neely, M., Guidi, M.: Parametric and nonparametric methods in population pharmacokinetics: experts'

discussion on use, strengths, and limitations. J. Clin. Pharmacol. **62**(2), 158–170 (2021–12). https://doi.org/10.1002/jcph.1993. https://www.ncbi.nlm.nih.gov/pubmed/34713491
7. Ishihara, N., Nishimura, N., Ikawa, K., Karino, F., Miura, K., Tamaki, H., Yano, T., Isobe, T., Morikawa, N., Naora, K.: Population pharmacokinetic modeling and pharmacodynamic target attainment simulation of piperacillintazobactam for dosing optimization in late elderly patients with pneumoni. Antibiotics **9**(3), 113 (2020). https://doi.org/10.3390/antibiotics9030113
8. Joe, S., Kuo, F.Y.: Remark on algorithm 659: Implementing Sobol's quasirandom sequence generator. ACM Trans. Math. Softw. **29**(1), 49–57 (2003). https://doi.org/10.1145/641876.641879. https://dl.acm.org/doi/10.1145/641876.641879
9. Joe, S., Kuo, F.Y.: Constructing Sobol sequences with better two-dimensional projections. SIAM J. Sci. Comput. **30**(5), 2635–2654 (2008). https://doi.org/10.1137/070709359. http://epubs.siam.org/doi/10.1137/070709359
10. Kolda, T.G., Lewis, R.M., Torczon, V.: Optimization by direct search: new perspectives on some classical and modern methods. SIAM Rev. **45**(3), 385–482 (2003). http://www.jstor.org/stable/25054427
11. Laboratory of Applied Pharmacokinetics and Bioinformatics: Pmetrics. https://github.com/LAPKB/Pmetrics/tree/dev (2024)
12. Leary, R.H.: Improved computational methods for statistically consistent and efficient pk/pd population analysis. In: PAGE 12. Verona, Italy (2003). www.page-meeting.org/?abstract=421
13. Lesperance, M.L.., Kalbfleisch, J.D.: An algorithm for computing the nonparametric MLE of a mixing distribution. J. Am. Stat. Assoc. **87**, 120–126 (1992)
14. Lindsay, B.: The geometry of mixture likelihoods: a general theory. Ann. Stat. **11**(1), 86–94 (1983)
15. Lindsay, B.G.: The geometry of mixture likelihoods: a general theory. Ann. Statist. **11**, 86–94 (1983)
16. Mallet, A.: A maximum likelihood estimation method for random coefficient regression models. Biometrika **73**(3), 645–656 (1986)
17. Neely, M.N., van Guilder, M.G., Yamada, W.M., Schumitzky, A., Jelliffe, R.W.: Accurate detection of outliers and subpopulations with pmetrics, a nonparametric and parametric pharmacometric modeling and simulation package for R. Ther. Drug Monit. **34**(4), 467–76 (2012). https://doi.org/10.1097/FTD.0b013e31825c4ba6. https://www.ncbi.nlm.nih.gov/pubmed/22722776
18. Schumitzky, A.: Nonparametric EM algorithms for estimating prior distributions. Appl. Math. Comput. **45**, 143–157 (1991)
19. Shah, N.R., Bulitta, J.B., Kinzig, M., Landersdorfer, C.B., Jiao, Y., Sutaria, D.S., Tao, X., Höhl, R., Holzgrabe, U., Kees, F., et al.: Novel population pharmacokinetic approach to explain the differences between cystic fibrosis patients and healthy volunteers via protein binding. Pharmaceutics **11**(6), 286 (2019)
20. Soraluce, A., Barrasa, H., Asín-Prieto, E., Sánchez-Izquierdo, J.Á., Maynar, J., Isla, A., Rodríguez-Gascón, A.: Novel population pharmacokinetic model for linezolid in critically ill patients and evaluation of the adequacy of the current dosing recommendation. Pharmaceutics **12**(1), 54 (2020). https://doi.org/10.3390/pharmaceutics12010054
21. Wang, X., Wang, Y.: Nonparametric multivariate density estimation using mixtures. Stat. Comput. **25**(1), 33–43 (2015)
22. Yamada, W.M., Neely, M.N., Bartroff, J., Bayard, D.S., Burke, J.V., van Guilder, M., Jelliffe, R.W., Kryshchenko, A., Leary, R., Tatarinova, T., Schumitzky, A.: An algorithm for nonparametric estimation of a multivariate mixing distribution with applications to population pharmacokinetics. Pharmaceutics **13**(1), 42 (2021). https://www.mdpi.com/1999-4923/13/1/42

Unrolling Deep Learning End-to-End Method for Phase Retrieval

Haiyan Cheng, Cristina Garcia-Cardona, Weihong Guo, Sara Hahner, Yuan Liu, Yifei Lou, Michela Marini, and Sui Tang

1 Introduction

In many imaging systems, such as X-ray diffraction, electron microscopy, or astronomical imaging, only the intensity of the wave can be directly measured, while the phase information is either lost or inaccessible. Phase retrieval (PR) is a computational technique employed to recover the phase information of a wave solely from intensity measurements. This process involves utilizing computational algorithms to recover the phase information from the recorded magnitude measurements. Phase retrieval finds applications in various fields, including physics, biology, materials science, and imaging technologies. For instance, PR is pivotal in coherent diffraction imaging (CDI) [41], image-based wavefront sensing [45], and radar/sonar sensing

H. Cheng
School of Computing and Information Sciences, Willamette University, Salem, OR, USA
e-mail: hcheng@willamette.edu

C. Garcia-Cardona (✉)
Los Alamos National Laboratory, Los Alamos, NM, USA
e-mail: cgarciac@lanl.gov

W. Guo
Department of Mathematics, Applied Mathematics and Statistics, Case Western Reserve University, Cleveland, OH, USA
e-mail: wxg49@case.edu

S. Hahner
Fraunhofer Institute for Scientific Computing and Algorithms (SCAI), Sankt Augustin, Germany

Y. Liu
Department of Mathematics, Statistics and Physics, Wichita State University, Kansas, KS, USA
e-mail: yuan.liu@wichita.edu

C. Garcia-Cardona, H. Lee (eds.), *Advances in Data Science*, Association for Women in Mathematics Series 37, https://doi.org/10.1007/978-3-031-87804-6_13

sectors [26]. It enables researchers to extract valuable information about structures and properties of samples without requiring specialized phase-sensitive detectors.

In this work, we consider a two-dimensional (2D) object that can be represented by a (column) vector $\mathbf{x} \in \mathbb{R}^{mn}$, where $m \times n$ is the dimension of the underlying 2D object. A physical observation model, denoted by $\mathcal{A}$, yields an idealized complex-valued observation $\mathcal{A}(\mathbf{x})$. We further assume the operator $\mathcal{A}$ is linear or can be approximated as linear. However, the detectors are limited to recording the magnitude, i.e., $\mathbf{y} = |\mathcal{A}(\mathbf{x})| + \boldsymbol{\epsilon}$, where $|\cdot|$ represents the element-wise magnitude, indicative of the photon flux measured by the detectors, and $\boldsymbol{\epsilon}$ signifies the observational noise matrix. PR aims at reconstructing $\mathbf{x}$ from the observed data $\mathbf{y}$, which is a highly ill-posed nonlinear inverse problem. Please refer to Sect. 3.1 for a more detailed description of the problem setting.

A critical regime within PR is the Fourier phase retrieval [6], where the linear operator $\mathcal{A}$ is related to the Fourier transform. This problem is fundamental in several fields, including X-ray crystallography [43], astronomy, coherent light microscopy, quantum state tomography, and remote sensing; please refer to [6] for more details.

Difficulties of PR The absence of phase information leads to non-unique solutions of PR. Identifiability, often hindered by intrinsic symmetries such as spatial translation, conjugate inversion, and constant global phase change (referred to as *trivial ambiguities*), exacerbates this challenge. These symmetries, long recognized in the PR literature, limit claims of uniqueness to modulo these trivial ambiguities. Consequently, in most applications, recovering any signal from the equivalent classes is deemed satisfactory. Empirically, it is observed that the complexity of a PR problem correlates directly with the number of its intrinsic symmetries. This relationship is particularly noticeable in Fourier PR, where these symmetries are more prevalent, thus amplifying the challenge.

On the other hand, the oversampling ratio serves as another indicator for determining the complexity of PR problems. This ratio is defined as

$$\tau = \frac{\text{number of (effective) measurements}}{\text{number of unknown pixels}}.$$

It contrasts the volume of measurements with the dimensionality of the signal to be recovered. A higher oversampling ratio implies a computationally less complex problem, indicating a data-rich environment relative to the signal parameters

Y. Lou
Department of Mathematics & School of Data Science and Society, University of North Carolina at Chapel Hill, Chapel Hill, NC, USA

M. Marini
Department of Mathematics, University of Houston, Houston, TX, USA

S. Tang
Department of Mathematics, University of California Santa Barbara, Santa Barbara, CA, USA

needing estimation. Consequently, τ is a vital metric for evaluating and managing the complexity of PR problems. The oversampling ratio $\tau = 4$ typically leads to a unique solution [3, 7]. Under this condition, the standard phasing algorithms empirically worked well. For Fourier phase retrieval, using dimension counting, [42] conjectured that $\tau = 2$ uniquely determines a unique phasing solution up to spatial shift, conjugate inversion, and global phase factor. A numerical verification of this conjecture is provided by the random phase illumination method [18].

1.1 Literature Review

Numerous algorithms have been developed to tackle various challenges within the PR paradigms. Regularization, by favoring certain solutions, enhances the robustness of the recovery process against ambiguities and perturbations. Most regularization approaches can be formulated as the following optimization problem:

$$\min_{\mathbf{x}\in\mathbb{C}^{mn}} \{L(\mathbf{x}, \mathbf{y}) + R(\mathbf{x})\}, \tag{1}$$

where the first term, $L(\mathbf{x}, \mathbf{y})$, enforces consistency with the intensity measurements $\mathbf{y}$ (the data-fidelity term). The second term, $R(\mathbf{x})$, penalizes unrealistic estimates to promote desirable properties in $\mathbf{x}$, which can be formulated either implicitly or explicitly. The relevant works can be broadly categorized as follows:

Projection Algorithms Early phase retrieval methods were pioneered by Gerchberg and Saxton (GS) [22], who developed an alternating projection approach to solve a nonlinear least square problem. This technique, which begins with random initialization, applies alternating time domain and Fourier magnitude constraints, often converging to a local minimum due to the interplay between convex and non-convex constraint sets. This tendency hampers the accuracy of the solution, even in the noiseless setting. Later, Fienup [20] introduced the Hybrid Input-Output (HIO) algorithm, which incorporates a time-domain correction step into the Gerchberg-Saxton (GS) algorithm to accelerate the convergence. However, HIO does not guarantee overall convergence and may occasionally result in local minima. Nevertheless, HIO and its variants remain widely used in optical phase retrieval, as explored in [4] and [37].

Gradient-Based Optimization To address the issue of getting stuck in local minima, gradient-based methods often integrate acceleration strategies, such as Nesterov acceleration and stochastic gradient descent. These algorithms have proven to be effective in solving phase retrieval problems [10, 30] with a variety of applications including ptychography, coded-diffraction imaging, and imaging from defocus [1, 5]. Key strategies for navigating non-convex landscapes include convex relaxation methods such as PhaseLift [11], maxcut [50], sketching methods [53], and phasemax [24]. A notable work in this category is the Wirtinger Flow

(WF) [10], which begins with an initial guess obtained from a spectral method, followed by gradient descent. It is proven to achieve the exact phase retrieval (up to trivial ambiguities) with independent Gaussian measurements, but it is hampered by high computational complexity. To speed up the process, truncated WF (TWF) [14] retains the original two-stage framework but improves efficiency through an adaptive gradient flow, providing a solution in linear time. While WF and TWF have demonstrated significant empirical success for certain types of phase retrieval problems that directly involve dealing with non-convex objectives, they come with limitations related to sensitivity to initialization, observation noise, and the need for careful parameter tuning. In particular, both WF and TWF can exhibit slow convergence, especially when the number of measurements is not significantly larger than the signal dimensionality. Additionally, the need to compute gradients in each iteration can pose scalability challenges for very large-scale problems, as the iterative nature of these algorithms can become computationally intensive. A comprehensive review is provided in [19].

Deep Learning Approach Recent advancements have positioned deep learning approaches, such as convolutional neural networks (CNNs) [29], graph neural networks (GNNs) [35], and attention-based transformers [49], as formidable tools in image analysis and natural language processing. These networks comprise convolutional layers that automatically learn hierarchical features from input images. Deep equilibrium model (DEQ) [2] is another promising trend, where the neural network can be viewed as "infinitely deep" with converging equilibrium points. DEQ can then be used to find these points directly via root finding. Gilton et al. [23] showed promising results when applying the DEQ model to linear inverse problems. Adapting deep learning methods to phase retrieval is generally challenging, as it is a nonlinear problem. Some early works begin with a denoising framework. For example, Plugging-in a Denoiser [12, 34, 39] leverages pre-trained CNNs as implicit regularizers, harnessing their inherent strength in image denoising. Specifically, Metzlter et al. [39] adopted the regularization by denoising (RED) approach [46] to phase retrieval, thus giving rise to the term PRred. Following the notation in (1), the regularizer in PRred has the form

$$R(\mathbf{x}) = \frac{\lambda}{2}\mathbf{x}^\top(\mathbf{x} - D(\mathbf{x})),$$

where the denoiser $D(\mathbf{x})$ can be arbitrary. PRred [39] utilizes a DnCNNs denoiser from [54] and employs a fast solver from [25] to address the associated minimization problem. Similarly, generative prior methods [8, 31, 48] constrain the solution using a neural network generator via generative adversarial networks (GANs), ensuring that the solution has an accurate representation. Additionally, Hu et al. [33] combined a transformer as an encoder and a CNN as a decoder to build a phase shift network, while a vision transformer (ViT) [16] was adapted for phase retrieval in ptychography [21], referred to as PtychoDV. Last but not least, the end-to-end approach represents a more radical shift since it involves training a neural network to directly approximate the inverse mapping or its proxies. Some pioneer works [27, 36, 40] have demonstrated promising outcomes.

Unrolling Algorithms Generally, the idea of unrolling [44] is built on iterative techniques that mimic traditional optimization methods but are structured as a fixed number of layers in a neural network. In particular, one takes these iterative processes and "unrolls" them into a finite sequence of operations, each corresponding to a layer in a neural network. By treating the iterations as layers, the entire process can be trained end to end using gradient-based optimization. This approach leverages the interpretability of classical algorithms while benefiting from the efficiency and adaptability of deep learning. Specifically for phase retrieval, unrolling networks are utilized in PtychoDV [21] for ptychography and in [17] for PR from coded diffraction patterns (CDP).

1.2 Our Contributions

Classical projection methods such as Fienup [20] yield satisfactory results when there are a sufficient number of clean measurements, i.e., without noise. However, these methods easily fail when there is noise and there are not enough measurements. To address the ill-posed nature of the PR problem, some approaches rely on a predefined regularization term, followed by an optimization algorithm to find the optimal solution. While such regularization terms encode the desired properties of the solution and are mathematically interpretable, they may only be effective for specific types of data. On the other hand, deep learning methods such as CNNs and GNNs, have been successful in many imaging processing tasks, but they often require a large amount of training data and lack interpretability.

In this chapter, we propose a novel algorithm for recovering phase information from noisy and mildly oversampled data by extending the work of Manifold and Graph Integrative Convolution Network (MAGIC) [52]. Originally devised for CT image reconstruction, MAGIC is applicable to linear inverse problems. In this chapter, we adapt the MAGIC framework for phase retrieval, a challenging nonlinear problem. Our success is built upon the alternating direction method of multipliers (ADMM) [9], which involves two subproblems, each with closed-form solutions. Following MAGIC, we construct a neural network by unrolling an optimization-based approach that exploits both local features encoded by a CNN and nonlocal features encoded by a graph convolutional network (GCN). Our contribution is threefold:

1. We develop an algorithm unrolling strategy that has built-in mathematical interpretability.
2. By this means, we make the unrolling of the nonlinear PR problem tractable via ADMM.

3. We demonstrate that, in some cases, a stand-alone unrolling-based post-processing network can improve phase retrieval results from some classical methods such as Fienup.

As a proof-of-concept study, we assess our approach by employing Fourier measurements of masked images. Comparative numerical experiments with traditional non-learning and learning-based methods reveal better performance in situations involving noisy measurements.

The rest of the chapter is organized as follows. Section 2 is dedicated to a brief review of the MAGIC framework. We then detail the proposed algorithms in Sect. 3, highlighting the adaptation from solving a linear CT problem to a nonlinear PR problem. We also incorporate a stand-alone enhancement model in Sect. 3 that utilizes the same network architecture while aiming to improve the image quality achieved by a traditional PR method. As a proof-of-concept study, Sect. 4 shows numerical experiments of a masked Fourier PR problem. We consider two datasets (dog and CT images) and two types of masks (binary masks and plus-minus 1 masks) under two settings: noiseless and noisy measurements. An ablation study is conducted in Sect. 4.3 to illustrate the influence of the key ingredients in the proposed workflow: CNN, GNN, and the improvement over the initial solution. Lastly, Sect. 5 concludes the chapter and points to some future directions.

2 MAGIC Review

The Manifold and Graph Integrative Convolution Network (MAGIC) [52] framework integrates regularizations with a learning-based method for CT image reconstruction. Specifically, MAGIC unrolls a gradient descent-based iterative scheme into a neural network, using a convolutional neural network (CNN) as a regularization term. The algorithm samples points from overlapped patches with a small size, constructs a graph, and applies a graph neural network (GNN) to extract low-dimensional nonlocal features. By combining CNN and GNN modules, MAGIC can capture information at both the pixel level and the topological structure to model the CT images.

Mathematically, MAGIC is built upon the so-called learned experts' assessment-based reconstruction network (LEARN) [13], to minimize the following objective function:

$$\hat{\mathbf{x}} = \arg\min_{\mathbf{x}} \frac{1}{2}\|A\mathbf{x} - \mathbf{y}\|_2^2 + \lambda R(\mathbf{x}), \tag{2}$$

where $\mathbf{x}$ is an image in a vector form that represents the attenuation coefficients, A denotes the Radon transform that projects $\mathbf{x}$ into the data measurements $\mathbf{y}$, $R(\cdot)$ denotes a regularization term applied to $\mathbf{x}$, and $\lambda > 0$ is a trainable weighting parameter. Note that the first term in (2) is a least squares to measure the data misfit. Instead of handcrafted regularization form, such as total variation, MAGIC and LEARN introduce a generalized regularization term, also known as Field of Expert (FoE) [47], which can be learned using deep learning techniques,

$$R(\mathbf{x}) = \sum_{l=1}^{N_f} \psi_l(\phi_l(\mathbf{x})), \tag{3}$$

where ϕ_l and ψ_l can be regarded as some linear convolution operators and nonlinear activation functions, respectively, to be learned using training datasets, and N_f denotes the total number of features to be considered.

Convolution and activation functions are fundamental components of CNNs. Convolution is used to learn features such as horizontal/vertical edges. Activation functions introduce non-linearity into the network, enabling it to learn complex patterns and relationships in the data. Different activation functions are used for different purposes. For instance, a rectified linear unit (ReLU) sets negative values to zero and passes positive values unchanged to help with faster convergence and mitigate the vanishing gradient problem. Sigmoid squashes the output between 0 and 1, which is useful for binary classification tasks. Softmax is used in the output layer of classification models to convert logits into probabilities, with each output representing the probability of a class.

Incorporating the generalized regularization (3) into the objective function (2) yields

$$\hat{\mathbf{x}} = \arg\min_{\mathbf{x}} \frac{1}{2}\|A\mathbf{x} - \mathbf{y}\|_2^2 + \sum_{l=1}^{N_f} \lambda_l \psi_l(\phi_l(x)), \tag{4}$$

with a set of weights λ_l that is associated with each feature. One step of the gradient-descent algorithm when minimizing (4) leads to

$$\mathbf{x}^{t+1} = \mathbf{x}^t - \alpha\left[A^T(A\mathbf{x}^t - \mathbf{y}) + \sum_{l=1}^{N_f} \lambda_l^t \phi_l^*(\psi_l'(\phi_l(\mathbf{x}^t))\right], \tag{5}$$

where t indexes the iteration number, $\alpha > 0$ is a step size, ϕ_l^* is the adjoint operator of ϕ_l, and ψ_l' is the derivative of ψ_l. The last term in (5) filters the image $\mathbf{x}^t$ spatially, which is replaced by a more general three-layer CNN module Φ [13], thus leading to

$$\mathbf{x}^{t+1} = \mathbf{x}^t - \alpha A^T(A\mathbf{x}^t - \mathbf{y}) + \Phi(\mathbf{x}^t) \text{ with } \Phi(\mathbf{x}^t) = \omega_3^t * \sigma\left(\omega_2^t * \sigma(\omega_1^t * \mathbf{x}^t)\right), \tag{6}$$

where $\{\omega_1, \omega_2, \omega_3\}$ is a set of convolution kernels to be learned, $*$ denotes the convolution operator, and $\sigma(\cdot)$ is the activation function applied elementwise after the hidden layers. With this setup, (6) can be viewed as a residual block with three parts: a skip connection, a data fidelity layer, and a spatial CNN module. By specifying the number of iterations, we can unroll (6) into a network with the same number of layers as the specified iteration count.

Here, the CNN module $\Phi(\mathbf{x}^t)$ in (6) is used to extract the local pixel-level features of an image $\mathbf{x}^t$. Consequently, CNN can be considered as a form of local regularization.

MAGIC also employs nonlocal regularization based on graph convolutional networks (GCNs), which are defined on the manifold of image patches. Denote a patch set $\mathcal{P}(\mathbf{x})$ as a collection of image patches of $\mathbf{x}$ with size $s_1 \times s_2$. For example, $p_{ij}(\mathbf{x})$ represents an image patch with pixel (i, j) at the top left corner of the image $\mathbf{x}$. We vectorize each image patch as a row vector stack row by row to form a matrix and denote such linear transform of $\mathbf{x}$ as $P(\mathbf{x}) \in \mathbb{R}^{p\times d}$, where d is the number of pixels in each patch and p is the number of patches. We construct a graph $\mathcal{G}(\mathcal{V}, \mathcal{E})$ with p nodes. We construct edges by using Euclidean distance to find neighbors. We define the adjacency matrix W as

$$W_{rq} = e^{-\frac{\|v_r - v_q\|_2^2}{\mu^2(\mathcal{V})}}, \tag{7}$$

where $v_r, v_q \in \mathcal{V}$ are two nodes (patches) in the graph $\mathcal{G}$ and $\mu(\mathcal{V})$, a function capturing the "typical" graph distances. Here we follow the original MAGIC method and use the median square distance between patches as $\mu^2(\mathcal{V})$. Adopting a renormalization trick, we let $\tilde{W} = I + W$ with the identity matrix I and define diagonal matrix $\tilde{D}$ with $\tilde{D}_{rr} = \sum_q \tilde{W}_{rq}$.

Now, nonlocal topological features from the low-dimensional patch manifold space are extracted by adding a GCN module Ψ acting on the image patches $P(\mathbf{x}^t)$ into (6), thus leading to

$$\mathbf{x}^{t+1} = \mathbf{x}^t - \alpha A^T(A\mathbf{x}^t - \mathbf{y}) + \Phi(\mathbf{x}^t) + \Psi(P(\mathbf{x}^t)), \tag{8}$$

where $\Psi(P(\mathbf{x}^t)) = \tilde{D}^{-\frac{1}{2}}\tilde{W}\tilde{D}^{-\frac{1}{2}}\sigma(\tilde{D}^{-\frac{1}{2}}\tilde{W}\tilde{D}^{-\frac{1}{2}}P(\mathbf{x}^t)\Theta_1^t)\Theta_2^t$, the GCN, with graph convolutional kernels $\Theta_1^t \in \mathbf{R}^{d\times p}$ and $\Theta_2^t \in \mathbf{R}^{p\times d}$ to-be-trained. In summary, MAGIC unrolls a fixed number of (8) into a neural network with respect to the parameters $\{\boldsymbol{\omega}_1^t, \boldsymbol{\omega}_2^t, \boldsymbol{\omega}_3^t, \Theta_1^t, \Theta_2^t\}$ when minimizing a loss function, defined by

$$\frac{1}{N_s}\sum_{i=1}^{N_s} \|\mathbf{x}_i - \hat{\mathbf{x}}_i\|_2^2, \tag{9}$$

where N_s denotes the number of the training samples, $\mathbf{x}_i$ is the predicted reconstruction from the neural network, and $\hat{\mathbf{x}}_i$ is corresponding label or ground-truth image. It was demonstrated in [52] that the effectiveness of both spatial CNN and topological GCN components leads to significant improvements compared to using only one of them.

3 The Proposed Algorithms

We describe two unrolling-based algorithms: one for phase retrieval and the other for image enhancement. Adapted from MAGIC to deal with the nonlinear inverse problem, both approaches share the same network architectures in terms of the unrolling techniques of an optimization method and two data-driven regularizations defined by CNN and GCN. As the unrolling framework requires an initial condition, we investigate the image enhancement model to offer an alternative that directly improves the image quality of the initial.

3.1 *Unrolling-Based Phase Retrieval*

We focus on a specific type of PR data that is a collection of Fourier-type measurements of masked images. The proposed methodology however works for any kind of PR data. We consider two types of masks in the experiments. One is called a binary mask in the sense that each mask matrix $M^{(j)} \in \mathbb{R}^{mn\times mn}$ is a diagonal matrix where the diagonal entries are either 1 or 0, with 1 at locations where the intensity of $\mathbf{x}$ is preserved and 0 where it is nullified. The other is called a plus-minus (PM) mask, where the diagonal entries are either $+1$ or -1, to avoid completely ignoring some pixels by nullifying them. Let $\mathcal{F} \in \mathbb{C}^{mn\times mn}$ denote the 2D Fourier transform matrix that acts on $\mathbf{x}$. In this setting, the forward model and the measurements can be succinctly represented as follows:

$$\mathcal{A} = \begin{pmatrix} \mathcal{F}M^{(1)} \\ \mathcal{F}M^{(2)} \\ \vdots \\ \mathcal{F}M^{(\ell)} \end{pmatrix} \in \mathbb{C}^{\ell mn\times mn}, \quad \mathbf{y} = \begin{pmatrix} \mathbf{y}^{(1)} \\ \mathbf{y}^{(2)} \\ \vdots \\ \mathbf{y}^{(\ell)} \end{pmatrix} \in \mathbb{R}^{\ell mn},$$

where ℓ denotes the number of masks.

Given the phaseless data $\mathbf{y} \in \mathbb{R}^{\ell mn}$, we aim to reconstruct the image $\mathbf{x} \in \mathbb{R}^{mn}$ by solving the following minimization problem

$$\min_{\mathbf{x}\in\mathbb{R}^{mn}} \frac{1}{2}\||\mathcal{A}\mathbf{x}| - \mathbf{y}\|_2^2 + \lambda R(\mathbf{x}), \tag{10}$$

where the data fidelity is measured by the least squares, R denotes a regularization term, and $\lambda > 0$ is a trainable weighting parameter. As the term $|\mathcal{A}\mathbf{x}|$ is not differentiable, we adopt the alternating direction method of multipliers (ADMM) [9] to minimize (10). The core idea of ADMM is to introduce auxiliary variables and decompose the problem into subproblems, each of which is easier to solve. In particular, we introduce an auxiliary variable $\mathbf{z} \in \mathbb{C}^{\ell mn}$ and convert (10) to an equivalent formulation

$$\min_{\mathbf{x},\mathbf{z}} \frac{1}{2}\||\mathbf{z}| - \mathbf{y}\|_2^2 + \lambda R(\mathbf{x}) \quad \text{s.t.} \quad \mathcal{A}\mathbf{x} = \mathbf{z}. \tag{11}$$

The corresponding augmented Lagrangian can be expressed by

$$\mathcal{L}_\rho(\mathbf{x}, \mathbf{z}; \mathbf{v}) = \frac{1}{2}\||\mathbf{z}| - \mathbf{y}\|_2^2 + \lambda R(\mathbf{x}) + \rho \mathrm{Re}\{\langle \mathbf{v}, \mathcal{A}\mathbf{x} - \mathbf{z}\rangle\} + \frac{\rho}{2}\|\mathcal{A}\mathbf{x} - \mathbf{z}\|_2^2, \tag{12}$$

where $\mathbf{v}$ is a Lagrangian multiplier or so-called dual variable and ρ is a positive penalty parameter. ADMM iterates as follows:

$$\begin{cases} \mathbf{x}^{t+1} = \arg\min_{\mathbf{x}} \mathcal{L}_\rho(\mathbf{x}, \mathbf{z}^t; \mathbf{v}^t) \\ \mathbf{z}^{t+1} = \arg\min_{\mathbf{z}} \mathcal{L}_\rho(\mathbf{x}^{t+1}, \mathbf{z}; \mathbf{v}^t) \\ \mathbf{v}^{t+1} = \mathbf{v}^t + (A\mathbf{x}^{t+1} - \mathbf{z}^{t+1}), \end{cases} \tag{13}$$

where t is an index of the iteration numbers.

We start with a closed-form solution for the $\mathbf{z}$-subproblem in (13), which is equivalent to

$$\min_{\mathbf{z}\in\mathbb{C}^{\ell mn}} \frac{1}{2}\||\mathbf{z}| - \mathbf{y}\|_2^2 + \frac{\rho}{2}\|\mathcal{A}\mathbf{x}^{t+1} - \mathbf{z} + \mathbf{v}^t\|_2^2. \tag{14}$$

Let $\mathbf{g} := \mathcal{A}\mathbf{x}^{t+1} + \mathbf{v}^t$. The closed form solution for $\mathbf{z}$ depends on two inputs, $\mathbf{y}$ and $\mathbf{g}$, which can be given by an operator G, i.e.,

$$G(\mathbf{y}, \mathbf{g}) = \begin{cases} \frac{\mathbf{y}+\rho|\mathbf{g}|}{1+\rho}\frac{\mathbf{g}}{|\mathbf{g}|} & \text{if } \mathbf{g} \neq 0 \\ \frac{\mathbf{y}}{1+\rho}c & \text{if } \mathbf{g} = 0, \end{cases} \tag{15}$$

where c is an arbitrary unit root. The derivation of (15) is based on the Wirtinger calculus [51] to deal with complex-valued $\mathbf{z}$ and find a stationary point of the objective function in (14).

Then we examine the $\mathbf{x}$-subproblem in (13), which can be equivalently expressed as

$$\min_{\mathbf{x}} \lambda R(\mathbf{x}) + \frac{\rho}{2}\|\mathcal{A}\mathbf{x} - \mathbf{z}^t + \mathbf{v}^t\|_2^2. \tag{16}$$

One step of gradient descent for minimizing (16) with respect to $\mathbf{x}$ with implicit regularization[1] yields

$$\mathbf{x}^{t+1} = \mathbf{x}^t - \alpha\left(\rho(\mathcal{A}^*\mathcal{A}\mathbf{x}^t - \mathcal{A}^*\mathbf{z}^t + \mathcal{A}^*\mathbf{v}^t)\right) + \Phi(\mathbf{x}^t) + \Psi(P(\mathbf{x}^t)), \tag{17}$$

[1] Implicit used here to denote that there is no assumption of an explicit relation between $R(\mathbf{x})$ and $\Phi(\mathbf{x}^t)$ or $\Psi(P(\mathbf{x}^t))$.

where $\alpha > 0$ is a step size, $\mathcal{A}^*$ denotes the complex conjugate of $\mathcal{A}$, and $P(\mathbf{x}^t)$ is a matrix obtained by collecting image patches from $\mathbf{x}^t$ as in (8). To expedite convergence, we choose to initialize using the results of a classical projection algorithm such as Fienup to set the value of $\mathbf{x}^0$.

Though the formula in (17) is defined in a vector form, all the computations can be implemented in the matrix formulation. We maintain the matrix representations of $\mathbf{x}$ and illustrate the calculation of $\mathcal{A}^*\mathcal{A}\mathbf{x}$ to be the sum of all 2D masks point-wise multiplied by the matrix version of $\mathbf{x}$.

The calculation of $\mathcal{A}^*\mathbf{z}$ is similar: note $\mathbf{z} = \begin{pmatrix} \mathbf{z}^1 \\ \mathbf{z}^2 \\ \vdots \\ \mathbf{z}^\ell \end{pmatrix}$ is a stacked vector, $\mathcal{A}^*\mathbf{z} =$

$\big((M^{(1)})^T\mathcal{F}^* \ (M^{(2)})^T\mathcal{F}^* \ \ldots \ (M^{(\ell)})^T\mathcal{F}^*\big) \begin{pmatrix} \mathbf{z}^1 \\ \mathbf{z}^2 \\ \vdots \\ \mathbf{z}^\ell \end{pmatrix} = \sum_j M^{(j)}\mathcal{F}^*(\mathbf{z}^j)$. Its matrix operation is the sum of the following terms: the jth mask pointwise multiplied by the inverse Fourier transform of the 2D matrix representation of $\mathbf{z}^j$.

Lastly, the calculation of $\mathcal{A}\mathbf{x}^{t+1}$ is also efficient: $\mathcal{A}x^{t+1} = \begin{pmatrix} \mathcal{F}M^{(1)} \\ \mathcal{F}M^{(2)} \\ \vdots \\ \mathcal{F}M^{(\ell)} \end{pmatrix} \mathbf{x}^{t+1}$, whose matrix representations are the l copies of Fourier transformation of $\mathbf{x}^{t+1}$ multiplied by l masks. In summary, the ADMM iterations (13) can be given by

$$\begin{cases} \mathbf{x}^{t+1} = \mathbf{x}^t - \alpha(\rho(\mathcal{A}^*\mathcal{A}x^t - \mathcal{A}^*\mathbf{z}^t + \mathcal{A}^*\mathbf{v}^t)) + \Phi(\mathbf{x}^t) + \Psi(P(\mathbf{x}^t)) \\ \mathbf{z}^{t+1} = G(\mathbf{y}, \mathcal{A}\mathbf{x}^{t+1} + \mathbf{v}^t) \\ \mathbf{v}^{t+1} = \mathbf{v}^t + \mathcal{A}\mathbf{x}^{t+1} - \mathbf{z}^{t+1} \end{cases} \tag{18}$$

Unlike a typical ADMM algorithm, which iterates until convergence, our objective is to determine the trainable parameters, such as those in the CNN and GCN modules. We iterate through the equations in (18) for a specified number of blocks (Fig. 1) to construct a neural network as shown in Fig. 2, with the updates of the ADMM variables as shown in Fig. 3. One can see that neural network is originated from an iterative algorithm, providing mathematical interpretability of the deep network.

Next, we describe the training procedure in detail. As the aforementioned neural network is differentiable with respect to these trainable parameters, we apply the stochastic gradient descent algorithm to find the optimal solutions. Given training data $\{(\mathbf{x}_i, \mathbf{y}_i, \mathbf{x}_i^0)\}_{i\in S_1}$ with $\mathbf{x}_i$ the ith ground-truth image, $\mathbf{y}_i$ the corresponding PR data, and $\mathbf{x}_i^0$ the initial guess, we adopt the algorithm unrolling scheme [28, 44] and let each $\mathbf{y}_i$ go through the iterative scheme outlined in Eqs. (18), propagating the

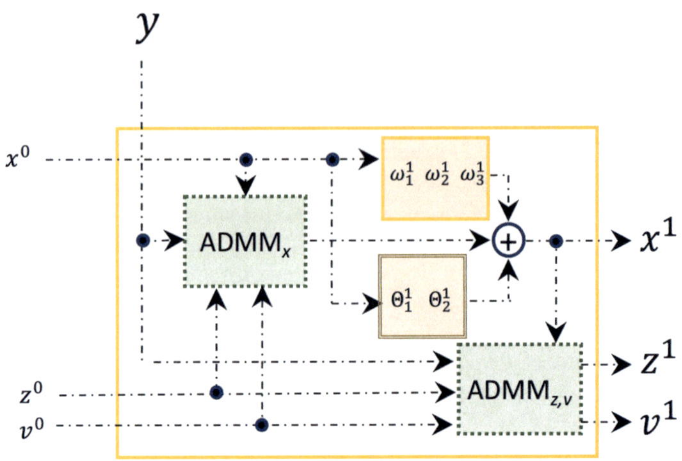

Fig. 1 MAGIC PR Block: one iteration of the ADMM-based algorithm includes updates for x, z, and v, as well as the block's CNN and GCN modules. During training, parameters $\{\omega_1, \omega_2, \omega_3\}$ of the CNN and $\{\Theta_1, \Theta_2\}$ of the GCN are learned

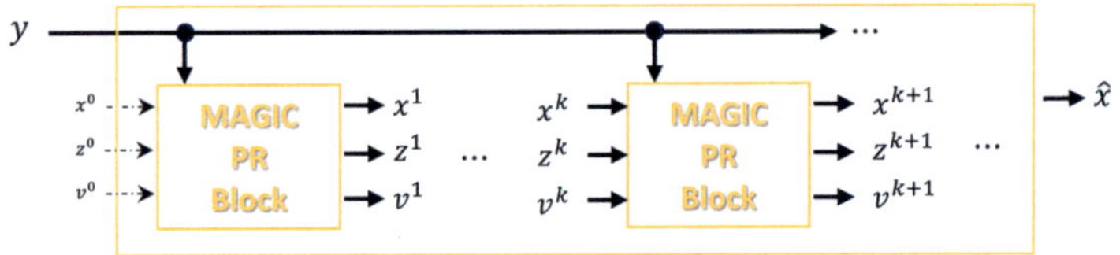

Fig. 2 The unrolled network is constructed by the composition of a predetermined number of MAGIC PR Blocks. The estimated reconstruction $\hat{x}$ corresponds to the x variable of the last block in the network

information forward in the unrolled neural network (Fig. 2). To solve for the model parameters ω_n's, Θ_n's in the neural networks and the gradient descent step-size α, we minimize the supervised mean squared error (MSE) loss function:

$$L = \frac{1}{N} \sum_i \|\mathbf{x}_i - \hat{\mathbf{x}}_i\|_2^2, \tag{19}$$

where $\mathbf{x}_i$ is the i-th given ground-truth image, $\hat{\mathbf{x}}_i$ is the output of the unrolled neural network to the i-th measurement, and N is the size of the training set.

3.2 *Unrolling-Based Image Enhancement*

In this section, we consider a stand-alone unrolling-based neural network to improve the quality from image denoising. For example, the input data, denoted by $\mathbf{g}$, could

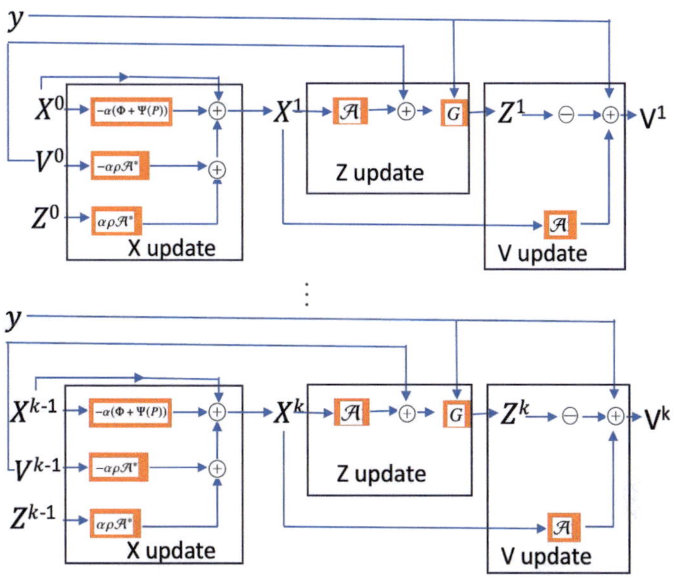

Fig. 3 Update of the ADMM variables in the unrolling algorithm (18), with each block representing one iteration and k blocks stacking a finite number of iterations to build a deep network architecture

be phase retrieval results obtained by using a handful of iterations of the Fienup algorithm. This can be treated as a post-processing step that sometimes works better than the unrolling-based phase retrieval method.

To improve the image quality from the input data $\mathbf{g}$, we consider to minimize the following energy function

$$\min_{\mathbf{x}} \frac{1}{2}\|\mathbf{x}-\mathbf{g}\|_2^2 + \lambda R(\mathbf{x}), \tag{20}$$

to find a solution $\mathbf{x}$ using a regularization function defined by R. We adopt the same strategies in the proposed PR pipeline, i.e., local and nonlocal regularization functions described by CNN and GCN, respectively.

Solving the minimization problem (20) using gradient descent, we get

$$\mathbf{x}^{t+1} = \mathbf{x}^t - \alpha\Big((\mathbf{x}^t - \mathbf{g})\Big) + \Phi(\mathbf{x}^t) + \Psi(P(\mathbf{x}^t)). \tag{21}$$

Adopting the unrolling scheme, we construct a neural network using the iterative scheme (21) and solve for parameters in the neural network with respect to MSE (19), where $\mathbf{x}_i$ is the given i-th ground-truth clean image, $\hat{\mathbf{x}}_i$ is the output of unrolled (21), and N is the size of the training set. Note that the denoising model (21) inherits the same network architecture, i.e., $\Phi(\cdot)$ and $\Psi(\cdot)$, as in the PR case (18). We include this model into an ablation study of the proposed methods, as outlined in Sect. 4.3.

4 Numerical Examples

Data Simulation We consider masked Fourier measurements to simulate phaseless data. Specifically, we start with an image, denoted by $\mathbf{x}$ in a vector form, which is regarded as the underlying ground truth, and three masks of the same dimension as the image. The justification for using three masks stems from the difficulty in achieving successful PR under an oversampling ratio of 2, and the theoretical assurance of a unique PR solution when the oversampling ratio is 4. We multiply the image with each mask elementwise, take the Fourier transform of each masked image, and only record the magnitude, leading to vectorized data denoted by $\mathbf{y}$. As a result, the number of the measurements in $\mathbf{y}$ is three times the number of the ground-truth image in $\mathbf{y}$ due to the use of three masks. We also add noise to test the robustness of our model. We use the same masks for each image in the dataset before splitting it into the training and testing sets.

Model Architecture For hyperparameters related to CNN and GCN, we adopt similar strategies as outlined in the original MAGIC paper [52]. For instance, for the CNN component, we use three convolutional layers, each one with 64 filters of size 3×3. We also use stride of 1 and padding of 1, but in contrast with the original paper that uses padding with zeros, we use padding mode with reflecting conditions, no bias in the convolutional layer and ReLU activation function after all the convolutional layers. Therefore, the number of parameters to learn per CNN component per block is 38,016 (first layer, $\boldsymbol{\omega}_1^k$, $3 \times 3 \times 1 \times 64$; second layer, $\boldsymbol{\omega}_2^k$, $3 \times 3 \times 64 \times 64$; and last layer, $\boldsymbol{\omega}_3^k$, $3 \times 3 \times 64 \times 1$). For the GCN component, we use two layers each with 64 neurons. To build the weighted graph, we use Eq. (7) and patches of size 6×6. The patches cover the image with stride 2, which yields a graph with 3844 vertices (i.e. a vertex per patch) for images of size 128×128. Therefore, the number of parameters to learn per GCN component per block is 4708 (first layer, Θ_1^k, $6 \times 6 \times 1 \times 64$; last layer, Θ_2^k, $64 \times 6 \times 6 \times 1 + 6 \times 6 \times 1$). In addition, for each ADMM block, we learn the parameter of α in Eq. (18 for $\mathbf{x}^{t+1}$ update. Briefly, this corresponds to a model with 42,725 parameters per block. We calculate the graph Laplacian twice with the first $k/2$ blocks using graph Laplacian calculated from the initial $\mathbf{x}^0$ and the last $k/2$ blocks using an updated one from the $\mathbf{x}^{k/2}$ iterate. A larger k implies more iterations and a deeper model that may not fit in memory or require greater data to train. In the experiments, we conduct a minimal exploration of the number of blocks k to use for each problem, running a few iterations for models with 10, 20, 30, and 50 blocks and greedily selecting the number of blocks yielding a smaller error. In one case (see sections below), we find that a larger number of blocks (120) produced better results.

Initialization and Training Setup We employ the Fienup algorithm, performing 50 iterations to obtain the starting point $\mathbf{x}^0$ in the iteration; see (18). As in the original MAGIC work, we use the ADAM optimizer and a decaying learning rate. The learning rate is set to decay by a multiplicative factor of 0.95 after each epoch. However, to capture the different expected influence of parameters, we group them in (i) neural network parameters (i.e., including CNN and GCN components) and (ii) ADMM parameters (including α), and we set two different initial learning rates for these two groups, $\mathrm{LR}^0_{\mathrm{NN}}$ and LR^0_{α}, respectively. Usually we set $\mathrm{LR}^0_{\mathrm{NN}} \ll \mathrm{LR}^0_{\alpha}$. We initialize the parameters for CNN and GCN with normal distributions with mean 0 and standard deviation equal to 0.001 or 0.0001, respectively. In most cases, we initialize the α parameters with 0.5. Nevertheless, we find it convenient to initialize them to 0.01 for the noisy data cases. We train our models for 50 epochs. During training, we make sure to clip the α parameters to the range $[10^{-7}, 0.9]$, since they correspond to gradient descent step sizes. Additionally, we use training, validation, and testing sets. The validation set is only used to track performance (MSE) and enact an early-stopping criterion: we stop training if the performance evaluated on the validation set does not improve for a span of a pre-specified number of epochs. This span of epochs, usually denoted as *patience*, was set to 25 in our experiments. In all cases, we report results over the testing set using the best model, i.e., the model that during training exhibited the lowest MSE in the validation set. This can be different from the model obtained in the last training epoch, which in turn can be obtained in less than 50 epochs if the early-stopping criterion is activated.

Comparison We compare the proposed approach with a classical projection approach (Fienup [20]), an advanced gradient-based method (TWF [14]), a gradient-based method (WF), and a deep learning approach (PRred [39]). We use the MATLAB codes for Fienup, TWF, and WF which are publicly available.[2] Standard metrics PSNR and SSIM (both using the dynamic range of the ground truth; see definitions in [32]) are used to quantitatively evaluate the performance of various competing methods. Additionally, we present some visual results for qualitative comparisons.

Datasets We consider two diverse datasets for our experimental study: images that contain dogs from ImageNet [15] and full-dose CT images [38]. We focus on dog images with the intention of learning features that are specific to dogs rather than other classes in ImageNet. We consider clean dog images (without noise) since most dog images contain fine details of fur and background, making the reconstruction susceptible to noise. Since CT images are generally piece-wise constant and can tolerate a certain amount of noise, we investigate both clean and noisy data to evaluate the robustness of the proposed method in handling noise.

4.1 Experiments on the Dog Dataset

We collect a dataset composed of 600 images of dogs in front of diverse backgrounds from ImageNet [15]. We resize each image to 128×128 pixels and split the data into 400 images for training, 100 images for validation, and 100 images for

[2] https://github.com/tomgoldstein/phasepack-matlab

testing. In this experiment, we use directly the phaseless simulated measurements—meaning that no additional noise is added to the measurement data.

We consider two types of masks. One consists of 50% of the value $+1$ and 50% of the value -1, referred to as a plus-minus (PM) mask. The other is a binary mask with 80% of value 1 and 20% of value 0. In both cases, the size of the measurements is triple that of the underlying image. However, the effective oversampling ratio for the binary mask is 2.4, taking into account the loss of information at locations where the mask value is 0.

For the PM mask, we train a 30-block unrolled MAGIC architecture, with the block configuration described before. This corresponds to a model with 1,281,750 parameters. We use a batch size of 64 and initial learning rates of $\mathrm{LR}_{\mathrm{NN}}^{0} = 0.01$ and $\mathrm{LR}_{\alpha}^{0} = 0.0001$. For the binary mask, we train a 120-block unrolled MAGIC architecture, with the block configuration described before. This corresponds to a model with 5,127,000 parameters. We use a batch size of 32 and initial learning rates of $\mathrm{LR}_{\mathrm{NN}}^{0} = 0.01$ and $\mathrm{LR}_{\alpha}^{0} = 0.0001$.

The results of the dog data are presented in Tables 1 and 2 for PSNR and SSIM, respectively. One can see that when there is no noise and PM mask is used, Fienup performs the best, with TWF the second best. For the two deep learning-based approaches, the proposed one is better than PRred and almost matches the performance of WF. On the other hand, in the case of binary masks, Fienup and TWF still perform well, although not as good as in the case of PM mask. In contrast, the WF method has a bad performance and is easily beaten by the proposed method. For the two deep learning-based approaches, the proposed one is, again, better than PRred.

Figures 4, 5, and 6 present visual results of the image reconstruction from phaseless data for the PM mask. Figures 7, 8, and 9 present visual results of the image reconstruction from phaseless data for the binary mask. Fienup and TWF achieve exact recovery of the underlying image for the PM mask, as expected due to the theoretical guarantees provided by an oversampling ratio of 3 in the PM case. However, these two methods lack a mechanism to fill in the missing information when binary masks are employed, causing dead pixels at the locations where the

Table 1 PSNR comparison on the clean dog dataset (without noise). Mean values over 100 testing images are reported with standard deviation in parenthesis

	Fienup	TWF	WF	PRred	Proposed
PM mask	83.31 (1.95)	60.55 (3.63)	35.44 (6.03)	15.88 (3.43)	33.32 (3.07)
Binary mask	25.19 (1.44)	26.76 (1.84)	13.57 (1.20)	8.98 (1.73)	23.65 (2.50)

Table 2 SSIM comparison on the clean dog dataset (without noise). Mean values over 100 testing images are reported with standard deviation in the parenthesis

	Fienup	TWF	WF	PRred	Proposed
PM mask	1.00 (0.00)	1.00 (0.00)	0.92 (0.15)	0.51 (0.10)	0.90 (0.07)
Binary mask	0.80 (0.05)	0.89 (0.04)	0.21 (0.03)	0.10 (0.02)	0.72 (0.09)

Fig. 4 Results for dog dataset with the PM mask. From left to right: ground truth, Fienup (PSNR, 81.84 dB; SSIM, 0.99), TWF (PSNR, 60.00 dB; SSIM, 0.99), WF (PSNR, 43.80 dB; SSIM, 0.99), PRred (PSNR, 13.51 dB; SSIM, 0.47), proposed (PSNR, 35.01 dB; SSIM, 0.94)

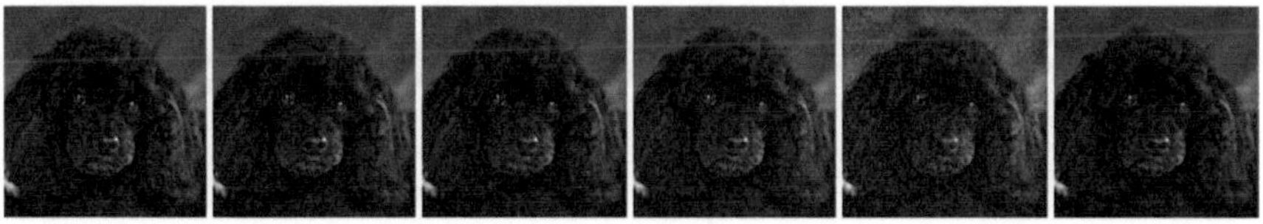

Fig. 5 Results for dog dataset with the PM mask. From left to right: ground truth, Fienup (PSNR, 90.59 dB; SSIM, 1.00), TWF (PSNR, 63.75 dB; SSIM, 0.99), WF (PSNR, 35.95 dB; SSIM, 0.93), PRred (PSNR, 26.06 dB; SSIM, 0.65), proposed (PSNR, 40.70 dB; SSIM, 0.97)

Fig. 6 Results for dog dataset with the PM mask. From left to right: ground truth, Fienup (PSNR, 80.92 dB; SSIM, 0.99), TWF (PSNR, 55.54 dB; SSIM, 0.99), WF (PSNR, 36.53 dB; SSIM, 0.97), PRred (PSNR, 10.98 dB; SSIM, 0.47), proposed (PSNR, 30.02 dB; SSIM, 0.89)

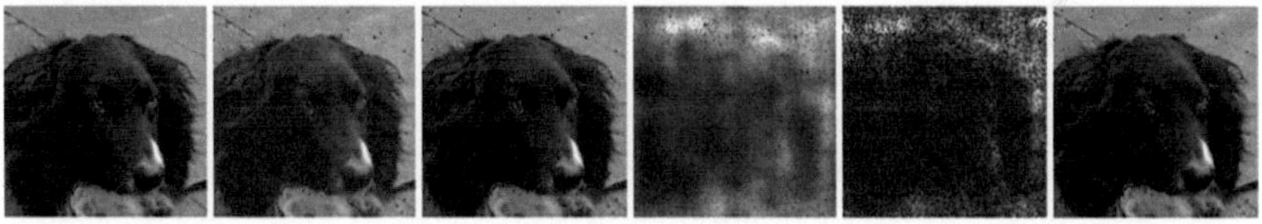

Fig. 7 Results for dog dataset with the binary mask. From left to right: ground truth, Fienup (PSNR, 28.29 dB; SSIM, 0.77), TWF (PSNR, 30.91 dB; SSIM, 0.95), WF (PSNR, 15.81 dB; SSIM, 0.22), PRred (PSNR, 12.71 dB; SSIM, 0.20), proposed (PSNR, 30.31 dB; SSIM, 0.87)

mask entry takes the value of 0. This loss of information impacts the performance of algorithms like Fienup and TWF and severely affects the reconstruction on the WF case. Additionally, our experimental setup falls outside the assumptions under which the theoretical guarantees for classical methods are typically established, which may explain the observed issues in this specific context. PRred relies on an image prior to guiding the image reconstruction but does not employ any information about the measurement operator. Results are noisy for the PM mask case and very distorted for the binary mask. The proposed approach utilizes both local and nonlocal features

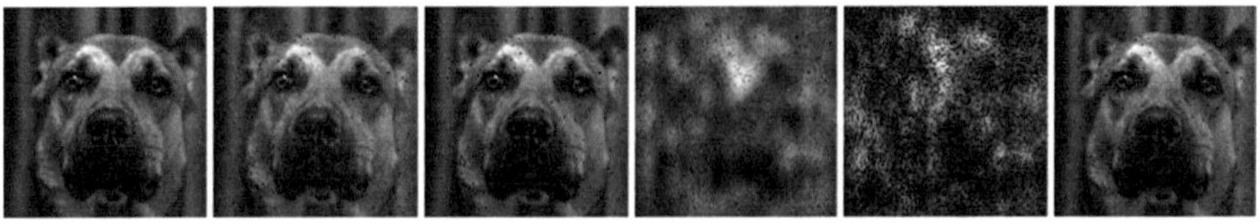

Fig. 8 Results for dog dataset with the binary mask. From left to right: ground truth, Fienup (PSNR, 28.40 dB; SSIM, 0.87), TWF (PSNR, 30.42 dB; SSIM, 0.93), WF (PSNR, 16.89 dB; SSIM, 0.30), PRred (PSNR, 13.55 dB; SSIM, 0.16), proposed (PSNR, 28.88 dB; SSIM, 0.88)

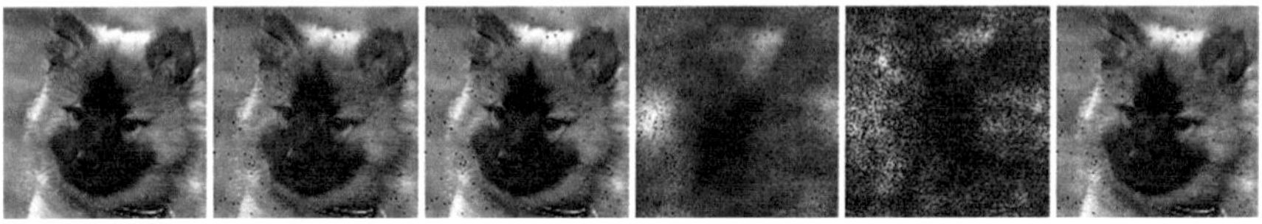

Fig. 9 Results for dog dataset with the binary mask. From left to right: ground truth, Fienup (PSNR, 26.13 dB; SSIM, 0.82), TWF (PSNR, 27.32 dB; SSIM, 0.88), WF (PSNR, 14.69 dB; SSIM, 0.24), PRred (PSNR, 10.62 dB; SSIM, 0.13), proposed (PSNR, 26.51 dB; SSIM, 0.82)

discovered by CNN and GCN for image reconstruction. While it produces metrics that are lower than Fienup and TWF, the corresponding visual results are appealing but still suffer from similar dead pixel artifacts than classical methods.

4.2 Experiments on the CT Dataset

The CT dataset contains a collection of CT images from ten different patients [38]. In our experiments, we use full-dose CT images from ten patients to simulate data for phase retrieval. We select eight patients for the training and validation sets and reserve two patients for testing. From the 8 training/validation patients, we randomly select 400 images to train the unrolling network and 100 images for validation. From the 2 reserved testing patients, we randomly select 100 images for testing. We consider both clean CT and noisy CT, with the latter simulated by adding Gaussian noise with mean 0 and standard deviation of 0.04 (i.e., about 4%) in the measurement domain (magnitude data). In other words, we add noise to each component of the vectorized data $\mathbf{y}$, clipping negative values to 10^{-5}, to guarantee positive measurements. This corresponds to about 22 dB PSNR and 5.3 dB SNR, between original $\mathbf{y}$ and corrupted signal $\mathbf{y}_n$ in the measurement domain. Although this seems like a mild noise level, it is enough to severely affect the performance of classical methods.

For the clean CT, we train a 20-block unrolled MAGIC architecture, with the block configuration described before. This corresponds to a model with 854,500 parameters. We set a batch size of 128 and initial learning rates of $\mathrm{LR}_{\mathrm{NN}}^{0} = 0.001$ and $\mathrm{LR}_{\alpha}^{0} = 0.0001$. For the noisy CT, we train a ten-block unrolled MAGIC architecture, but we modify the CNN component to include more layers and filters.

Table 3 PSNR comparison using the CT dataset (standard deviation is provided in parenthesis)

	Fienup	TWF	WF	PRred	Proposed
Clean CT	88.09 (1.24)	64.63 (2.73)	40.96 (6.72)	13.29 (3.02)	38.60 (1.63)
Noisy CT	12.10 (0.91)	11.78 (0.90)	11.33 (0.90)	12.61 (1.54)	22.98 (0.69)

Table 4 SSIM comparison using the CT dataset (standard deviation is provided in parenthesis)

	Fienup	TWF	WF	PRred	Proposed
Clean CT	1.00 (0.00)	1.00 (0.00)	0.92 (0.15)	0.31 (0.05)	0.93 (0.02)
Noisy CT	0.08 (0.02)	0.08 (0.02)	0.07 (0.01)	0.16 (0.03)	0.40 (0.03)

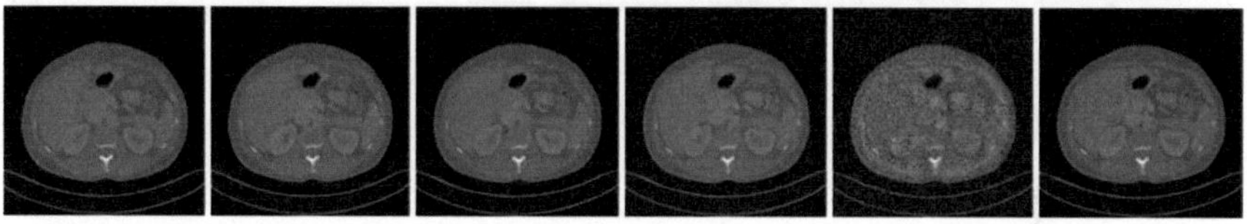

Fig. 10 Results for CT dataset. From left to right: ground truth, Fienup (PSNR, 87.13 dB; SSIM, 0.99), TWF (PSNR, 61.19 dB; SSIM, 0.99), WF (PSNR, 39.11 dB; SSIM, 0.94), PRred (PSNR, 11.62 dB; SSIM, 0.31), proposed (PSNR, 39.94 dB; SSIM, 0.95)

We use 5 convolutional layers, each one with 128 filters of size 3×3. We also use a stride of 1 and padding of 1, padding mode with reflecting conditions, no bias in the convolutional layer, and ReLU activation function in all the layers. Therefore, the number of parameters to learn per CNN component per block is 444,672 (first layer, $\boldsymbol{\omega}_1^k$, $3 \times 3 \times 1 \times 128$; second, third, and fourth layers ($\boldsymbol{\omega}_2^k$, $\boldsymbol{\omega}_3^k$, and $\boldsymbol{\omega}_4^k$, respectively), $3 \times 3 \times 128 \times 128$ (each); and last layer, $\boldsymbol{\omega}_5^k$, $3 \times 3 \times 128 \times 1$). The other layers remain as before (i.e., Θ_1^t, Θ_2^t corresponds to 4708 parameters per GCN component per block and 1 α parameter per block). Thus, this corresponds to a model with 449,381 per block, for a total of 4,493,810 parameters. We use a batch size of 64 and initial learning rates of $\text{LR}_{\text{NN}}^0 = 0.1$ and $\text{LR}_{\alpha}^0 = 0.0001$.

The PSNR and SSIM values of the reconstructed images are reported in Tables 3 and 4. One can see the same pattern in clean CT as in the dog data under the PM mask, albeit, the proposed method's performance is improved, probably due to the higher homogeneity of the CT dataset. However, in the presence of noise, the performance of Fienup, TWF, and WF drops dramatically, resulting in significantly poorer results compared to the proposed approach. Considering learning-based methods, our approach largely outperforms PRred.

We present CT image results for phase retrieval without noise in Figs. 10, 11, and 12 and for phase retrieval with noise in Figs. 13, 14, and 15. In the noise-free case, Fienup and TWF achieve the perfect reconstruction with no visual difference from the ground truth and are nearly followed by WF. PRred exhibits noisy results, whereas the proposed approach generates very good visual results with SSIMs that are very close to WF. In Figs. 13, 14, and 15, when the noise is present, it appears that noise propagates throughout the entire images recovered by Fienup,

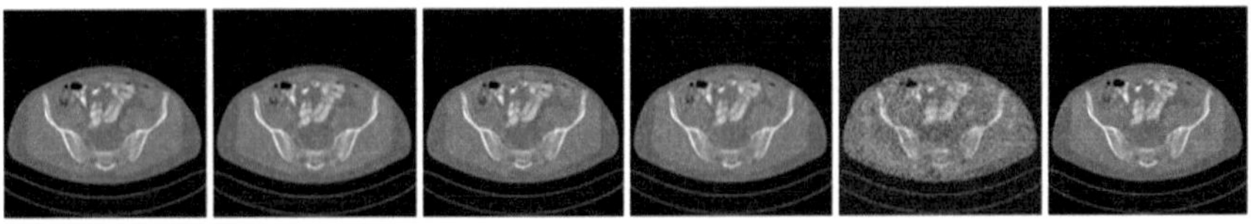

Fig. 11 Results for CT dataset. From left to right: ground truth, Fienup (PSNR, 86.85 dB; SSIM, 0.99), TWF (PSNR, 63.58 dB; SSIM, 0.99), WF (PSNR, 43.56 dB; SSIM, 0.98), PRred (PSNR, 11.57 dB; SSIM, 0.30), proposed (PSNR, 38.43 dB; SSIM, 0.94)

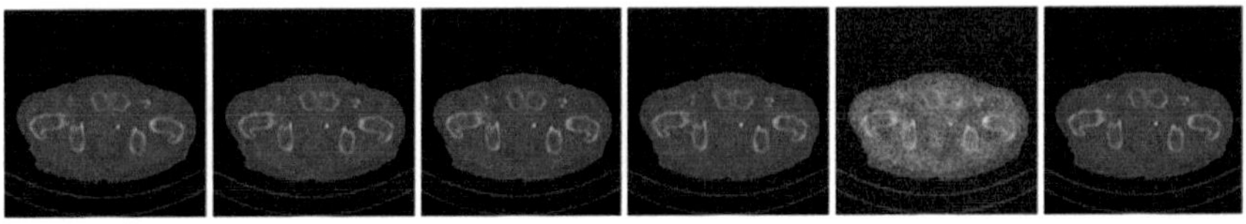

Fig. 12 Results for CT dataset. From left to right: ground truth, Fienup (PSNR, 91.46 dB; SSIM, 1.00), TWF (PSNR, 66.53 dB; SSIM, 0.99), WF (PSNR, 47.32 dB; SSIM, 0.99), PRred (PSNR, 18.98 dB; SSIM, 0.41), proposed (PSNR, 42.32 dB; SSIM, 0.96)

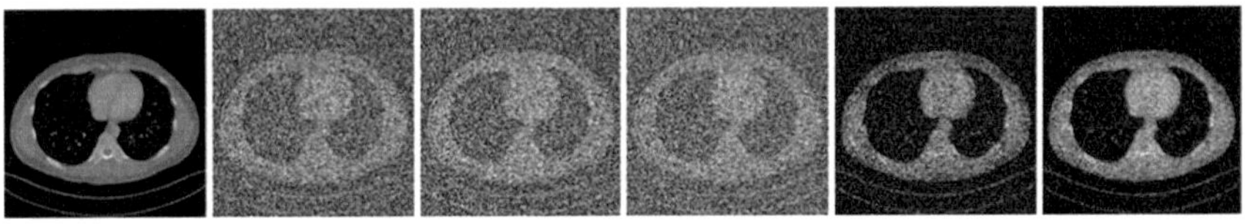

Fig. 13 Results for CT dataset for measurements with additive Gaussian noise. From left to right: ground truth, Fienup (PSNR, 12.37 dB; SSIM, 0.12), TWF (PSNR, 12.25 dB; SSIM, 0.12), WF (PSNR, 11.84 dB; SSIM, 0.11), PRred (PSNR, 12.96 dB; SSIM, 0.25), proposed (PSNR, 22.52 dB; SSIM, 0.47)

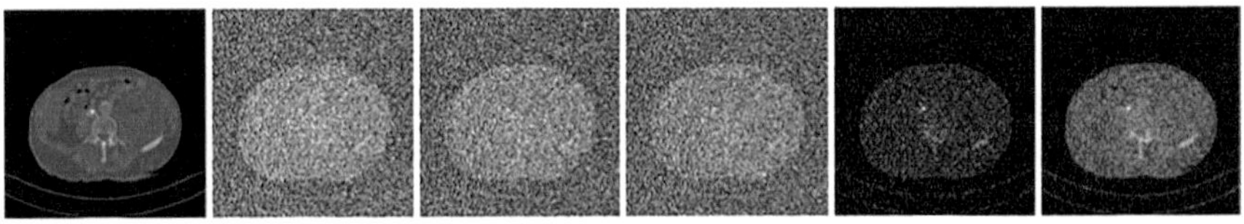

Fig. 14 Results for CT dataset for measurements with additive Gaussian noise. From left to right: ground truth, Fienup (PSNR, 13.34 dB; SSIM, 0.07), TWF (PSNR, 13.13 dB; SSIM, 0.06), WF (PSNR, 12.69 dB; SSIM, 0.05), PRred (PSNR, 16.83 dB; SSIM, 0.16), proposed (PSNR, 24.19 dB; SSIM, 0.38)

TWF, and WF, resulting in poor reconstruction performance. PRred is able to reduce the amount of noise, overpassing the performance of the classical methods. The proposed approach produces the best visual result for phase retrieval from noisy data among all the methods compared.

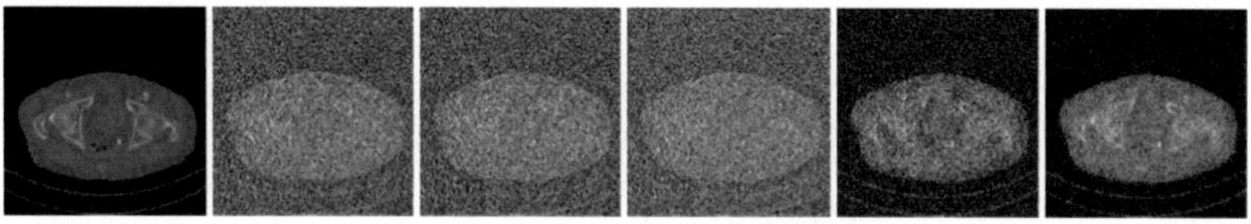

Fig. 15 Results for CT dataset for measurements with additive Gaussian noise. From left to right: ground truth, Fienup (PSNR, 14.29 dB; SSIM, 0.09), TWF (PSNR, 14.02 dB; SSIM, 0.08), WF (PSNR, 13.34 dB; SSIM, 0.07), PRred (PSNR, 14.50 dB; SSIM, 0.16), proposed (PSNR, 24.87 dB; SSIM, 0.42)

4.3 Ablation Study of MAGIC-Based Approach

We provide an ablation study of the proposed MAGIC-based approach for phase retrieval based on the noisy CT dataset. In the training stage, we have the option to exclude either the CNN component (i.e., no CNN) or the GCN component (i.e., no GCN) in order to investigate the influence of each component on the overall performance. As the proposed workflow (18) requires an initial condition $\mathbf{x}^0$, we additionally incorporate an experimental study on a post-processing enhancement module, aiming to improve the image quality by learning the CNN and GCN components, as elaborated in Sect. 3.2. We refer to this approach as the simpler denoising method.

We carry out four experiments: the first one employs our complete approach; the second uses the simpler enhancement formulation, i.e., only enhances the initial solution (which is computed via a handful of iterations of Fienup); the third experiment does not use graph-based regularization; and the fourth omits the CNN-based regularization. We denote these experiments as full-model, only-denoise, no-GCN, and no-CNN, respectively.

In terms of parameters, the full-model has 449,381 parameters per block as explained before (444,672 parameters per CNN component per block, 4708 parameters per GNN component per block, and one parameter of α per block); the only-denoise has the same number of parameters per block as full-model; the no-GCN has 444,673 parameters per block (444,672 parameters per CNN component per block and one parameter of α per block); and the no-CNN has 4709 parameters per block (4708 parameters per GCN component per block and one α parameter per block).

In all cases, we set a batch size of 64, initial learning rates of $\mathrm{LR}^0_{\mathrm{NN}} = 0.1$ and $\mathrm{LR}^0_{\alpha} = 0.0001$, and we initialize α to 0.01. As in the other cases, we train our models for 50 epochs using the ADAM optimizer with a decaying learning rate (set to decay by a multiplicative factor of 0.95 after each epoch). During training, we clip the α parameters to the range $[10^{-7}, 0.9]$ along with an early-stopping criterion. We report results estimated over the testing set using the best model.

We compare all the MAGIC-inspired variants in terms of PSNR and SSIM in Table 5 and show visual results in Figs. 16, 17, and 18. From these results, it is clear

Table 5 Ablation study of MAGIC-inspired variants on noisy CT dataset (standard deviation is provided in parenthesis)

	Full-Model	Only-Denoise	No-GCN	No-CNN
PSNR [dB]	22.98 (0.69)	24.44 (0.77)	24.08 (0.70)	18.10 (0.86)
SSIM	0.40 (0.03)	0.59 (0.03)	0.45 (0.03)	0.25 (0.04)
Number of parameters	4,493,810	4,493,810	4,446,730	47,090

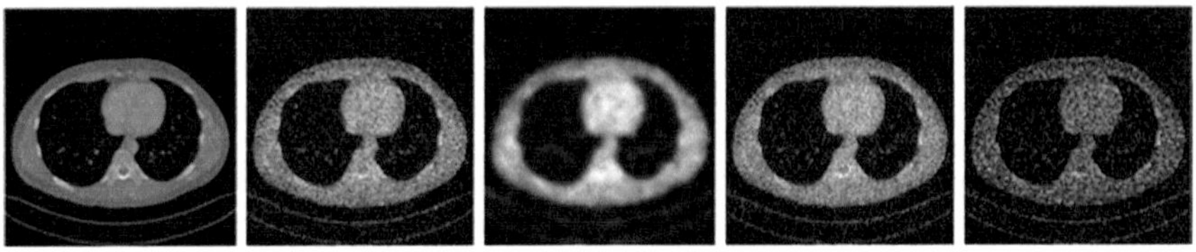

Fig. 16 Results for CT dataset for measurements with additive Gaussian noise. From left to right: ground truth, full model (PSNR, 22.52 dB; SSIM, 0.47), only denoise (PSNR, 23.44 dB; SSIM, 0.63), no GNN (PSNR, 23.38 dB; SSIM, 0.51), no CNN (PSNR, 18.69 dB; SSIM, 0.35)

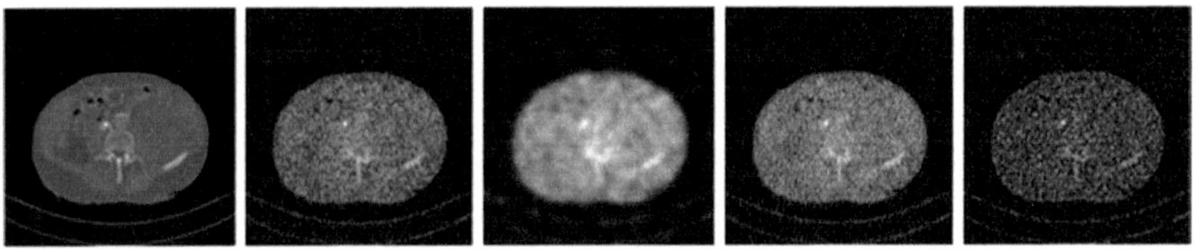

Fig. 17 Results for CT dataset for measurements with additive Gaussian noise. From left to right: ground truth, full model (PSNR, 24.19 dB; SSIM, 0.38), only denoise (PSNR, 25.99 dB; SSIM, 0.62), no GNN (PSNR, 25.31 dB; SSIM, 0.43), no CNN (PSNR, 19.60 dB; SSIM, 0.24)

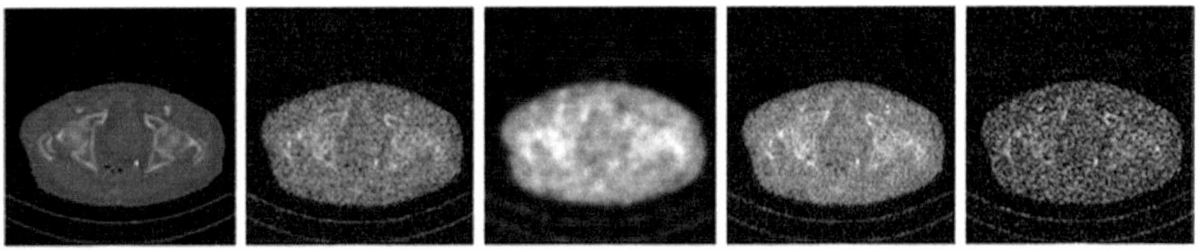

Fig. 18 Results for CT dataset for measurements with additive Gaussian noise. From left to right: ground truth, full model (PSNR, 24.87 dB; SSIM, 0.42), only denoise (PSNR, 26.62 dB; SSIM, 0.64), no GNN (PSNR, 26.11 dB; SSIM, 0.48), no CNN (PSNR, 20.12 dB; SSIM, 0.25)

that removing the CNN component, as in no-CNN, leads to a significant degradation in performance. This implies that the local features learned by CNN are more crucial than the nonlocal features learned by GCN. We also observe that the only-denoise model outperforms the proposed unrolling-based PR one, exhibiting the best metrics over all the methods. Nevertheless, the only-denoise visual results tend to be blurry. This variant may also have a more limited scope due to its high dependency on the initial condition. The no-GCN model yields better metrics than the full-model, reducing a bit more of the noise and capturing somewhat better the structure than the full-model without as much blurring as only-denoise. Note that we experiment with several hyperparameter configurations.[3] Although we do not claim the optimal configuration is guaranteed, in most cases, the resulting reconstructions follow the same trend, which are reported here. However, some configurations lead to unstable training (with loss reduction only in early epochs) or progress much more slowly than others. A more rigorous exploration of the hyperparameter configuration is out of the scope of this work. Overall, using the local operator helps reduce blurry artifacts in the reconstruction (as the only-denoise model). However, balancing this against the local and nonlocal smoothing components (CNN or GCN) is challenging within the additive context of $\mathbf{x}^{t+1}$ update in Eq. (18). This may be even more problematic when training with limited or noisy measurement data.

5 Conclusion and Future Work

We proposed an unrolling-based deep learning approach for phase retrieval. While the proposed method is applicable in a general sense, this chapter specifically focused on phase retrieval from Fourier measurements of masked images. We adopt data adaptive local and nonlocal regularization based on CNN and GCN. The proposed algorithm outperforms state-of-the-art methods in recovering phase information from noisy measurements. Future work includes the extension to general PR settings and more realistic PR applications.

The exploration of deep learning methods for phase retrieval is still ongoing, particularly in understanding their performance in noiseless scenarios and when the number of measurements is small—conditions where most classical methods tend to fail. While deep learning methods have shown great potential, especially in large-scale problems, realizing this potential often requires careful parameter tuning and significant engineering efforts, such as collecting more data or better balancing parameter initialization and learning rate initialization and decay. We believe that with further numerical studies and optimizations, deep learning approaches can achieve performance levels comparable to classic methods, such as Fienup, WF, and TWF, in noiseless cases; however, more research is needed to fully validate this.

Acknowledgments The authors would like to acknowledge the support from the Women in Data Science and Mathematics Research Workshop (WiSDM) hosted by IPAM at UCLA from August 7 to 11, 2023, which initiated the collaboration. C. Garcia-Cardona was funded by the Los Alamos National Laboratory LDRD Program Director's Initiatives (DI) project 20230771DI. Y. Liu is partially supported by NSF 2213436. Y. Lou is partially supported by NSF CAREER 2414705. S. Tang is partially supported by NSF 2111303 and NSF CAREER 2340631.

[3] We narrowly explore different k, as in the other data cases, but also change learning rate initialization, learning rate decay, and parameter initialization.

Competing Interests The authors have no conflicts of interest to declare that are relevant to the content of this chapter.

References

1. Allen, L., Oxley, M.: Phase retrieval from series of images obtained by defocus variation. Opt. Commun. **199**(1–4), 65–75 (2001)
2. Bai, S., Kolter, J.Z., Koltun, V.: Deep equilibrium models. In: Advances in Neural Information Processing Systems, vol. 32 (2019)
3. Balan, R., Casazza, P., Edidin, D.: On signal reconstruction without phase. Appl. Comput. Harmon. Anal. **20**(3), 345–356 (2006)
4. Bauschke, H.H., Combettes, P.L., Luke, D.R.: Phase retrieval, error reduction algorithm, and fienup variants: a view from convex optimization. JOSA A **19**(7), 1334–1345 (2002)
5. Beltran, M.A., Paganin, D.M., Uesugi, K., Kitchen, M.J.: 2d and 3d x-ray phase retrieval of multi-material objects using a single defocus distance. Opt. Express **18**(7), 6423–6436 (2010)
6. Bendory, T., Beinert, R., Eldar, Y.C.: Fourier phase retrieval: uniqueness and algorithms. In: Compressed Sensing and its Applications: Second International MATHEON Conference 2015, pp. 55–91. Springer, Berlin (2017)
7. Bendory, T., Edidin, D.: Algebraic theory of phase retrieval. Not. AMS **69**(9), 1487–1495 (2022)
8. Bohra, P., Pham, T.a., Dong, J., Unser, M.: Bayesian inversion for nonlinear imaging models using deep generative priors. IEEE Trans. Comput. Imaging **8**, 1237–1249 (2022)
9. Boyd, S., Parikh, N., Chu, E., Peleato, B., Eckstein, J., et al.: Distributed optimization and statistical learning via the alternating direction method of multipliers. Found. Trends® Mach. Learn. **3**(1), 1–122 (2011)
10. Candes, E.J., Li, X., Soltanolkotabi, M.: Phase retrieval via wirtinger flow: theory and algorithms. IEEE Trans. Inform. Theory **61**(4), 1985–2007 (2015)
11. Candes, E.J., Strohmer, T., Voroninski, V.: Phaselift: exact and stable signal recovery from magnitude measurements via convex programming. Commun. Pure Appl. Math. **66**(8), 1241–1274 (2013)
12. Chang, X., Bian, L., Zhang, J.: Large-scale phase retrieval. eLight **1**(1), 1–12 (2021)
13. Chen, H., Zhang, Y., Chen, Y., Zhang, J., Zhang, W., Sun, H., Lv, Y., Liao, P., Zhou, J., Wang, G.: Learn: learned experts' assessment-based reconstruction network for sparse-data ct. IEEE Trans. Med. Imaging **37**(6), 1333–1347 (2018)
14. Chen, Y., Candes, E.: Solving random quadratic systems of equations is nearly as easy as solving linear systems. In: Advances in Neural Information Processing Systems, vol. 28 (2015)
15. Deng, J., Dong, W., Socher, R., Li, L.J., Li, K., Fei-Fei, L.: Imagenet: a large-scale hierarchical image database. In: 2009 IEEE Conference on Computer Vision and Pattern Recognition, pp. 248–255 (2009). https://doi.org/10.1109/CVPR.2009.5206848
16. Dosovitskiy, A., Beyer, L., Kolesnikov, A., Weissenborn, D., Zhai, X., Unterthiner, T., Dehghani, M., Minderer, M., Heigold, G., Gelly, S., et al.: An image is worth 16×16 words: transformers for image recognition at scale (2020). arXiv preprint arXiv:2010.11929
17. Estupiñán, J., Jerez, A., Bacca, J., Arguello, H.: Deep unrolled phase retrieval approach from coded diffraction patterns. In: 2021 XXIII Symposium on Image, Signal Processing and Artificial Vision (STSIVA), pp. 1–4. IEEE, Piscataway (2021)
18. Fannjiang, A., Liao, W.: Phase retrieval with random phase illumination. JOSA A **29**(9), 1847–1859 (2012)
19. Fannjiang, A., Strohmer, T.: The numerics of phase retrieval. Acta Numerica **29**, 125–228 (2020)
20. Fienup, J.R.: Phase retrieval algorithms: a comparison. Appl. Opt. **21**(15), 2758–2769 (1982)

21. Gan, W., Zhai, Q., McCann, M.T., Cardona, C.G., Kamilov, U.S., Wohlberg, B.: Ptychodv: vision transformer-based deep unrolling network for ptychographic image reconstruction. IEEE Open J. Signal Process. **5**, 539–547 (2024)
22. Gerchberg, R.W.: Phase determination from image and diffraction plane pictures in the electron microscope. Optik **34**, 275–284 (1971)
23. Gilton, D., Ongie, G., Willett, R.: Deep equilibrium architectures for inverse problems in imaging. IEEE Trans. Comput. Imaging **7**, 1123–1133 (2021)
24. Goldstein, T., Studer, C.: Phasemax: convex phase retrieval via basis pursuit. IEEE Trans. Inform. Theory **64**(4), 2675–2689 (2018)
25. Goldstein, T., Studer, C., Baraniuk, R.: A field guide to forward-backward splitting with a fasta implementation (2014). arXiv preprint arXiv:1411.3406
26. Gonzalez, H.A., Liu, C., Vogginger, B., Kumaraveeran, P., Mayr, C.G.: Doppler disambiguation in MIMO FMCW radars with binary phase modulation. IET Radar Sonar Navigation **15**(8), 884–901 (2021)
27. Goy, A., Arthur, K., Li, S., Barbastathis, G.: Low photon count phase retrieval using deep learning. Phys. Rev. Lett. **121**(24), 243902 (2018)
28. Gregor, K., LeCun, Y.: Learning fast approximations of sparse coding. In: International Conference on Machine Learning (2010). https://api.semanticscholar.org/CorpusID:13333501
29. Gu, J., Wang, Z., Kuen, J., Ma, L., Shahroudy, A., Shuai, B., Liu, T., Wang, X., Wang, G., Cai, J., et al.: Recent advances in convolutional neural networks. Pattern Recognit. **77**, 354–377 (2018)
30. Guizar-Sicairos, M., Fienup, J.R.: Phase retrieval with transverse translation diversity: a nonlinear optimization approach. Opt. Express **16**(10), 7264–7278 (2008)
31. Hand, P., Leong, O., Voroninski, V.: Phase retrieval under a generative prior. In: Advances in Neural Information Processing Systems, vol. 31 (2018)
32. Hore, A., Ziou, D.: Image quality metrics: PSNR vs. SSIM. In: 2010 20th International Conference on Pattern Recognition, pp. 2366–2369. IEEE, Piscataway (2010)
33. Hu, K., Sun, D., Zhao, Y.: Enhanced single-frame interferometry via hybrid conv-transformer architecture for ultra-precise phase retrieval. Opt. Express **32**(17), 30226–30241 (2024)
34. Kamilov, U.S., Bouman, C.A., Buzzard, G.T., Wohlberg, B.: Plug-and-play methods for integrating physical and learned models in computational imaging: theory, algorithms, and applications. IEEE Signal Process. Mag. **40**(1), 85–97 (2023)
35. Kipf, T.N., Welling, M.: Semi-supervised classification with graph convolutional networks (2016). arXiv preprint arXiv:1609.02907
36. Manekar, R., Tayal, K., Kumar, V., Sun, J.: End to end learning for phase retrieval. In: ICML Workshop on ML Interpretability for Scientific Discovery (2020)
37. Marchesini, S.: Invited article: a unified evaluation of iterative projection algorithms for phase retrieval. Rev. Sci. Instrum. **78**(1) (2007)
38. McCollough, C.: TU-FG-207A-04: overview of the low dose ct grand challenge. Med. Phys. **43**(6Part35), 3759–3760 (2016). https://doi.org/10.1118/1.4957556
39. Metzler, C., Schniter, P., Veeraraghavan, A., Baraniuk, R.: prdeep: robust phase retrieval with a flexible deep network. In: International Conference on Machine Learning, pp. 3501–3510. PMLR (2018)
40. Metzler, C.A., Heide, F., Rangarajan, P., Balaji, M.M., Viswanath, A., Veeraraghavan, A., Baraniuk, R.G.: Deep-inverse correlography: towards real-time high-resolution non-line-of-sight imaging. Optica **7**(1), 63–71 (2020)
41. Miao, J., Charalambous, P., Kirz, J., Sayre, D.: Extending the methodology of x-ray crystallography to allow imaging of micrometre-sized non-crystalline specimens. Nature **400**(6742), 342–344 (1999)
42. Miao, J., Sayre, D.: On possible extensions of x-ray crystallography through diffraction-pattern oversampling. Found. Crystallogr. **56**(6), 596–605 (2000)
43. Millane, R.P.: Phase retrieval in crystallography and optics. JOSA A **7**(3), 394–411 (1990)

44. Monga, V., Li, Y., Eldar, Y.C.: Algorithm unrolling: interpretable, efficient deep learning for signal and image processing. IEEE Signal Process. Mag. **38**(2), 18–44 (2021). https://doi.org/10.1109/MSP.2020.3016905
45. Roddier, C., Roddier, F.: Wave-front reconstruction from defocused images and the testing of ground-based optical telescopes. JOSA A **10**(11), 2277–2287 (1993)
46. Romano, Y., Elad, M., Milanfar, P.: The little engine that could: regularization by denoising (red). SIAM J. Imaging Sci. **10**(4), 1804–1844 (2017)
47. Roth, S., Black, M.J.: Fields of experts. Int. J. Comput. Vision **82**, 205–229 (2009)
48. Uelwer, T., Oberstraß, A., Harmeling, S.: Phase retrieval using conditional generative adversarial networks. In: 2020 25th International Conference on Pattern Recognition (ICPR), pp. 731–738. IEEE, Piscataway (2021)
49. Vaswani, A.: Attention is all you need (2017). arXiv preprint arXiv:1706.03762
50. Waldspurger, I., d'Aspremont, A., Mallat, S.: Phase recovery, maxcut and complex semidefinite programming. Math. Program. **149**, 47–81 (2015)
51. Wirtinger, W.: Zur formalen theorie der funktionen von mehr komplexen veränderlichen. Mathematische Annalen **97**(1), 357–375 (1927)
52. Xia, W., Lu, Z., Huang, Y., Shi, Z., Liu, Y., Chen, H., Chen, Y., Zhou, J., Zhang, Y.: Magic: manifold and graph integrative convolutional network for low-dose ct reconstruction. IEEE Trans. Med. Imaging **40**(12), 3459–3472 (2021)
53. Yurtsever, A., Udell, M., Tropp, J., Cevher, V.: Sketchy decisions: convex low-rank matrix optimization with optimal storage. In: Artificial intelligence and statistics, pp. 1188–1196. PMLR (2017)
54. Zhang, K., Zuo, W., Chen, Y., Meng, D., Zhang, L.: Beyond a gaussian denoiser: residual learning of deep CNN for image denoising. IEEE Trans. Image Process. **26**(7), 3142–3155 (2017)

Performance Analysis of MFCC and wav2vec on Stuttering Data

Venera Adanova and Maksat Atagoziev

1 Introduction

Over the last decade, speech recognition systems have evolved dramatically. Thanks to advances in machine and deep learning, we can witness automatic recognition systems (ASR) such as Alexa, Siri, or Google with astonishing performance. However, these systems are trained on fluent speech and fail to recognize speech with disorders, such as stuttering.

Stuttering, aka stammering, is a complex speech disorder that negatively affects the communication ability of 1% of the population. Persons who stutter (PWS) often know what they want to say; however, the speech is interrupted by involuntary pauses and word or sound repetitions. Identification of stuttering in a speech is a challenging problem involving multiple disciplines such as pathology, psychology, acoustics, and signal processing.

The majority of studies conducted on stuttering data aim to detect and identify the dysfluency types in audio recordings. These types of dysfluency are generally defined as blocks, prolongations, sound/word/phrase repetitions, and interjections [15, 19]. Blocks are defined as involuntary pauses before words. Prolongations are elongated syllable, like *I am s[sss]ory*. Repetitions involve sound, word, or phrase repetitions. For example, *I made [made] dinner* represents a word repetition. In order to avoid above-defined stuttering types, a person who stutters learns to use filler words like “*um*, *uh*, *you know*, etc.” These filler words are

V. Adanova (✉)
TED University, Ankara, Turkey
e-mail: venera.adanova@tedu.edu.tr

M. Atagoziev
Ostim Technical University, Ankara, Turkey
e-mail: maksat.atagoziev@ostimteknik.edu.tr

C. Garcia-Cardona, H. Lee (eds.), *Advances in Data Science*, Association for Women in Mathematics Series 37, https://doi.org/10.1007/978-3-031-87804-6_14

known as interjections. Note that the dysfluency types might be named differently in different studies.

The ability to differentiate between stuttering types enables improvements in the design of assistive speech technologies. For instance, speech recognition software could be optimized to handle prolongations differently from repetitions. Knowing the type of dysfluency allows developers to fine-tune how technology interacts with stuttered speech.

It would also be very useful for speech therapists. Currently, speech therapists record audio of their patients while they speak and then manually annotate the stuttering types observed in the speech. Based on the frequency of stuttering types, the severity of speech disorder is identified. The improvements in patient's speech after the therapy are also identified by the same process. This manual intervention limits treatment to the confines of the therapist's office. Automatic detection and identification ability can significantly enhance treatment strategies by providing real-time, objective analysis of speech patterns. Automated systems can quickly identify and categorize dysfluencies, allowing speech therapists to track progress more precisely and adjust interventions dynamically based on detailed data. This automation also enables the development of personalized therapy tools, such as mobile apps that provide instant feedback, and improves speech recognition technologies by making them more adaptive to dysfluent speech. Additionally, automatic detection can accelerate research by providing large-scale, consistent data for studying stuttering.

Though ASR systems have evolved, studies involving stuttering detection and identification are scarce. The main reason for the deficiency of studies in stuttered speech is the lack of data. Many studies in this field use in-house datasets, which are small, manually labelled datasets. These types of datasets are not publicly available. Some publicly available datasets, such as UCLASS, are not labelled. Even if the datasets are publicly available and labeled, it is highly imbalanced where more than half of the dataset contains fluent data and the other half is shared among different dysfluency types. Just like any speech-related problem, detecting stuttering requires lots of data for accurate learning.

Typically, works conducted on stuttering detection and identification are done based on some datasets, either in-house or public, learn to classify between fluent and dysfluent speech (stuttering detection), and distinguish dysfluency types (identification).

The works that use in-house datasets [1, 8, 9, 11, 17] typically use small self-labeled data, which is not shared publicly.

There is only a handful number of publicly available stuttering datasets. The very first and also the smallest one is the UCLASS dataset [10]. It contains 457 audio recordings of monologues, conversations, and readings, and only small amount of them has transcriptions. The dataset is not labeled according to dysfluency types.

The FluencyBank dataset [18] contains audio and video files with transcriptions for the interviews conducted for 32 adults and children who stutter. However, the dataset is not labeled.

The scarcity of labeled data led to the creation of synthetic dataset, LibriStutter [13], which consists of 50 speakers (approximately 20 hours). The dataset was generated by injecting random stuttering to LibriSpeech dataset, which consists of fluent speech. The audio signals were segmented into four-second windows, and for every window, either one of the stuttering events, as sound, word, and phrase repetitions, prolongations, and interjections, was injected or left untouched.

Bayerl et al. [4] suggest their own dataset, namely, Kassel State of Fluency (KSoF), which consists of 5500 clips of stuttered speech in German. The clips were labeled with the six stuttering event types: blocks, prolongations, sound/word/phrase repetitions, interjections, and speech modifications. The last type is therapy specific and indicates whether the speaker's speech is modified after the therapy. The dataset also has some metadata, like the gender of a speaker, therapy status, type of microphone used, etc.

The largest dataset, *Stuttering Events in Podcasts* (SEP-28k), was released recently by Lea et al. [15]. SEP-28k is the first publicly available annotated dataset. It contains about 28,000 3-second clips from podcast recordings. The SEP28k corpus also has 4144 3-second annotated clips from the FluencyBank dataset. Bayerl et al. [6] subsequently introduced an extended SEP-28k, which contains also the gender and speaker information. Along with the extended data, they proposed possible partitioning ways of data into train and test set.

Typically, studies [2, 14, 20] choose Mel-frequency cepstral coefficients (MFCC) as the feature representation for audio clips. Lately, some of the studies [6, 7] proposed using features extracted from pretrained wav2vec 2.0 [3] network. Wav2vec network is learned on a large amount of fluent speech, takes raw data as an input, and produces a feature vector describing each audio data.

In this work, we perform our experiments on the subset of SEP28k dataset and compare predictive powers of MFCC and wav2vec feature representations. We use simple Siamese network having a single task model, which learns to differentiate between stuttering types, and then gradually convert the model into multitask learner with multiple heads. We then observe improvements that these transformations bring into the classification task.

2 Dataset

In our study, we utilize a subset of the SEP28k dataset. The audio recordings were sourced from eight different shows, with each episode being segmented into 3-second clips. The dataset comprises a total of 28,177 clips, equivalent to approximately 23.5 hours of audio. Each clip was annotated by three annotators, with annotations categorized into two types: stuttering and non-stuttering. Stuttering types encompass various dysfluency forms such as prolongation, block, interjection, and sound/word repetition, as well as instances where no stuttering occurred. Non-stuttering types include unintelligible speech, natural pauses, uncertainty, background music, and poor audio quality. Our primary focus lies on analyzing the stuttering types within the dataset.

While SEP28k is the only available large dataset, it is indeed a very challenging one. One of the challenges comes from the fact that it is very imbalanced. More than half of the data contains fluent speech, and approximately 10% is given for a particular dysfluency type. Second, each clip might have several annotations. Thus, a single clip might contain both prolongation and block dysfluency types while being also annotated by one of the annotators as a fluent speech. Lastly, it is also imbalanced in terms of speaker. Thus, host speech dominates 60% of the data. Also, the distribution of shows is imbalanced. The number of clips for one of the shows, Women Who Stutter, form 33% of overall clips. Considering that Women Who Stutter and He Stutters share the same host (Pamela Mertz), a large amount of clips is dominated by the speech of a single person.

The labels given by annotators can be misleading. Williams and Kent [21] reported on the results of a study where college students listen to a recording of an adult speaker imitating various types of dysfluency. On one occasion, they were instructed to mark all "stuttered" interruptions on a transcript of the recording, and on another presentation of the same recording, they were told to mark all "normal" interruptions. It was observed that people tend to hear what they were instructed to listen for. Hence, those interruptions that were marked as stuttered under one set of instructions were marked as normal interruptions under the other. The authors called this phenomena as "confusion." Based on this study, we decided to narrow the size of our dataset by including only the annotations that were agreed upon all three annotators. We construct a smaller subset from SEP28k, which we call confidence list. It consists of clips that were assigned to the same type by all three annotators. However, the interjection type never was annotated alone, as it also belongs to no stuttered word type. Hence, we also include the clips for which three annotators selected both no stuttered words and interjections. Moreover, the clips where all three selected both no stuttered words and natural pause are also included as fluent speech. We also form confidence list for FluencyBank. However, only dysfluency typed clips were included to the dataset, as the percentage of fluent clips already form the large portion of the dataset. There are overall 3901 clips in our dataset, and the distribution of different types is illustrated in Fig. 1.

3 Proposed Framework

Given the audio clips, we initially extract their MFCCs and wav2vec representations. The extracted MFCCs and wav2vecs are further fed to our baseline model, by training which we learn new embeddings (features) for the clips. The baseline model that we use to extract embeddings is the Siamese network with contrastive loss shown in Fig. 3 (shaded area). The choice of this network is not random. It was shown in [12] that Siamese networks learn well under the scenarios with a small number of representatives from each class. This is indeed the case for our data.

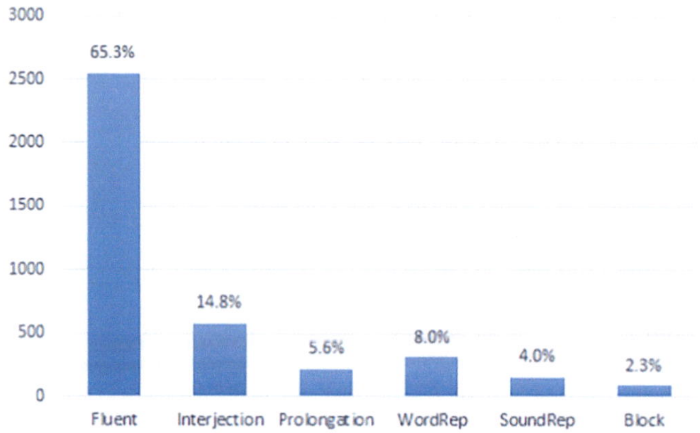

Fig. 1 The distribution of stuttering types in our dataset. Observe that fluent data form 65% of the dataset

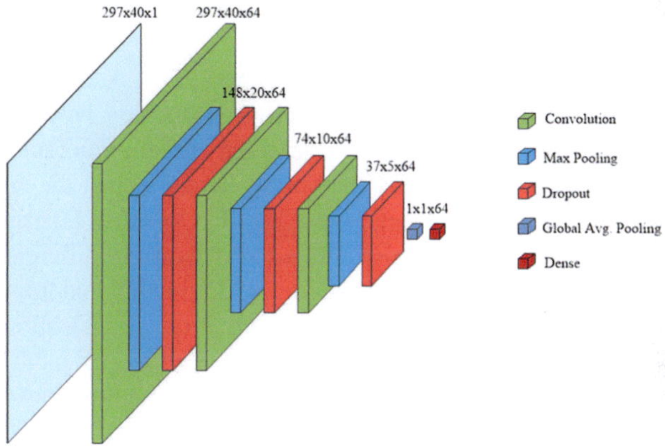

Fig. 2 The architecture of our subnetwork

There are two inputs to the network and two identical subnetworks that share the weights. For each pair, the two subnetworks produce embeddings that are then used to compute the Euclidean distance between a pair of inputs. The main goal of the network is not to learn to classify different stuttering types but to differentiate between them.

The subnetwork consists of three blocks. Each block contains a convolutional layer followed by max pooling and dropout layers. The last layer does global averaging, which returns the desired 64×1 dimensional vector. The details on input and output dimensions are illustrated in Fig. 2.

As a loss function for the baseline model, we use contrastive loss, which is defined as the following:

$$L_{baseline} = y \cdot d^2 + (1 - y) \cdot \max(margin - d, 0)^2 \tag{1}$$

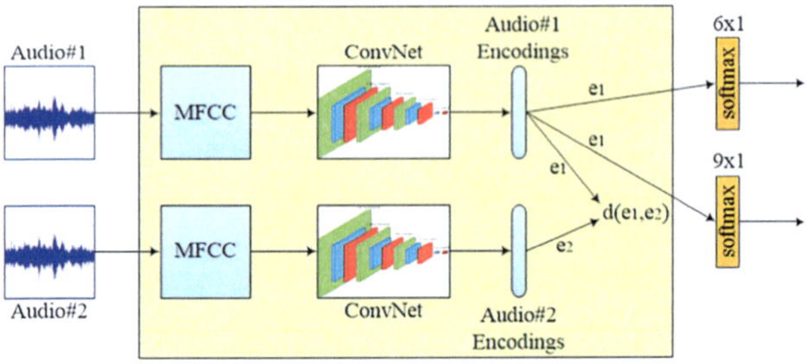

Fig. 3 Proposed framework. The model within the yellow box is the baseline model. An example of converting the baseline model to MTL by adding two classification heads is illustrated. The first classification head learns to categorize stuttering types; the second head learns to differentiate between different shows

where y is a true label, 1 if the audio pairs are of the same class and 0 otherwise, and d is the Euclidean distance between the outputs of twin network embeddings. Margin is 1.

Our baseline model does a single-task learning (STL). Adding additional auxiliary heads to the network can increase the generalization power of the model [5, 20, 22]. During our experiments, we extend the model to do multitask learning (MTL). Thus, we add a classification head to the model so that the model can also learn to classify between the six stuttering types. In our other experiment, we add another classification head, which also forces the network to differentiate between different shows. A Siamese network along with possible additional auxiliary classification heads is shown in Fig. 3, where each of the classification heads is fed with the new features computed for the first input of the baseline model. For both of these classification heads, we use sparse categorical cross-entropy loss function.

Hence, the overall loss of an MTL model is given by:

$$L = \lambda_{baseline} \cdot L_{baseline} + \lambda_{auxiliary} \cdot L_{auxiliary} \tag{2}$$

As was mentioned before, our model learns new features (embeddings) for the audio data, which are subsequently fed to machine learning models for classification purposes.

4 Experimental Results

4.1 Features

We first compute MFCC and wav2vec features for every audio clip, which are then fed to our models. All clips are read with the sampling frequency of 16,000.The MFCC features are computed using *speechpy* library, using the frame length of

0.025, frame stride of 0.01, and number of filters of 40. For every 3-second-long clip, using these parameters, we produce 297×40 features. wav2vec features are obtained from the last layer of wav2vec network, which produces 149×768 features.

4.2 Data Augmentation

The data is highly imbalanced, so we produce new samples using augmentation techniques. The augmented data is used only during the training process. Since fluent speech already takes up to 65% of the data, we only augmented the clips with dysfluency labels.

Data augmentation is done using $audiomentations$ library. For every dysfluent clips in the training set, we add Gaussian noise with a minimum amplitude of 0.001 and a maximum amplitude of 0.015, time stretch up/down to 25%, and shift pitch up/down 4 semitones.

4.3 Data Splitting

The data is divided into three: train, validation, and test sets. We experiment with two different splitting techniques. In the first split type, which we call $frequency$ $split$, the train set contains clips of 10 most frequent speakers, plus the dysfluency clips from FluencyBank, which gives us overall 2434 clips. What is left is divided between validation and test sets. Thus, validation set consists of 734 clips of next most-frequent speakers, and the test set contains 733 clips of less-frequent speakers. The second split type is just a random split, where 20% of the dataset is given to both validation and test sets by random assignment.

4.4 Training

We train four different models: baseline model, baseline model with stuttering classification head, baseline model with stuttering and show classification heads, and baseline model with stuttering, show, and binary classification heads. We call them BM, BSM, BSSM, and BSSBM, respectively. The latter three are MTL models.

We use Adam optimizer with $learning\_rate = 0.001$, and the batch size is 8. A larger batch size could be used for MFCC features, but for wav2vec, it is impossible due to its size and our GPU limitations. Hence, we kept the batch size equally small for both kinds of features. We use the early stopping technique with a patience of 5, monitoring the loss function on the validation set. The number of epochs is kept 100 for both representations. In our experiments, early stopping was triggered after around 10 epochs for MFCC feature for all model types. For wav2vec features, it was triggered after around 30 epochs.

For the first MTL model, the weights of losses are equal; thus, $\lambda_{baseline} = 0.5$, and $\lambda_{stuttering} = 0.5$. For the second MTL model, we pay less importance to the show classification head as it is used more like regularization; hence, the weights are distributed as $\lambda_{baseline} = 0.4$, $\lambda_{stuttering} = 0.4$, and $\lambda_{show} = 0.2$. Lastly, the fourth model classification heads are distributed as $\lambda_{baseline} = 0.3$, $\lambda_{stuttering} = 0.3$, $\lambda_{show} = 0.1$, and $\lambda_{binary} = 0.3$.

4.5 Results

After training the aforementioned four models, we extract new features from the data and perform classification using K-nearest neighbor (KNN) with k= 7. The choice of a large k is to ensure the stability of classification results. The reported results illustrate the performance of KNN on test data, unless otherwise stated. We consider different classification scenarios. First, we consider how KNN classifies all types: dysfluent plus fluent speech. Second, we observe how it handles the binary case: when we combine all dysfluent types into one class and label them as the non-fluent class and observe the performance of the model on fluent versus non-fluent classification. Lastly, we consider only dysfluent types and observe how KNN performs when fluent speech is excluded.

F1 score results for the classification of all stuttering types, when the frequency split is used, are given in Table 1. Observe that the results are the worst for the block type. This is because it has the fewest observations. It is natural to have high results for the fluent type as 65% of data consists of fluent speech. Note that while MFCC has better prediction power on the interjection type, wav2vec is better in predicting prolongations.

Table 2 shows F1 scores for the random split case. The classification results are much better for prolongations and sound and word repetitions for this case. Note that while in frequency split case the maximum F1 score for prolongations is 0.28

Table 1 F1 score for stuttering classification for frequency split. (P, Prolongation; B, block; SR, sound repetition; WR, word repetition; I, interjection; F, fluent; BM, baseline model; BSM, baseline with stuttering classification head; BSSM, BSM with show classification head; BSSBM, BSSM with binary classification head)

	F1 Score MFCC/wav2vec					
Model	P	B	SR	WR	I	F
BM	0.11/0.08	0.04/0.00	0.06/0.03	0.08/0.13	0.19/0.15	0.69/0.74
BSM	0.28/0.17	0.23/0.04	0.16/0.13	0.13/0.12	0.38/0.20	0.77/0.71
BSSM	0.24/0.24	0.00/0.08	0.10/0.07	0.10/0.13	0.39/0.15	0.80/0.72
BSSBM	0.26/0.32	0.00/0.04	0.06/0.12	0.15/0.14	0.32/0.22	0.77/0.78

Table 2 F1 score for stuttering classification for random split. (P, Prolongation; B, block; SR, sound repetition; WR, word repetition; I, interjection; F, fluent; BM, baseline model; BSM, baseline with stuttering classification head; BSSM, BSM with show classification head; BSSBM, BSSM with binary classification head)

	F1 Score MFCC/wav2vec					
Model	P	B	SR	WR	I	F
BM	0.19/0.19	0.04/0.04	0.09/0.21	0.06/0.16	0.15/0.18	0.69/0.74
BSM	0.34/0.43	0.17/0.04	0.19/0.27	0.19/0.19	0.37/0.24	0.77/0.74
BSSM	0.34/0.43	0.10/0.16	0.09/0.21	0.12/0.23	0.35/0.26	0.75/0.77
BSSBM	0.34/0.25	0.11/0.11	0.10/0.18	0.17/0.15	0.34/0.23	0.74/0.70

Table 3 Accuracy results for both MFCC and wav2vec features

	Accuracy		
Model	MFCC	/	wav2vec
BM	0.49	/	0.55
BSM	0.62	/	0.57
BSSM	0.57	/	0.61
BSSBM	0.57	/	0.52

Table 4 F1 score for binary classification

	F1-Score MFCC/wav2vec	
Model	Fluent	Non-fluent
BM	0.66/0.73	0.43/0.49
BSM	0.77/0.73	0.51/0.51
BSSM	0.74/0.76	0.52/0.51
BSSBM	0.72/0.68	0.47/0.46

for all model and feature types, in the random split case, it improves up to 0.43. Since the classification results for the random split are better, we will proceed with the random split data in our further analysis. MFCC produces the best classification results for all disfluency types when BSM model is used. For wav2vec, the best results are obtained for the BSSM model. This is also true for accuracy results, as shown in Table 3.

When the embeddings from the models are used to perform binary classification, i.e., fluent versus all dysfluency types, the results are slightly better for MFCC representations for both fluent and non-fluent types in MTL models, as shown in Table 4. The STL model gives better results with wav2vec.

Table 5 shows the F1 scores when only the embeddings of dysfluent types are classified. In this case, the fluent type was discarded. Again, we observe that MFCC representations perform well in predicting the interjection type and wav2vec representations are better at predicting prolongations. Sound and word repetitions are better predicted with MFCC, and for block types, both are equally poor.

To visualize the embeddings learned by the models, we use the t-stochastic neighborhood embedding technique (tSNE) introduced by van der Maaten and

Table 5 F1 score for dysfluency type classification. (P, Prolongation; B, block; SR, sound repetition; WR, word repetition; I, interjection; BM, baseline model; BSM, baseline with stuttering classification head; BSSM, BSM with show classification head; BSSBM, BSSM with binary classification head)

Model	F1 Score MFCC/wav2vec				
	P	B	SR	WR	I
BM	0.28/0.35	0.05/0.06	0.36/0.26	0.28/0.36	0.42/0.35
BSM	0.45/0.62	0.18/0.15	0.45/0.43	0.46/0.32	0.54/0.50
BSSM	0.43/0.55	0.19/0.23	0.24/0.24	0.48/0.41	0.61/0.49
BSSBM	0.48/0.39	0.15/0.12	0.24/0.43	0.43/0.28	0.60/0.45

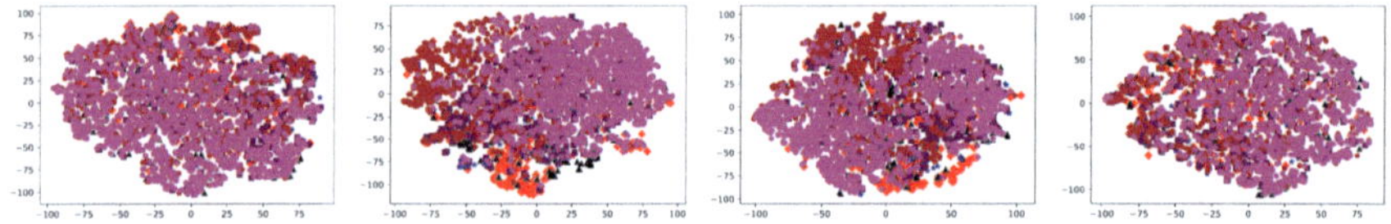

(a) (b) (c) (d)

Fig. 4 tSNE results of the embeddings extracted for the training set from different models learnt using MFCCs. (**a**) BM, (**b**) BSM, (**c**) BSSM, (**d**) BSSBM

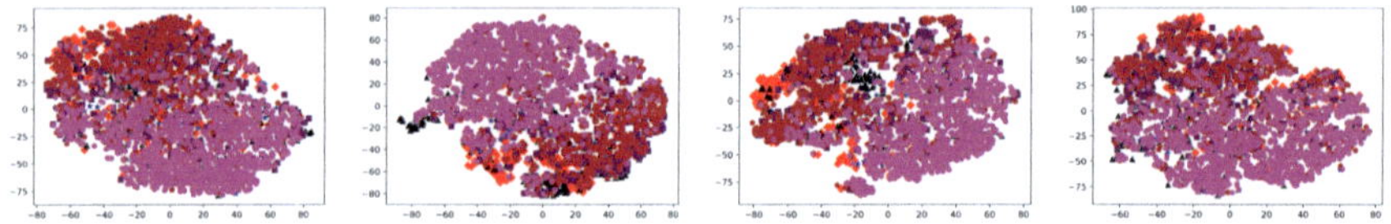

(a) (b) (c) (d)

Fig. 5 tSNE results of the embeddings extracted for the training set from different models learnt using wav2vecs. (**a**) BM, (**b**) BSM, (**c**) BSSM, (**d**) BSSBM

Hinton [16]. The train set embeddings are reduced to two dimensions and are colored according to their stuttering type. The embeddings learned for the training data with MFCCs as feature representation are illustrated in Fig. 4. Observe that the STL model (Fig. 4a) is not able to learn the stuttering categories. When the stuttering classification head is included in the network, it begins to learn the embeddings that represent different stuttering types (Fig. 4b). However, as we increase the number of auxiliary classification heads, the clusters start to merge due to the regularization introduced by the extra classification heads (Fig. 4d).

The train set embeddings with wav2vec results are shown in Fig. 5. Observe that in BSSBM model, the cluster boundaries start to fade. This is also clear form of the classification results on training set given in Table 6. We can see a significant performance decrease with the extra classification head, especially for wav2vec results.

Table 6 F1 score for stuttering classification of training set. (P, Prolongation; B, block; SR, sound repetition; WR, word repetition; I, interjection; F, fluent; BM, baseline model; BSM, baseline with stuttering classification head; BSSM, BSM with show classification head; BSSBM, BSSM with binary classification head)

	F1 Score MFCC/wav2vec					
Model	P	B	SR	WR	I	F
BM	0.60/0.61	0.50/0.48	0.45/0.53	0.55/0.57	0.63/0.66	0.71/0.79
BSM	0.79/0.71	0.63/0.52	0.60/0.63	0.61/0.58	0.78/0.65	0.81/0.80
BSSM	0.76/0.73	0.57/0.52	0.59/0.61	0.63/0.61	0.78/0.69	0.80/0.82
BSSBM	0.74/0.62	0.62/0.52	0.58/0.48	0.63/0.57	0.77/0.58	0.81/0.75

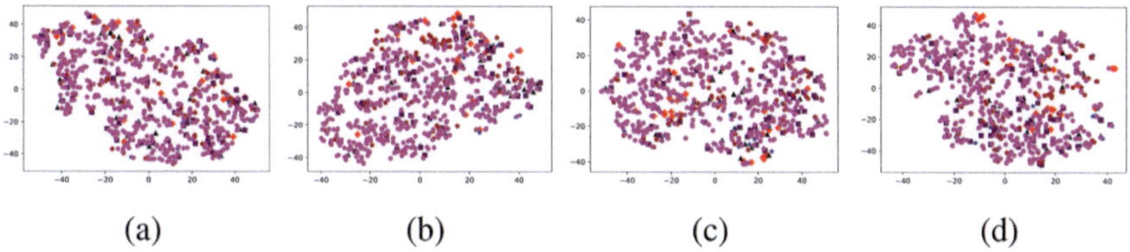

Fig. 6 tSNE results of the embeddings extracted for the test set from different models learnt using MFCCs. (**a**) BM, (**b**) BSM, (**c**) BSSM, (**d**) BSSBM. The observations are colored based on the stuttering type

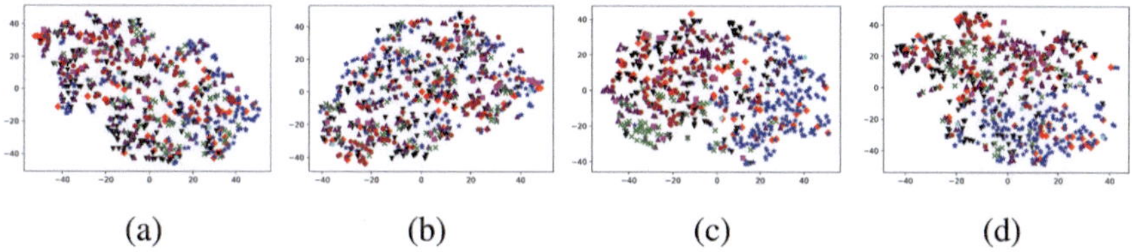

Fig. 7 tSNE results of the embeddings extracted for the test set from different models learnt using MFCCs. (**a**) BM, (**b**) BSM, (**c**) BSSM, (**d**) BSSBM. The observations are colored based on the show type

While for the training data the embeddings form clusters according to the stuttering types, the same cannot be said for the test data. Observe the tSNE results for the embeddings extracted for the test set shown in Fig. 6. No clusters can be observed, just a magenta class, which represents the fluent class, covering all over the space. However, if we color the observations in Fig. 6 according to the show that they were taken from, then we can see some patterns of clustering as shown in Fig 7. Somehow, the test data is greatly influenced by the show type rather than the stuttering type. Seems like during training the data, our models also capture the specifics of the shows, like gender of a speaker, music on the background, etc.

For comparison of our stuttering identification results, we refer to the results presented by [20]. In this work, the authors investigate how multitask learning (MTL) and adversarial learning (ADV) models perform on the stuttering detection

Table 7 Comparison of stuttering classification results of our models with the models in [20]. (P, Prolongation; B, block; R, average of sound repetition and word Repetition; I, interjection; F, fluent; MTL, multitask learning; ADV, adversarial learning; BSM, baseline with stuttering classification head; BSSM, BSM with show classification head)

		F1 Score				
Model	Representation	P	B	R	I	F
MTL [20]	MFCC	0.36	0.21	0.34	0.54	0.67
ADV [20]	MFCC	0.37	0.20	0.35	0.58	0.66
BSM	MFCC	0.34	0.17	0.19	0.37	0.77
BSM	wav2vec	0.43	0.04	0.23	0.24	0.74
BSSM	MFCC	0.34	0.10	0.11	0.35	0.75
BSSM	wav2vec	0.43	0.16	0.22	0.26	0.77

problem. Their models are evaluated on the SEP-28k dataset. Recall that in our experiments, we use only a small fraction of the original dataset, as we included only those clips for which all annotators agreed on the annotation. The work in [20] used almost the entire dataset with some cleaning. The clips are represented using MFCCs. Since the work combined sound repetitions with word repetitions and considered them as a single repetition type, we report the average results for these two under one repetition type. The results are given in Table 7. Our models, built on a smaller dataset, perform better in classifying prolongations and fluent types.

5 Discussion

The SEP-28k dataset, as discussed earlier, is a challenging dataset. It is highly imbalanced, with more than 60% of the dataset consisting of fluent speech, while the remainder is distributed among disfluent types. As suggested by Lea et al. [15], interjections are the easiest type to recognize, which is also evident from our results. Blocks are difficult to detect because the gasp for breath or pause is often inaudible and may require visual accompaniment. Additionally, there are very few samples of blocks. The results shown in Table 7 illustrate how challenging this type is. Sound repetitions are also difficult to detect because syllables can vary in duration, count, style, and articulation. However, since the percentage of speech with repetitions is much higher than that of blocks, the identification results for repetitions are much better.

Obviously, we need more labeled data to train models with high performance. SEP-28k is not a small dataset. However, its annotations are unreliable for most of the audio clips. The curators of SEP-28k [15] reported inter-annotator agreement measurements for different disfluency types. The results show that word repetitions, interjections, sound repetitions, and no disfluencies are more consistent (0.62, 0.57, 0.40, 0.39), while blocks and prolongations had only fair or slight agreement (0.25,

0.11). This unreliability is also reflected in Table 7. Although the work in [20] used almost the entire SEP-28k dataset, their performance results for block types are not significantly better than ours. Furthermore, the identification of prolongations is better with our models.

6 Summary and Conclusion

In this work, we performed our experiments on a subset of the SEP-28k dataset and compared the predictive powers of MFCC and wav2vec feature representations. The performance of both feature representations was observed to be almost the same, with MFCC having a slight edge. We found that wav2vec performs better on prolongations, sound repetitions, and word repetitions. While wav2vec generally is slightly inferior to MFCC, it also has the disadvantage of being computationally expensive. Its large size requires more memory and makes it slower to train. Therefore, MFCC appears to be a better option.

We also observed that although our models have learned embeddings to represent different stuttering types, they do not generalize well. When visualizing the test data embeddings, no distinct clusters according to stuttering type are apparent. However, when we color the visualization according to show types, some groups become noticeable. We believe this issue arises from additional data in the audio clips, such as the speaker's gender and background music. Therefore, techniques to suppress metadata in the dataset should be developed.

Competing Interests This study was funded by TEDU BAP Grant No. T-22-B2010-90108.

References

1. Amruth, V., Lavanya, K., Manoj, N., Umme, H., Deepika, M.B.: A novel approach for stutter speech recognition and correction. Int. J. Res. Appl. Sci. Eng. Technol. **8**, 544–547 (2020)
2. Arjun, K.N., Karthik, S., Kamalnath, D., Chanda, P., Tripath, S.: Automatic correction of stutter in dysfluent speech. Proc. Comput. Sci. **171**, 1363–1370 (2020)
3. Baevski, A., Zhou, Y., Mohamed, A., Auli, M.: wav2vec 2.0: A framework for self-supervised learning of speech representations. In: Advances in Neural Information Processing Systems, pp. 12449–12460 (2020)
4. Bayerl, S., von Gudenberg, A.W., Hönig, F., Nöth, E., Riedhammer, K.: KSoF: The kassel state of fluency dataset – a therapy centered dataset of stuttering. In: Proceedings of the Language Resources and Evaluation Conference LREC, pp. 1780–1787 (2022)
5. Bayerl, S.P., Gerczuk, M., Batliner, A., Bergler, C., Amiriparian, S., Schuller, B., Nöth, E., Riedhammer, K.: Classification of stuttering – the compare challenge and beyond. Comput. Speech Lang. **81** (2023)
6. Bayerl, S.P., Wagner, D., Nöth, E., Bocklet, T., Riedhammer, K.: The influence of dataset partitioning on dysfluency detection systems. In: Text, Speech, and Dialogue (2022)
7. Bayerl, S.P., Wagner, D., Nöth, E., Riedhammer, K.: Detecting dysfluencies in stuttering therapy using wav2vec 2.0. In: Interspeech 2022 (2022)

8. Dash, A., Subramani, N., Manjunath, T., Yaragarala, V., Tripathi, S.: Speech recognition and correction of a stuttered speech. In: 2018 International Conference on Advances in Computing, Communications and Informatics (ICACCI), pp. 1757–1760 (2018)
9. Heeman, P., Lunsford, R., McMillin, A., Yaruss, J.S.: Using clinician annotations to improve automatic speech recognition of stuttered speech. In: Interspeech, pp. 2651–2655 (2016)
10. Howell, P., Davis, S., Bartrip, J.: The UCLASS archive of stuttered speech. J. Speech Lang. Hear. Res. **52**, 556–596 (2009)
11. Howell, P., Sackin, S.: Automatic recognition of repetitions and prolongations in stuttered speech. In: Proceedings of the First World Congress on Fluency Disorders, pp. 372–374 (1995)
12. Koch, G.R.: Siamese neural networks for one-shot image recognition (2015). https://api.semanticscholar.org/CorpusID:13874643
13. Kourkounakis, T., Hajavi, A., Etemad, A.: Detecting multiple speech dysfluencies using a deep residual network with bidirectional long-short-term memory. In: IEEE International Conference on Acoustics, Speech and Signal Processing, ICASSP 2020, pp. 6089–6093 (2020)
14. Kourkounakis, T., Hajavi, A., Etemad, A.: Fluentnet: End-to-end detection of stuttered speech dysfluencies with deep learning. IEEE/ACM Trans. Audio Speech Lang. Process. **29**, 2986–2999 (2021)
15. Lea, C., Mitra, V., Joshi, A., Kajarekar, S., Bigham, J.P.: Sep-28k: A dataset for stuttering event detection from podcasts with people who stutter. In: IEEE International Conference on Acoustics, Speech and Signal Processing, ICASSP 2021, pp. 6798–6802 (2021)
16. van der Maaten, L.J.P., Hinton, G.E.: Visualizing high-dimensional data using t-SNE. Mach. Learn. Res. **9**, 2579–2605 (2008)
17. Nöth, E., Niemann, H., Haderlein, T., Decher, M., Eysholdt, U., Rosanowski, F., Wittenberg, T.: Automatic stuttering recognition using hidden markov models. In: INTERSPEECH (2000)
18. Ratner, N.B., MacWhinney, B.: Fluency bank: A new resource for fluency research and practice. J. Fluency Disorders **56**, 69–80 (2018)
19. Sheikh, S.A., Sahidullah, M., Hirsch, F., Ouni, S.: Machine learning for stuttering identification: review, challenges and future directions. Neurocomputing **514**, 385–402 (2022)
20. Sheikh, S.A., Sahidullah, M., Hirsch, F., Ouni, S.: Robust stuttering detection via multi-task and adversarial learning. In: European Signal Processing Conference (EUSIPCO) (2022)
21. Williams, D.E., Kent, L.R.: Listener evaluations of speech interruptions. J. Speech Hear. Res. **1**, 124–131 (1958)
22. Zagoruyko, S., Komodakis, N.: Wide residual networks (2016). arXiv preprint arXiv:1605.07146

Part V
Data Science and Higher Education

Active Learning for Reducing Gender Gaps in Undergraduate Computing and Data Science

Philip S. Chodrow, Harlin Lee, Natalie Lao, and Vincent Monardo

1 Introduction

This report describes the experience of two instructors (Philip Chodrow and Harlin Lee, the first and second authors) of the course "PIC 16A: Python Programming with Applications" at the University of California, Los Angeles, in the period 2020–2022, as well as our efforts to measure and assess the success of our course design. We joined UCLA as Hedrick Visiting Assistant Adjunct Professors of Mathematics, Chodrow in 2020 and Lee in 2021. For both of us, this was our first academic position after completing our PhDs, and PIC 16A was our first opportunity to teach a course as instructor of record. The Program in Computing (PIC) of the UCLA Department of Mathematics offers a range of courses in applied computing for students who are not majoring in computer science or engineering disciplines. Students who take a prescribed selection of PIC courses, as well as

The first two authors contributed equally.

P. S. Chodrow (✉)
Middlebury College, Middlebury, VT, USA
e-mail: pchodrow@middlebury.edu

H. Lee
School of Data Science and Society, University of North Carolina, Chapel Hill, NC, USA
e-mail: harlin@unc.edu

N. Lao
App Inventor Foundation, Alamo, CA, USA
e-mail: natalie@appinventorfoundation.org

V. Monardo
Massachusetts Institute of Technology, Cambridge, MA, USA
e-mail: monardo@mit.edu

C. Garcia-Cardona, H. Lee (eds.), *Advances in Data Science*, Association for Women in Mathematics Series 37, https://doi.org/10.1007/978-3-031-87804-6_15

several computing-oriented courses in their major, can add a "Specialization in Computing" to their bachelor's degree. At the time in which Chodrow and Lee taught in the PIC program, common majors of enrolled students included cognitive science, neuroscience, economics, applied mathematics, and biology.

PIC 16A: Python Programming with Applications requires only a single pre-requisite course in introductory programming in C + +. Because of this, PIC 16A is a very common choice for PIC students to take as a second or third course in computing. Due to enrollment pressures, the majority of students in PIC 16A during the period in which we taught it were juniors and seniors. PIC 16A has been (and continues to be) offered many times by many instructors at UCLA. Offerings of the course prior to Chodrow's arrival in 2020 primarily focused on Python coding constructs, including relatively little content related to data science or machine learning.

In preparation for his first offering, Chodrow administered an informal entrance survey approximately one week before the beginning of the fall 2020 term. This entrance survey indicated significant discomfort and discouragement with programming among many students, with this discomfort appearing especially prevalent among female students. Chodrow's initial design of PIC 16A therefore reflected an explicit ambition to increase overall confidence and interest in computing and to narrow gender gaps along these axes. Chodrow approached this ambition through both content and format decisions. In choosing course content, he hypothesized that reorienting much of the course around data science and machine learning would connect more effectively to student interests and help hesitant students gain confidence. In the course format, Chodrow decided to emphasize active learning, project-based activities, and learning communities of practice, all with the aspiration of promoting growth mindset.

Chodrow offered PIC 16A a total of four times across four UCLA 10-week terms: Fall 202X, Winter 202X, and Spring 202X. In winter 2022, Lee designed a version of PIC 16A, which adopted several of Chodrow's content choices and format interventions while also implementing changes to the course format (Fig. 1). This report describes our experience teaching this course, as well as what we learned from a series of entrance and exit surveys, which we administered in three different terms. Chodrow administered surveys in winter 2021 (W21) and spring 2021 (S21), while Lee administered surveys in fall 2022 (F22). In Sect. 2 we describe our curriculum, our format decisions, and their rationale. We also

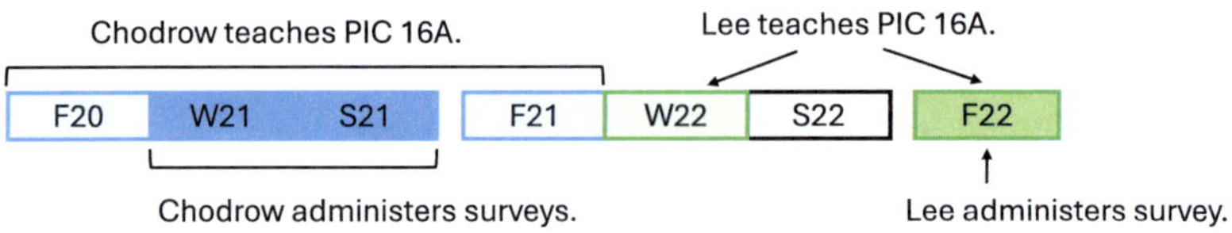

Fig. 1 Timeline of PIC 16A offerings by Chodrow and Lee. Three boxes colored with dark blue and dark green are the three course offerings discussed in this chapter. Lee did not teach PIC 16A in S22

note (Sect. 2.3) the most important differences between Chodrow's and Lee's respective implementations. In Sect. 3, we describe the survey instruments that we administered in each term, the results of which we describe in Sect. 4. We close in Sect. 5 with reflections on our findings and how our experience teaching PIC 16A continues to inform our pedagogy.

2 Course Design

The course Program in Computing: Python with Applications I (PIC 16A) is a quarter-long (10-week) course in Python programming and data science at the University of California, Los Angeles (UCLA). UCLA is a large, flagship university in the public University of California system. As of fall 2024, 22% of UCLA undergraduate students are Hispanic, 6.5% are Black, and 35.1% are Asian or Pacific Islander.[1] Women comprise 60% of the undergraduate student body.

2.1 Content

PIC 16A has a single prerequisite: one course in C++ programming, in which students are introduced to fundamental programming constructs like variables, data types, control flow, functions, and object-oriented programming. In the first 3–4 weeks of the course, PIC 16A reintroduces and reinforces these concepts in Python. The remainder of the course is primarily dedicated to data science applications in Python. The core sequence includes numerical programming, data visualization, tabular data, exploratory data analysis, and machine learning. Under the broad heading of machine learning, students studied data acquisition, data cleaning, feature selection, cross-validation, model training, model evaluation, and basic auditing. The primary software for this course sequence included the `numpy`, `matplotlib`, `pandas`, and `scikit-learn` packages for Python (Table 1).

A major feature of the data science sequence is a sustained series of lectures and activities using the Palmer Penguins dataset [7]. Through this series, students clean the data, visualize various features, implement several approaches to feature selection, apply classification and clustering algorithms, and assess the performance of their models using confusion matrices and decision regions. This series also provides a smooth segue into a cumulative project, which serves as the primary summative assessment for the course. In Chodrow's W21 and S21 offerings, this cumulative project involved further development and synthesis of a complete

[1] Figures from https://www.ucla.edu/about/facts-and-figures as accessed on September 5, 2024.

Table 1 PIC 16A data science and machine learning modules in weeks 4 through 9. First three weeks cover standard programming concepts including object-oriented programming

Data science topics	Data	Tools
File I/O	Tabular, Text	`urllib`, `csv`
Matrix and array operation	Numerical	`numpy`
Image processing	Image	`numpy`
Data visualization	Tabular	`matplotlib`
Project management	Code	`git`, command line
Data wrangling	Tabular	`pandas`
Machine learning		
Overview	Tabular	`scikit-learn`
Supervised learning	Tabular, Image	
Overfitting	Tabular, Numerical	
Clustering	Tabular	
ML Ethics		

analysis for the Palmer Penguins dataset, while in Lee's F22 offering, the cumulative project was open and proposed by the students. Lee's project assignment also encouraged students to learn about best practices for code management, collaboration, and communication via `git` and GitHub.

2.2 *Organization*

2.2.1 Basics

The weekly course format included five contact hours: three 50-minute Lecture periods and two 50-minute Discussion (lab) periods.[2] In class and on assignments, students interacted with Python via the Jupyter Notebook app provided by the Anaconda Python distribution. Videoconferencing via Zoom was used for many remote Lecture periods and remote office hours, depending on UCLA policy at the time. Typical sections of PIC 16A contained 50–70 students and were staffed by one instructor and either one or two graduate teaching assistants (TAs).

Although our emphasis here is not on remote learning methodologies, it should be noted that initial development of this course offering took place during the fall term of 2020 at the height of the COVID-19 pandemic.

[2] Our capitalization convention is that Lecture and Discussion refer to specific weekly class periods, while the uncapitalized "lecture" describes any content presentation by the instructor.

2.2.2 Learning Communities of Practice

PIC 16A was initially developed for an online environment during the height of the COVID-19 pandemic. Forming a cohesive learning community [1] for a large class via remote instruction appeared challenging. In order to promote student belonging and form functional support networks, the course design instead encouraged small learning communities of practice [20]. These communities of practice were groups of three, which worked together on Discussion activities as described below. Chodrow assigned students to groups based on their responses in an informal non-anonymous pre-course entrance survey, which described their interests and confidence in computing as well as their prior experience.[3] Based on advice from a more experienced colleague, Chodrow tried to create groups that had similar interests (e.g., data science, software development) but varying levels of confidence or experience; the intention was that more confident students would, with structure, support the experience of less confident students. Chodrow also made an effort to avoid groups in which a single female student would work with two male students; this was aimed against the risk of men dominating a conversation to the exclusion of the single female group member. Groups therefore contained either zero, two, or three female students, but never exactly one. On the other hand, Lee's groups were assigned randomly. Groups were persistent throughout the term except in rare cases of conflict or strong dysfunction between group members. These groups worked twice weekly on Discussion activities and also worked together on an end-of-semester data science project in Chodrow's sections.

2.2.3 Active Learning

Active learning—as opposed to traditional lecture-based formats—is known to improve student performance and narrow achievement gaps [2, 6, 12, 19]. For this reason, our design of PIC 16A emphasizes the Discussion period, rather than the Lecture period, as centers of the student learning experience. In a typical Discussion section, students worked in groups of three on a scaffolded Jupyter notebook. This notebook usually began with several exercises that helped students reinforce their learning at the "Remember" and "Understand" stages of Bloom's taxonomy [11]. Progressive exercises in the notebook encouraged students to the higher "Apply" and "Analyze" stages. See Fig. 2 for an example.

During the Discussion period, students were typically supported by both graduate teaching assistants (TAs) and undergraduate learning assistants (LAs). LAs are undergraduate students who receive training in inclusive pedagogy through UCLA's Center for Education Innovation and Learning in the Sciences. Most LAs had taken a prior offering of PIC 16A. LAs act as near-peer mentors [18], with their primary role being to encourage equitable participation among all group members and to help

[3] This survey was distinct from the more formal, anonymous entrance survey described in Sect. 3.

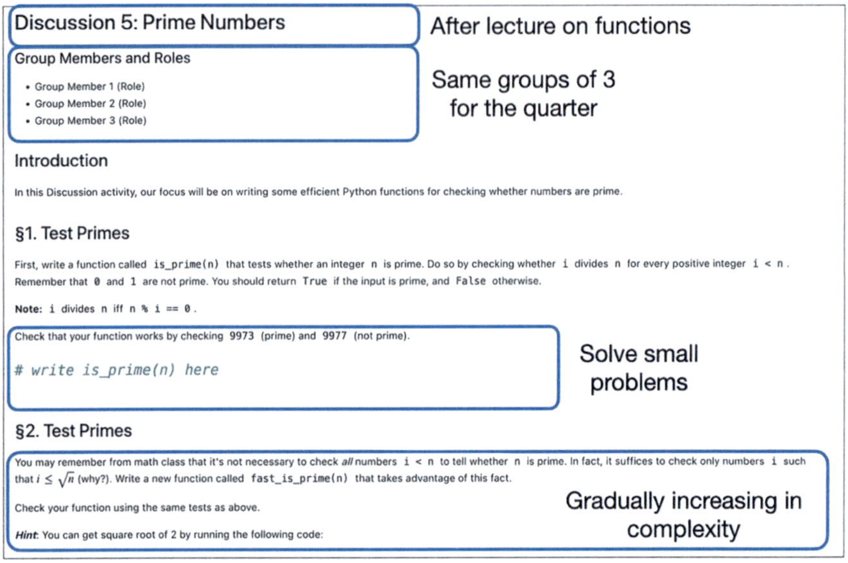

Fig. 2 Example of active learning employed in PIC 16A Discussions. Students work on worksheets like this in small groups, which builds learning communities that are especially critical in large institutions post-pandemic

stuck groups find their way to the next part of the assignment. The instructor was not usually present during Discussion sections, although in some terms, Chodrow "swapped" with his TA and managed one of the two weekly Discussion periods, while the TA managed one of the optional Lecture periods.

Within each three-person group, students took on one of three roles inspired by the pair-programming paradigm. As a learning methodology, pair-programming may offer modest benefits for student learning and may offer a particular learning benefit for female students [8]. For the three-person groups of PIC 16A, three roles were used: a Driver who writes code and makes low-level implementation decisions, a Proposer who guides the Driver at a high level, and a Reviewer who gives feedback on the solutions that the Proposer and Driver have crafted. Roles rotated each Discussion day, ensuring that each group member inhabited each role with approximately equal frequency. This put a lower bound on the extent to which students could withhold participation in their groups. Attendance in these Discussion periods was mandatory; failure to attend would result in failure to get credit for the Discussion assignment unless the student received explicit permission from the instructor to complete the assignment independently in response to an emergency.

Lecture delivery also emphasized student activity and engagement. In the W21 and S21 offerings of PIC 16A described here, lecture content was delivered via pre-recorded videos, which students watched asynchronously, outside of the scheduled

Lecture period. Students could then choose whether to attend[4] scheduled Lecture periods, which typically included a mix of question-and-answer, supplementary topics, and support on current assignments.

In the F22 offering, lectures were delivered in-person during scheduled Lecture periods. In all cases, the most common lecture format was live coding in a notebook. Students were encouraged to download the notebook prior to class and code along with the instructor throughout the Lecture period in order to build experience and muscle memory writing common code constructs. The lectures were recorded and uploaded to a class Web site afterward. We note that in both versions, lecture attendance was not mandatory.

2.2.4 Nurturing Growth Mindset

Some students described themselves using fixed-mindset language in the initial entrance survey, ascribing to themselves intrinsically less ability to excel at programming when compared to their peers. Growth mindset, in which students view their ability to achieve as malleable and able to be improved, is the opposite of fixed mindset [5]. Growth mindset is known to promote learning, especially among lower-achieving students [21]. Several studies have suggested that project-based learning experiences, in which students grapple with multi-part challenges motivated by real-world problems, can promote growth mindset [9, 17], although a formal causal link remains elusive [10]. In order to nurture growth mindset, most substantive assignments from the course constitute scaffolded mini-projects, with a satisfying product or insight waiting for the student upon completion (Table 2).

In one example of a Discussion activity from the Palmer Penguins sequence, students begin by writing a function to efficiently compute aggregated summary

Table 2 Example topics from homework and Discussion activities in different stages of PIC 16A

Weeks	Topic	Assignment examples
1–3	Data structures	Markov language models
	Iteration	PageRank
4–6	Array programming	Image manipulation
	Data wrangling	Data visualization
7–10	Machine learning	Logistic regression
	Impact and bias	Reproducing [14]

[4] Chodrow was explicit with students that the primary purpose of the scheduled Lecture period was to provide a space for students to ask questions and receive support. As a result, many students who were confident in the material or who did not desire additional course engagement chose not to attend these sessions; typical attendance rates were around 20%–40%. This approach in part reflected departmental policy, which prohibited instructors from requiring attendance at both Lecture and Discussion sections. Chodrow's choice to prioritize the active learning Discussion sections therefore necessarily de-emphasized attendance during scheduled Lecture.

statistics from the Palmer penguins data using the pandas package. In the following part, they use this function to explore the data and search for features that seem to distinguish different species of penguins. Next, they use findings from these tables to manually implement a shallow, axis-aligned decision tree using if-statements. They then evaluate this decision tree against the data. By the time students complete the activity, they have used data manipulation and basic control flow in order to explore the idea of prediction. This activity leads to an upcoming introduction of machine learning algorithms that automate the process of fitting predictors to data.

2.3 *Implementation Differences*

The implementations of the PIC 16A model by Chodrow (W21 and S21) and by Lee (F22) shared the above common features. There were, however, important differences in these implementations as well. Three such differences were especially large and offer important context for our findings in the following section.

2.3.1 Delivery

Chodrow's offerings of PIC 16A in W21 and S21 were fully remote, with all course activities taking place in Zoom meetings. In contrast, Lee's offering in F22 was fully in-person, with no remote instruction per pandemic-related departmental policy.

In Chodrow's offerings, all students completed the same cumulative project using Palmer Penguins, while Lee's students were permitted to choose their own project topic. The Palmer Penguins project was still provided as an option, and students with lower confidence who sought more guidance and scaffolding were encouraged to take that route. While it was not required that the projects had to be about data science and machine learning, fifteen out of seventeen student groups chose to do so. Of these fifteen, seven groups worked on classification, three on clustering (including two on Palmer Penguins), three on regression, and two on data analysis and visualization. The remaining two worked on building a scientific tool and an interactive game. Based on feedback from her W22 section students (not included in this study), Lee's students in F22 were also allowed to choose their own project partners as opposed to working with their discussion groups.

2.3.2 Disruption

The W21 and S21 offerings by Chodrow took place during the height of the COVID-19 pandemic in the USA. Although this was a difficult time for students and faculty alike, these offerings were not significantly disrupted beyond the necessity of ongoing remote instruction. The class was largely able to follow its course as intended by the instructor. In contrast, Lee's F22 offering was disrupted by a strike among graduate students across the UC system demanding a living wage. Because graduate TAs were the primary staffing for Discussion sections, the last month of Discussion sections was cancelled. The cancelled sections included many of the activities in the data science sequence.

2.3.3 Instructor Identity

Our expressed identities may also have played roles in student perceptions of the course. Chodrow is a white man whose first language is English, while Lee is a Korean woman whose first language is Korean and who speaks English fluently as a second language. Student evaluations of teaching (SETs) are known to show bias against female instructors [3], with some additional evidence of intersectional bias against women of color [4]. Although the questions in our entrance and exit surveys are not SETs as they do not ask students to evaluate teaching, these questions do ask students to reflect on the depth of their learning and their experience in various aspects of the course. It is possible that student biases in response to the different identities of the instructors may have played a role in their responses.

3 Methods

As described above, we designed PIC 16A with the hypothesis that social, active learning and project-based learning would help increase confidence and interest in computing while reducing gender gaps. In this section, we describe how we assessed the effectiveness of our designs using entrance and exit surveys that we collected during the course. Students completed entrance surveys during the first two weeks of each course and completed exit surveys during the final week of classes and the final exam period. On the survey, we asked questions related to comfort social learning environments; confidence with respect to programming; and interest in programming and machine learning. Students were incentivized to complete these surveys through assignments that conferred participation credit or extra credit, typically on the order of 0.25% toward their final average in the course. We emphasize that the initial purpose of these surveys was to inform future course improvement, rather than to perform formal evaluation or produce publishable work. We decided to write this article using this data after the described course offerings were complete. We especially note that we do not support the practice of using grades to incentivize student participation in research studies, and would not have done so had we planned to write this report from the outset. Additionally, because the surveys were initially intended to inform course improvement, the questions were designed by each instructor primarily in response to their pedagogical interests. As a result, the survey questions are similar but not identical across the three course offerings. Our data collection was retroactively considered by Institutional Review Boards at our present institutions (Middlebury College and UNC Chapel Hill),

Table 3 Counts of student respondents by gender to each entrance and exit survey in the three studied course offerings of PIC 16A. A small number of self-identified nonbinary students are not shown due to identifiability concerns

Term	Instructor	Gender	Entrance	Exit
W21	Chodrow	F	35	31
		M	36	32
S21	Chodrow	F	31	25
		M	25	17
F22	Lee	F	34	22
		M	29	17

which both determined that the publication of this report with the collected data did not constitute human subjects research and therefore did not require IRB approval.

Our surveys yielded response rates ranging between 80% and 95%, with generally lower response rates in exit surveys. Table 3 shows the number of students responding to each survey. Although small numbers of nonbinary students also participated in class and took the entrance and exit surveys, we exclude them from reporting due to identifiability concerns.

We collected both entrance and exit surveys through UCLA's learning management system (LMS). In Chodrow's W21 and S21 offerings, the LMS used was a Moodle-based app called Common Collaborative Learning Environment (CCLE). In W21 and S21, students were asked to respond to a series of statements on a four-point scale: "1: Strongly Disagree"; "2: Disagree"; "3: Agree"; and "4: Strongly Agree." These statements related to students' interest, comfort, and confidence in programming. Full text of each statement is displayed in Tables 4, 5 and 6. For the purposes of analysis, both "3: Agree," and "4: Strongly Agree," responses are considered to indicate agreement with prompts.

In Lee's F22 offering, the LMS used was Canvas (CCLE was decommissioned during the 2022–2023 academic year). Students were asked to respond to a series of prompts on a 5-point scale. Some prompts were statements, where students were asked to rate their level of agreement with the statement, for example, "1: Disagree a lot"; "2: Disagree a little"; "3: Neither agree nor disagree"; "4: Agree a little"; and "5: Agree a lot." Others were questions, where students were asked to rate their level of interest, ranging from "1: Not at all interested"; "2: Not very interested"; "3: Neutral"; "4: Somewhat interested"; and "5: Very interested." Lastly, when asked about the effectiveness of having learning partners, the choices were "1: Hindered a lot"; "2: Hindered a little"; "3: Neither helped nor hindered"; "4: Helped a little"; and "5: Helped a lot." Unlike in the previous offerings, this survey round included neutral responses. In analysis, only responses at levels "4" and "5" were considered to demonstrate agreement or interest. Lee's design also included several prompts, posed only in the exit survey, related to confidence and student learning from working with partners. The prompts for each offering are shown in Tables 4, 5, and 6 in Sect. 4.

4 Results

Table 4 summarizes survey findings for Chodrow's W21 and S21 sections. We show the proportion of responses indicating agreement ("3, Agree," or "4, Strongly Agree") for each question, term, survey round, and gender. There are 14 questions asked in both terms and two additional questions that were asked only in S21. Segmenting by gender and term, we have a total of 60 pairs of entrance and exit survey responses. To test for the significance of changes in agreement rates, we used a Mann-Whitney U test, a nonparametric test for difference of distributions. We did two sets of such tests. In the first set, we tested the null hypothesis that the distribution of responses in the exit survey was the same as that of the entrance survey, with bolded proportions indicating that this hypothesis was rejected.[5] Of the 60 pairs of entrance and exit surveys, 32 showed a significant difference in the distribution of responses at the 95% confidence level. All of these significant differences show increased rates of agreement from entrance to exit surveys. Most of the significant changes relate confidence in programming and computational thinking and comfort in classroom settings, with fewer significant changes related to career confidence or connections between programming and other interests. The smaller number of changes in interests may be due in part to the already-high rates of agreement on these questions on the entrance survey.

We additionally tested the null hypothesis that the distribution of responses of female and male students were the same, on each of the entrance and exit surveys separately. Again, this was done using the Mann-Whitney U test. We considered a gender gap to be present on a given question if the null hypothesis is rejected. We mark the presence of a gap on the entrance and exit surveys in the final two columns of Table 4. We consider a gap to have closed if a significant difference on the entrance survey is no longer significant on the exit survey; we consider a gap to have opened if a significant difference on the exit survey is not present on the entrance survey; and we consider a gap to have persisted if a significant difference is present on both surveys. Of the 30 total questions across both terms, we observed six gaps that closed, 3 gaps that opened, and 3 gaps that persisted. In each of the gaps that opened, agreement rates increased among both groups, with an increase for male students that left their agreement rates at 94% or above.

Table 5 shows the same hypothesis-testing approach for Lee's F22 offering. Eight questions were asked at both entrance and exit surveys, leading to 16 pairs of responses to be tested via the Mann-Whitney U test. One of them rejected the null hypothesis that the entrance and exit survey responses have the same distributions at $p < 0.05$. This supports the alternative hypothesis that the female students' self-perceived, "ability to explain machine learning and data science concepts to [...] peers" have changed after the quarter. We repeat the gender gap testing on F22 responses as well. We observed gender gaps in four out of eight questions at

[5] We emphasize that, although an agreement rate is shown, the hypothesis test is performed on the full, uncompressed distribution of responses.

Table 4 Summary hypothesis testing for W21 and S21. We show the proportion of students who agreed (3, "Agree," or 4, "Strongly Agree") on each question. **Bold** values in the first two "Exit" columns indicate a statistically significant difference in the distribution of responses between entrance and exit surveys at the 95% confidence level among students of the specified gender. We also show significance levels for difference in distribution on the entrance and exit surveys. A 1 indicates the presence of a statistically significant gap. A gender gap has closed if a significant difference on the entrance survey is no longer significant on the exit survey. The final two questions were asked only in S21

		Female		Male		Gap	
Question	Term	Entrance	Exit	Entrance	Exit	Entrance	Exit
I usually feel comfortable asking questions in class.	W21	46%	**90%**	69%	**91%**		
	S21	29%	**96%**	56%	**94%**	1	
I usually feel comfortable attending office hours.	W21	66%	**97%**	69%	**91%**		
	S21	65%	**100%**	72%	94%		
I usually feel comfortable asking for help from my peers.	W21	66%	94%	75%	**97%**		
	S21	68%	**88%**	72%	**94%**		
I usually feel comfortable when explaining my thought process to others.	W21	54%	**94%**	81%	**100%**	1	
	S21	52%	76%	76%	94%	1	1
I usually feel comfortable working in groups.	W21	71%	**94%**	72%	**94%**		
	S21	58%	**88%**	72%	**88%**		
I usually feel comfortable writing about how my code works.	W21	63%	**97%**	86%	**97%**		
	S21	74%	84%	76%	**94%**		1
Other people can learn from how I approach problems.	W21	80 %	90%	83%	**97%**		
	S21	71%	76%	80 %	94%	1	1
I can usually understand what task a program achieves by reading the code.	W21	51%	**90%**	69%	**94%**	1	
	S21	61%	**100%**	76%	**100%**		
Computer programming is fun.	W21	86%	**97%**	89%	**100%**		
	S21	74%	92%	88%	88%		
I can see connections between programming and my hobbies.	W21	69%	**87%**	92%	**94%**	1	
	S21	71%	76%	88%	94%		
I can see connections between programming and my academic interests.	W21	91%	100%	94%	100%	1	
	S21	97%	96%	96%	88%		
I can see connections between programming and my long-term career goals.	W21	89%	97%	97%	97%	1	
	S21	87%	80 %	92%	94%		
If I wanted to, I could eventually have a career that involved programming.	W21	60 %	81%	83%	88%	1	1
	S21	77%	72%	84%	82%		
If I wanted to, I could eventually have a career that involved data science or machine learning.	W21	66%	87%	83%	**94%**		1
	S21	81%	72%	92%	88%		

(continued)

Table 4 (continued)

		Female		Male		Gap	
Question	Term	Entrance	Exit	Entrance	Exit	Entrance	Exit
I can often improve code by removing inefficiencies redundancies.	S21	58%	84%	48%	**88%**		
I am able to solve problems that interest me using programming.	S21	39%	**68%**	60 %	**94%**		1

Table 5 Summary hypothesis testing for F22. We show the proportion of students who agreed (4, "Agree/Somewhat interested," or 5, "Strongly Agree/Very interested") on each question for display purposes only. **Bold** values in the first two "Exit" columns indicate a statistically significant difference in the distribution of responses between entrance and exit surveys at the 95% confidence level among students of the specified gender. We also show significance levels for difference in distribution on the entrance and exit surveys. A 1 indicates the presence of a statistically significant gap. A gender gap has closed if a significant difference on the entrance survey is no longer significant on the exit survey

		Female		Male		Gap	
Question	Term	Entrance	Exit	Entrance	Exit	Entrance	Exit
How interested are you in machine learning and data science from a career/work perspective?	F22	26%	24%	45%	47%	1	
How interested are you in machine learning and data science from a personal/hobby perspective?	F22	24%	31%	45%	29%	1	
How interested are you in understanding machine learning and data science theory?	F22	15%	10 %	41%	29%		
I have the ability to explain machine learning and data science concepts to my peers.	F22	3%	**14%**	5%	35%		
I have the ability to implement machine learning and data science projects.	F22	18%	10 %	27%	35%		
I have the ability to learn machine learning and data science concepts.	F22	18%	28%	64%	53%	1	
I like machine learning and data science.	F22	24%	17%	41%	50 %		
I think machine learning and data science is interesting.	F22	41%	62%	68%	71%	1	

Table 6 Exit only questions for F22. We show the proportion of students who agreed—4, "Agree/Helped a little," or 5 "Strongly Agree/Helped a lot"—on each question

		Female	Male	Both
Question	Term	Exit	Exit	Exit
How would you describe your final project partners' impact on your learning?	F22	93%	80%	88%
During discussions, how would you generally describe your partners' impact on your learning?	F22	96%	88%	93%
I will be able to complete a machine learning or data science project (of a similar level and scale to projects from this class) on my own.	F22	67%	69%	67%
I am proud of what I was able to accomplish in my final project.	F22	81%	81%	81%
I feel that I was successful in this class.	F22	89%	63%	79%

entrance surveys, all of which closed at exit surveys. These four questions were about the students' interest in machine learning and data science and their self-perceived ability to learn related concepts.

Several questions were asked only in the exit survey (Table 6). These questions asked students about their confidence and about the effectiveness of partners in supporting their learning. Four out of five exit-only prompts elicited stronger positive response from female students than male students. The greatest observed difference between male and female students on exit-only prompts was in response to the prompt, "I feel that I was successful in this class."

The decrease in agreement rates for the career-oriented confidence questions in S21 and a few interest questions in F22 for male students is a concerning observation about which we regrettably do not have follow-up data.

5 Discussion

Our quantitative results show mixed evidence regarding the effectiveness of our interventions with respect to our goal of reducing gender gaps in data science and computing. We observed a number of closed or narrowed gender gaps along axes of intellectual confidence, interest, and comfort in a social classroom environment. There were different patterns across the sections; Chodrow's W21 section tended to have greater improvements for female students than did his S21 section, resulting in more closed gaps. Lee's F22 section showed several gender gaps closing, although in several cases this was due to a decrease in interest or enthusiasm among male students (who also tended to feel that they were less successful in the class than their female peers). As noted in Sect. 2.3, this offering was severely disrupted by a graduate student strike, resulting in the cancellation of many Discussion activities in the section of the course focusing on data science techniques. We hypothesize that this disruption contributed to reduced interest and enthusiasm among this section.

Furthermore, as alluded to in Sect. 2.3, the difference in instructor identities could have been a confounding variable as well.

We take our findings to reflect partial success toward the overall course aim of reducing gender gaps in undergraduate data science and computing, especially relating to intellectual confidence, interest, and comfort in social classroom environments. We additionally hypothesize that these findings partially reflect the project-based Discussion and homework activities, in which students repeatedly experience the feeling of achieving a meaningful, satisfying task through their programming efforts.

Analytical Limitations

We also acknowledge several limitations to our analysis. The lack of a comparison group makes it difficult to rule out alternative mechanisms that may promote student comfort, including natural acclimatization during the course, instructor persona, and current events such as the COVID-19 pandemic, the US election, or ongoing protests concerning police violence against Black Americans. We also note that our analysis does not include any consideration of student performance in the class; it is possible that students who built high levels of confidence and interest in the course may not have in fact demonstrated the greatest learning of the course material. A related consideration is course attendance. Chodrow's offerings especially allowed students to choose whether or not to attend scheduled Lecture periods in order to access additional practice, content, and support. Chodrow did not track attendance during the scheduled lecture periods. Although attendance at the scheduled Lecture periods may be reasonably expected to impact student attitude and performance, we unfortunately do not have data with which to support this. As noted previously, differences in delivery model, course environment, and instructor identity may have contributed to differences in results between W21 and S21 on the one hand and F22 on the other.

Another important caveat in our analysis relates to students who chose to drop PIC 16A. As shown in Table 3, there were fewer exit survey respondents than entrance survey respondents in all sections, reflecting in part the handful of students who drop the course after the entrance survey is conducted. If, as seems likely, the population of students who dropped tended to have lower rates of confidence, interest, or comfort with the course design, then our descriptions of entrance-to-exit survey improvements may be artificially inflated.

Reflections and Future Work

We take our findings to offer tentative support to the idea the interventions we implemented in PIC 16A can help increase student comfort in social learning environments, intellectual confidence in programming, and interest in data science and computing while (in some cases) narrowing gender gaps along these axes. Especially in light of the limitations described above, we consider this evidence to be promising but ultimately inconclusive.

As noted above, PIC 16A was for each of us our very first opportunity to teach as an instructor of record. At the time we designed the course, neither of us were experienced in course design, active learning techniques, or evidence-based pedagogy broadly construed. The design we describe reflects our well-intentioned hypotheses about how to implement active, project-based, and social learning experiences for students, but there is much of which we were not aware. For example, in the time that we designed PIC 16A, we were not aware of the Peer Instruction protocol [13] or its successful deployment in many computing courses [16]. As another important point, we note a specific aspect that was missing from our efforts to encourage growth mindset. As emphasized in [15], an important catalyst for growth mindset is metacognition, the reflection of students on the process of their own learning. Our course design offered students a few opportunities to reflect on their learning, including several free-response questions on mid-semester surveys and a group contributions statement as part of final projects. With the benefit of hindsight and experience, however, we also feel that we missed several opportunities to encourage a sustained practice of metacognition as part of our course designs. It would have been especially efficient to implement structured reflective writing as part of the lab submission process, which we largely did not do.

Since our time teaching PIC 16A, we have moved on to different institutions in which we continue to grow as instructors. Several of the core ideas from our design of PIC 16A live on in our pedagogical practice.

Chodrow I am now a faculty in computer science at Middlebury College, a selective small liberal arts school in Vermont. My teaching portfolio includes introductory discrete mathematics, machine learning, and network science. My encouraging experience with PIC 16A informed many of my course designs at Middlebury. My offering of introduction to computing emphasizes collaborative and active group work, including both a weekly lab and one day a week of in-class practice time. My offering of discrete mathematics is fully flipped. This course also includes a weekly lab that emphasizes applications. This course emphasizes metacognition through standards-based learning, reflection prompts on lab assignments, and a structured revision process for assignments. My offering of machine learning is fully project-based and also emphasizes interactive computing in Jupyter notebooks. This course also emphasizes reflection through writing prompts at regular intervals in the course, as well as a portfolio-based assessment process. Because I am offering many of these courses for the first time at the time of writing, I haven't yet administered surveys for these courses of the kind I administered for PIC 16A. That said, my interest in data-informed course design remains strong. I am especially interested in incorporating discussion of the social impacts of computing technologies into his classrooms and in studying whether such discussions are experienced differently by students with different identities.

Lee Since 2023, I have been a faculty at the University of North Carolina at Chapel Hill, School of Data Science and Society. I am teaching a large undergraduate class on introduction to data science, which assumes no prerequisites in math or programming beyond high school algebra. Based on the positive feedback from PIC 16A, I have adapted the weekly format (i.e., in-person Lectures + group-based Discussions) and final group project to this brand-new course. This class emphasizes statistics and non-technical aspects of data science (e.g., data storytelling, data life cycle) more compared to PIC 16A, but we make best effort to still teach those concepts using active learning and project-based learning. At the time of writing this reflection, this course is being offered for the first time, so it has a lot of room for improvement, but I was able to start from a solid baseline. I am grateful for the UCLA colleagues, students, TAs, and LAs who made the PIC 16A experience a relative success.

Acknowledgments Michael Perlmutter (Boise State University) also contributed to the design of PIC 16A during his time at UCLA. Kirill Gura (UCLA) and Erin George (UCLA) made major contributions as graduate teaching assistants to the design of PIC 16A.

Course Repository Materials from the most recent offering of Chodrow's PIC 16A may be accessed at https://www.philchodrow.prof/PIC16A/syllabus/. Available materials include lecture notes, homeworks, and most Discussion activities. Please be aware that this Web site is no longer maintained.

Competing Interests The authors have no conflicts of interest to declare that are relevant to the content of this chapter.

Ethics Approval Proposed analysis of de-identified survey data (Reference ID 405500) was reviewed by the Office of Human Research Ethics at University of North Carolina at Chapel Hill, which has determined that this submission does not constitute human subjects research as defined under federal regulations [45 CFR 46.102 (e or l) and 21 CFR 56.102(c)(e)(l)] and does not require IRB approval. Similarly, the IRB of Middlebury College has determined that this submission does not constitute human subjects research as defined under federal regulations [45 CFR 46.102 (e or l) and 21 CFR 56.102(c)(e)(l)] and does not require IRB approval (Middlebury IRB Protocol #258).

References

1. Andrade, M.S.: Learning communities: examining positive outcomes. J. College Student Retention: Res. Theory Practice **9**(1), 1–20 (2007). https://doi.org/10.2190/E132-5X73-681Q-K188
2. Ballen, C.J., Wieman, C., Salehi, S., Searle, J.B., Zamudio, K.R.: Enhancing diversity in undergraduate science: self-efficacy drives performance gains with active learning. CBE—Life Sci. Educ. **16**(4), ar56 (2017)
3. Boring, A.: Gender biases in student evaluations of teaching. J. Public Econ. **145**, 27–41 (2017). https://doi.org/10.1016/j.jpubeco.2016.11.006
4. Chávez, K., Mitchell, K.M.: Exploring bias in student evaluations: gender, race, and ethnicity. Political Sci. Politics **53**(2), 270–274 (2020). https://doi.org/10.1017/S1049096519001744
5. Dweck, C.S.: Mindset: The New Psychology of Success, 1st ed edn. Random House, New York (2006)
6. Freeman, S., Eddy, S.L., McDonough, M., Smith, M.K., Okoroafor, N., Jordt, H., Wenderoth, M.P.: Active learning increases student performance in science, engineering, and mathematics. Proc. Natl. Acad. Sci. **111**(23), 8410–8415 (2014). https://doi.org/10.1073/pnas.1319030111
7. Gorman, K.B., Williams, T.D., Fraser, W.R.: Ecological sexual dimorphism and environmental variability within a community of Antarctic penguins (genus Pygoscelis). PLoS One **9**(3), e90081 (2014)
8. Hanks, B., Fitzgerald, S., McCauley, R., Murphy, L., Zander, C.: Pair programming in education: a literature review. Comput. Sci. Educ. **21**(2), 135–173 (2011). https://doi.org/10.1080/08993408.2011.579808

9. Iwamoto, D.H., Hargis, J., Vuong, K.: The effect of project-based learning on student performance: an action research study. Int. J. Scholar. Technol. Enhanced Learn. **1**(1), 24–42 (2016)
10. Kokotsaki, D., Menzies, V., Wiggins, A.: Project-based learning: a review of the literature. Improving Schools **19**(3), 267–277 (2016)
11. Krathwohl, D.R.: A Revision of bloom's taxonomy: an overview. Theory Into Practice **41**(4), 212–218 (2002). https://doi.org/10.1207/s15430421tip4104_2
12. Lorenzo, M., Crouch, C.H., Mazur, E.: Reducing the gender gap in the physics classroom. Am. J. Phys. **74**(2), 118–122 (2006)
13. Mazur, E.: Peer instruction: Getting students to think in class. In: AIP Conference Proceedings, pp. 981–988. IOP Institute of Physics Publishing (1997)
14. Obermeyer, Z., Powers, B., Vogeli, C., Mullainathan, S.: Dissecting racial bias in an algorithm used to manage the health of populations. Science **366**(6464), 447–453 (2019)
15. Petrik, R.L., Vega, J., Vindas-Meléndez, A.R.: A reflection on growth mindset and meritocracy. J. Hum. Math. **12**(1), 408–421 (2022)
16. Porter, L., Bailey Lee, C., Simon, B.: Halving fail rates using peer instruction: a study of four computer science courses. In: Proceeding of the 44th ACM Technical Symposium on Computer Science Education, pp. 177–182 (2013)
17. Reid, K.J., Ferguson, D.M.: Do design experiences in engineering build a "growth mindset"in students? In: 2014 IEEE Integrated STEM Education Conference, pp. 1–5. IEEE, Piscataway (2014)
18. Tenenbaum, L.S., Anderson, M.K., Jett, M., Yourick, D.L.: An innovative near-peer mentoring model for undergraduate and secondary students: STEM focus. Innov. Higher Educ. **39**(5), 375–385 (2014). https://doi.org/10.1007/s10755-014-9286-3
19. Theobald, E.J., Hill, M.J., Tran, E., Agrawal, S., Arroyo, E.N., Behling, S., Chambwe, N., Cintrón, D.L., Cooper, J.D., Dunster, G., Grummer, J.A., Hennessey, K., Hsiao, J., Iranon, N., Jones, L., Jordt, H., Keller, M., Lacey, M.E., Littlefield, C.E., Lowe, A., Newman, S., Okolo, V., Olroyd, S., Peecook, B.R., Pickett, S.B., Slager, D.L., Caviedes-Solis, I.W., Stanchak, K.E., Sundaravardan, V., Valdebenito, C., Williams, C.R., Zinsli, K., Freeman, S.: Active learning narrows achievement gaps for underrepresented students in undergraduate science, technology, engineering, and math. Proc. Natl. Acad. Sci. **117**(12), 6476–6483 (2020). https://doi.org/10.1073/pnas.1916903117
20. Wenger, E.: Communities of practice: a brief introduction (2011)
21. Yeager, D.S., Hanselman, P., Walton, G.M., Murray, J.S., Crosnoe, R., Muller, C., Tipton, E., Schneider, B., Hulleman, C.S., Hinojosa, C.P., Paunesku, D., Romero, C., Flint, K., Roberts, A., Trott, J., Iachan, R., Buontempo, J., Yang, S.M., Carvalho, C.M., Hahn, P.R., Gopalan, M., Mhatre, P., Ferguson, R., Duckworth, A.L., Dweck, C.S.: A national experiment reveals where a growth mindset improves achievement. Nature **573**(7774), 364–369 (2019). https://doi.org/10.1038/s41586-019-1466-y

Quantifying and Documenting Gender-Based Inequalities in the Mathematical Sciences in the United States

Ron Buckmire, Carrie Diaz Eaton, Joseph E. Hibdon Jr., Jakini Auset Kauba, Drew Lewis, Omayra Ortega, José L. Pabón, Rachel Roca, and Andrés R. Vindas-Meléndez

1 Introduction

The kinds of problems mathematics and data science can be used to solve are extremely varied, running the gamut from theoretical with no foreseen applications to those that are immediately applicable to important real-world phenomena like climate change, epidemiology, and social networks. There is a rapidly growing body

R. Buckmire (✉)
Mathematics Department, School of Computer Science and Mathematics, Marist University, Poughkeepsie, NY, USA
e-mail: ron.buckmire@marist.edu

C. D. Eaton
Digital & Computational Studies, RIOS Institute, Bates College, Lewiston, ME, USA

J. E. Hibdon Jr.
Department of Mathematics, Northeastern Illinois University, Chicago, IL, USA

J. A. Kauba
School of Mathematical and Statistical Sciences, Clemson University, Clemson, SC, USA

D. Lewis
Independent Consultant, Los Gatos, CA, USA

O. Ortega
Department of Mathematics & Statistics, Sonoma State University, Rohnert Park, CA, USA

J. L. Pabón
Department of Mathematical Sciences, Newark, NJ, USA

R. Roca
Department of Computational Mathematics, Science, and Engineering, Michigan State University, East Lansing, MI, USA

A. R. Vindas-Meléndez
Department of Mathematics, Harvey Mudd College, Claremont, CA, USA

C. Garcia-Cardona, H. Lee (eds.), *Advances in Data Science*, Association for Women in Mathematics Series 37, https://doi.org/10.1007/978-3-031-87804-6_16

of work using tools from the mathematical sciences to analyze the mathematical sciences itself that has recently been described as the "mathematics of mathematics" or "MetaMath" [6]. This term was chosen as an allusion to the broader field of "science of science" [11] or "SciSci" that uses scientific and mathematical tools to analyze science as a whole as well as its individual disciplines.

In this chapter, we present a contribution to the mathematics of mathematics to introduce readers to the kinds of questions that can be asked and the types of tools and techniques that can be used in the emerging MetaMath field. We are inspired by the work of Wapman, Zhang, Larremore, and Clauset [37] published in *Nature* in 2022 that quantified hierarchy in faculty hiring and retention in a wide array of academic disciplines in the United States by analyzing a very large dataset of nearly 300,000 faculty members in over 10,000 departments at almost 400 PhD-granting institutions from 2011 to 2020. The research we present here builds on and leverages the ideas and analyses presented in Wapman et al. [37] but focuses on specific disciplines in the mathematical sciences that are available in their dataset [36], namely, mathematics, statistics, and operations research. This results in a dataset that includes nearly 10,000 faculty members in well over 200 departments that granted PhDs in the mathematical sciences from 2011 to 2020.

Using mathematical tools from network science [15], Wapman et al. produced a "prestige ranking" for every department at every PhD-granting institution in their dataset and quantified hierarchy in faculty hiring and retention in a large number of academic disciplines. In particular, Wapman et al. constructed a directed graph of PhD-granting departments with an edge from department A to department B corresponding to a faculty member who earned a PhD at department A being hired into department B. Unlike other rankings like those published by *U.S. News & World Report* that purport to assign prestige based on an arbitrary selection of attributes, Wapman et al.'s prestige rankings are based upon the characteristics of their directed graph of departments. It is assumed that "prestigious" departments will prefer to hire faculty from other "prestigious" departments, and then the overall hierarchy of departments is inferred from the topology of the network using the SpringRank algorithm [9]. We will refer to the definition of prestige from Wapman et al. [37] as "Wapman-prestige."

We utilize Wapman's quantification of prestige to produce a definition of "elite," which we will use to help us quantify and document gender-based inequality in the mathematical sciences at PhD-granting institutions in the United States. We define "elite" institutions as those that are within the top quartile as defined by Wapman-prestige.

Because of the nature of the dataset [36], which contains gender data but not other demographic data (and in particular no data on race or ethnicity), we are only able to conduct analyses using gender and not other identity characteristics. However, by combining this dataset with publicly available data from the National Science Foundation (NSF) on awards made by the Division of Mathematical Sciences (DMS) [27], which is the primary funder of mathematical sciences research in the United States, we can conduct a MetaMath-related investigation of relationships among the variables of percentage of women in a mathematics department, Wapman-prestige, and NSF DMS funding.

Our thesis is that we can quantify and document inequality in mathematical sciences departments at PhD-granting institutions in the United States. Specifically, in this chapter, we will address the following research questions:

- RQ1: What is the relationship between the percentage of women in a mathematical sciences department and that department's Wapman-prestige?
- RQ2: What is the relationship between Wapman-prestige and funding received from the National Science Foundation's Division of Mathematical Sciences?
- RQ3: What is the relationship between the percentage of women in a mathematics department and funding received from the National Science Foundation's Division of Mathematical Sciences?

The rest of this chapter is organized as follows. In Sect. 2, we provide some examples of recent research that uses data science and mathematics to analyze the mathematics and science communities. We describe the data and methods used in our research in Sect. 3. Specifically, in this section, we describe the processing of the Wapman et al. data and the NSF funding data that is required so that we can investigate our research questions that support the thesis of this chapter, i.e., that gender-based inequality exists (and can be documented and quantified) in the mathematical sciences in the United States. We provide details about the statistical data analysis performed to establish the existence of quantifiable relationships between our variables of interest, gender percentage of faculty in mathematical sciences departments at PhD-granting institutions, amount of funding received from NSF from 2011 to 2020, and departmental Wapman-prestige. The results of the data analysis and discussion of these results are given in Sects. 4 and 5, respectively. We end the chapter by discussing in Sect. 6 some limitations of the work presented here and recognizing that there is a lot more work that can (and should) be done to quantify and document inequality in the mathematical sciences.

2 Existing Work on Inequality and Hierarchy in Mathematics

In this section, we provide a short survey of selected recent work that uses tools, topics, and techniques from mathematics and data science to describe, document, and discuss inequality in the mathematical sciences. We organize our summary of the literature in this area into three topics: (1) analysis of the (lack of) diversity in the mathematical sciences; (2) existence of hierarchy in mathematics and other academic disciplines; and (3) evidence of inequality in the mathematical sciences. For a longer survey of the areas discussed here, as well as the broader field of the mathematics of mathematics, we refer the reader to the recent paper by Buckmire et al. [6].

2.1 The Demographics and Diversity of the Mathematics Community

It is well documented that women are underrepresented at all levels in the mathematics community and that their representation declines as they progress through the academic system [22]. Women made up approximately 42.6% of recipients of bachelor's degrees in mathematics between 2013 and 2018 [3]. However, only 29% of doctorate recipients in mathematical sciences were women (in 2017–2018) [12]. Women were 28% of hires in doctoral-granting mathematical sciences departments in 2019 [14].

Similarly, underrepresentation in the mathematical sciences based on race and ethnicity is profound and persistent [23]. For example, from 2013 to 2018, the racial and ethnic makeup of undergraduate math graduates ranged from 64.9% White, 7.5% Latino/Hispanic, and 5.0% Black to 52.5% White, 9.9% Latino/Hispanic, and 4.2% Black [3]. Between 1993 and 2002, less than 5% of those who earned doctoral degrees in mathematics were Black, Latino/Hispanic, or Indigenous, even though those communities made up a quarter of the general population of the United States at that time [17].

Vitulli [35] examined the representation of women being hired by mathematics departments, based upon data from annual surveys conducted by the American Mathematical Society (AMS). Prior to 2012,[1] the AMS reported this data by dividing departments into three groups based on the reputational rankings in the 1995 (or previously, 1982) National Research Council report on doctoral departments [24, 25]. Group I contained the highest rated 25.9% of the departments. Group II was the next highest 30.3%, while Group III contained the remaining departments. Vitulli found that from 1991 to 2011, 20.5% of the faculty hired by Group I departments were women, while 26.3% of the faculty hired by the remaining departments were women.

We acknowledge that the research discussed in this section is just a small sample of the literature analyzing the demographics and diversity of the mathematical sciences community.

2.2 The Existence of Hierarchy in Mathematics and Science

Recently, researchers have used available data on faculty positions at institutions of higher education in the United States to document the existence of hierarchies in faculty-hiring networks in academia. These hierarchical structures [15] in mathematics and science demonstrate that some institutions have greater influence

[1] The AMS changed how they report this data in 2012, as the newest National Research Council report no longer provided a total ordering of departments, instead reporting multiple measures for each department.

on faculty hiring than others [21]; this is the essential characteristic of what we call Wapman-prestige. Institutions nearer the top of the hierarchy that are "more prestigious" (i.e., they have greater Wapman-prestige) are more likely to have graduating doctoral students go on to obtain faculty positions at institutions that are "less prestigious" and lower in the hierarchy. Clauset et al. [8] demonstrated the existence of hierarchy in faculty hiring in a study involving departments in computer science, business, and history. Wapman et al. [37] expanded this analysis to cover 295,089 faculty in 10,612 departments at 368 PhD-granting institutions and all academic disciplines for the years 2011–2020. FitzGerald et al. [10] complements Wapman et al.'s research and partially replicated their results by using data from the Mathematics Genealogy Project to restrict their analysis to mathematics faculty. These results confirm that hierarchies exist in faculty-hiring networks in the mathematical sciences.

2.3 *Evidence of Inequality in Mathematics and Science*

There are multiple research articles that use quantitative tools and techniques to analyze and highlight examples of inequality in the mathematics community. Topaz et al. [34] analyzed the editorial boards of 435 mathematical science journals and found that women accounted for a mere 8.9% of editorial positions. Editorial positions play important gatekeeping roles and represent status in the mathematics community, so the underrepresentation of women in this area demonstrates inequality based on gender in the area of power over knowledge production. Brisbin and Whitcher [2] found that women are underrepresented as authors among papers in the mathematical sciences uploaded to the arXiv preprint repository and that there are certain subfields (particularly concentrated within "pure" or theoretical mathematics) with even larger discrepancies. Researchers have analyzed data describing different aspects of academic activity and demonstrated myriad ways that gender can negatively mediate opportunity for advancement, participation, and achievement in science and mathematics [16, 18, 30, 33]. Schlenker [32] notes that fields with applications to the social or physical sciences such as numerical analysis, mathematical modeling, or statistics (i.e., fields seen as being in applied mathematics) seem to be viewed by some as having low status in the wider mathematical community. A large study investigating class backgrounds in academia by Morgan et al. [19] found that faculty are much more likely than the general population to have a parent with a PhD, with the effect being even more pronounced at institutions in the top quintile of *U.S. News & World Report* rankings.

3 Data and Methods

In this section, we will describe the data and explain the methodology used to obtain our results. Our data and code are publicly available at [4]. We utilized two datasets, one sourced from Wapman et al. [36] and the second from awards made by the Division of Mathematical Sciences (DMS) at the US National Science Foundation (NSF) between 2011 and 2020 [26].

The Wapman et al. dataset required some nuance to interpret. The dataset consisted of a census of tenured or tenure-track faculty employed at *all* PhD-granting institutions in the United States from the years 2011–2020. Faculty were only included in this sample if they were employed in the majority of the years under review. This dataset is centered on departments, rather than on faculty, and these departments are each assigned to a field such as "Mathematics." In particular, a department may be accounted for in multiple fields; a "Department of Mathematics and Statistics" would have its faculty included twice, in the fields of "Mathematics" and "Statistics."

Because our goal is to quantify and document inequality in the mathematics community as a whole, we choose to define the mathematical sciences as broadly as possible (see [5]). This choice results in a reduction of Wapman et al.'s original dataset of 295,089 faculty in all academic disciplines offering PhDs in the United States to 9814 faculty that are distributed among the fields of mathematics, statistics, or operations research (Table 1). We adopt the convention of capitalizing these three terms when referring to these fields present in the data throughout the rest of this chapter.

In our analysis, we incorporate the department prestige rankings from Wapman et al. [36], which we refer to as Wapman-prestige. Because of the way these are computed (a department has greater Wapman-prestige if its PhD graduates are hired by departments that have greater Wapman-prestige), some departments do not have a Wapman-prestige ranking. This could happen if the department does not have a PhD program in one of the three fields of mathematics, statistics, or operations research (recall the Wapman dataset began with *institutions* that grant PhDs) or if none of its PhD graduates were hired by departments with Wapman-prestige rankings. In mathematics, for example, there are 223 departments listed, but only 161 of these have a Wapman-prestige ranking. Departments without Wapman-prestige rankings were not included in the data analyses involving Wapman-prestige below but were included in the analysis of NSF funding later in the chapter. Figure 1

Table 1 Faculty present in the Wapman et al. dataset in the fields of mathematics, statistics, and operations research

Field	Departments	Faculty members	Percentage of women
Mathematics	223	7328	16.8%
Statistics	122	2576	20.9%
Operations Research	51	1034	19.3%

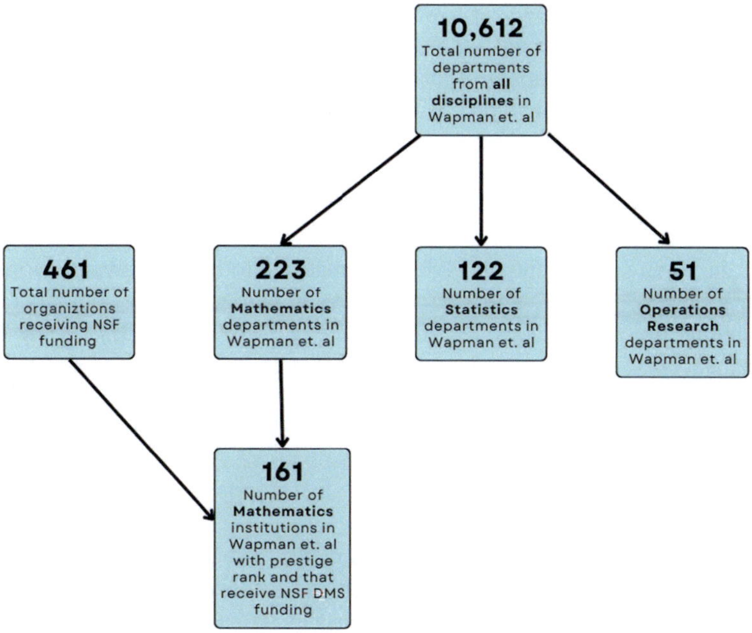

Fig. 1 Flow diagram depicting number of organizations and institutions used for data analysis

displays the various sizes of the datasets used in our analysis in this chapter that are also presented in Table 1.

Since the Group I departments—in the groupings used by the AMS historically to divide departments by perceived quality (see Sect. 2.1 above)—constituted about 25% of departments, in our analysis, we consider the upper quartile (in terms of Wapman-prestige rankings) as "elite" departments and compare this group to the remaining 75% of departments, which we refer to as "non-elite."

Unfortunately, the Wapman et al. dataset only included gender as a binary variable (male/female). In fact, gender is self-reported for a small percentage (6%) of faculty in their initial dataset. They then attempted to infer the gender of the remaining faculty based on their names and the use of software that claims to assign gender based on names relatively accurately; ultimately, binary gender was ascribed to a total of 85% of the faculty listed in their dataset. We include only these faculty who were ascribed a binary gender in our analyses, which eliminates roughly 15% of the total due to the inability to accurately ascribe a gender to these data entries.

We separately obtained data from the NSF publicly accessible database about awards made by the Division of Mathematical Sciences (DMS) in the Directorate of Mathematical and Physical Sciences (MPS) from 2011 to 2020 [26]. These data were aggregated by institution. Since the institution names in the Wapman et al. dataset typically did not match the formal organization names listed by the NSF, we manually aligned these in order to compare the two datasets. On average, DMS awarded \$235 million per year toward achieving its mission to support "a wide range of research in mathematics and statistics aimed at developing and exploring the properties and applications of mathematical structures" [27]. We removed from this dataset awards made to individuals (e.g., fellowships for post-docs and graduate students) as well as awards through the Mathematical Sciences Research Institutes program, which support research institutes separate from mathematics departments (though most, but not all, are hosted at PhD-granting institutions with mathematics departments).

Of the awards to institutions, 80% were matched to the departments of interest in the following analysis. The NSF, and DMS in particular, is the primary source of funding for mathematics research in the United States [27]. While some mathematicians (such as several authors of this chapter) receive funding from NSF divisions other than DMS and directorates other than MPS (e.g., the Division of Undergraduate Education and the Directorate for STEM Education) as well as from other federal agencies (e.g., the National Institutes of Health), we consider DMS funding a reasonable measure for overall financial support of mathematics by the federal government.

There are two major caveats to our use of the NSF DMS funding data. First, NSF awards are made to institutions rather than specifically to departments, which is the unit of analysis provided by the Wapman et al. data. Second, since faculty in Statistics and Operations Research typically have more varied sources of funding, we only considered the field of mathematics in our analyses of funding discussed below.

4 Results

In this section, we present the results of our research into the distribution of faculty and funding at mathematical sciences PhD-granting institutions in the United States. In Fig. 2, the percentage of women in the fields of mathematics, statistics, and operations research from 2011 to 2020 is given. We note that the percentage of women in mathematics lags behind operations research and statistics throughout the time period of the dataset, mirroring the lower percentage of women that earn PhDs in mathematics versus the other two fields [12]. We further note the percentage of women in mathematics, statistics, and operations research is far below the percentage of women in Academia as a whole for the time period covered by the dataset.

Next, we computed the percentages of faculty in each department inferred to be women and plotted these according to Wapman-prestige rank in the fields of mathematics, statistics, and operations research in Fig. 3. To compute this percentage, we used as a denominator the total number of faculty in a department for which a gender was inferred, in effect removing from our sample any faculty members whose gender could not be inferred. In Fig. 3, the blue circles are clustered in the lower-left corner of all three subfigures; this corresponds to the data demonstrating

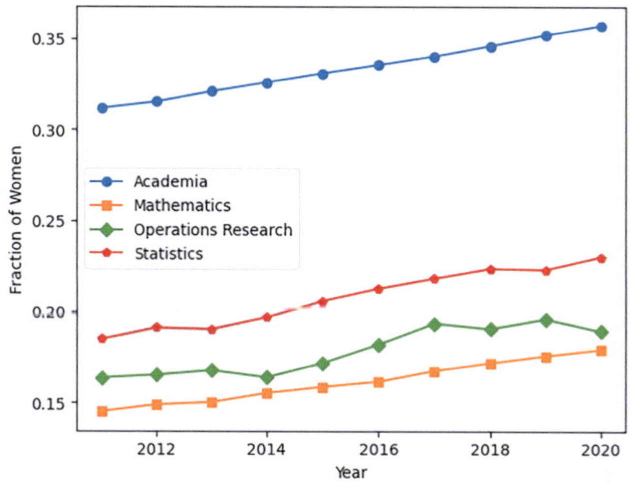

Fig. 2 Fraction of women in the fields of mathematics, statistics, and operations research, as well as academia as a whole, over the 10-year period of the dataset

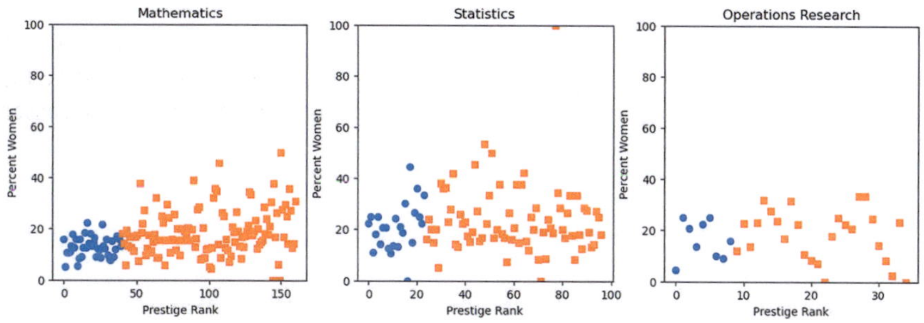

Fig. 3 Percentage of women by department in the fields of mathematics, statistics, and operations research. Color and shape distinguish the upper quartile of Wapman-prestige (blue circles) from the lower three quartiles (orange squares)

that elite departments (in the top quartile of Wapman-prestige) also have a low percentage of women (below 20% for mathematics).

In order to address RQ1, we then considered the elite institutions as a group and calculated the percentage of faculty at these institutions that are women (Table 2); in each case, we see that the percentage of women among these elite institutions is lower than among non-elite institutions. A chi-squared test for each field was conducted, finding only the difference in mathematics to be significant ($p < 0.001$). We also conducted a Kendall tau test to determine if there is an association between the percentage of women and Wapman-prestige rank of mathematics departments; we found a significant negative association between Wapman-prestige rank and rank by percentage of women ($\tau = -0.23$, $p < 0.001$). In other words, higher Wapman-

Table 2 The percentage of faculty at elite and non-elite institutions who are women in each field

Field	Percentage of women among elite institutions	Percentage of women among non-elite institutions
Mathematics	12.5%	18.1%
Statistics	21.3%	21.6%
Operations research	17.0%	18.7%

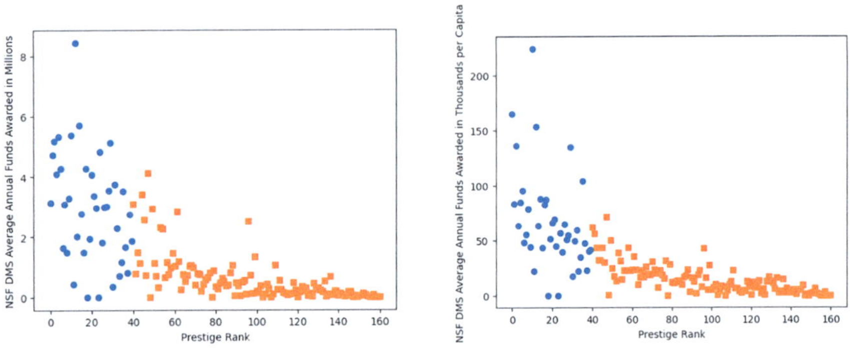

(a) Total funding

(b) Funding per faculty

Fig. 4 Average annual grant funding for mathematics departments by the prestige rank of the department, displayed by total funding to the department (**a**) and on a per capita basis accounting for the varying number of faculty in each department (**b**)

prestige of a Mathematics department is associated with having a lower percentage of women.

To address RQ2, we explored the distribution of DMS funding to the 161 mathematics departments with Wapman-prestige rankings from 2011 to 2020. The elite institutions (i.e., the upper quartile by Wapman-prestige) were awarded in aggregate \$119M per year in grant funding, while the non-elite institutions, of which there are three times as many, were awarded only \$70M in aggregate of NSF money per year. We plotted this funding by Wapman-prestige of each department in Fig. 4 on both a total and per-faculty basis.

We also computed the total amount of funding received by elite (top quartile by Wapman-prestige) and non-elite (lower three quartiles by Wapman-prestige) departments over the time period in question. Elite departments received 64.7% of the total funding in our dataset, compared to 35.3% for the non-elite departments (despite these being thrice as numerous). Here we also conducted a Kendall tau test, finding a significant positive relationship between Wapman-prestige and DMS funding ($\tau = 0.68$, $p < 0.001$). In addition, we conducted a Kendall tau test for a relationship between Wapman-prestige and DMS funding per faculty to account for variance in department sizes; this was also significant ($\tau = 0.70$, $p < 0.001$). In other words, higher Wapman-prestige is associated with more DMS funding.

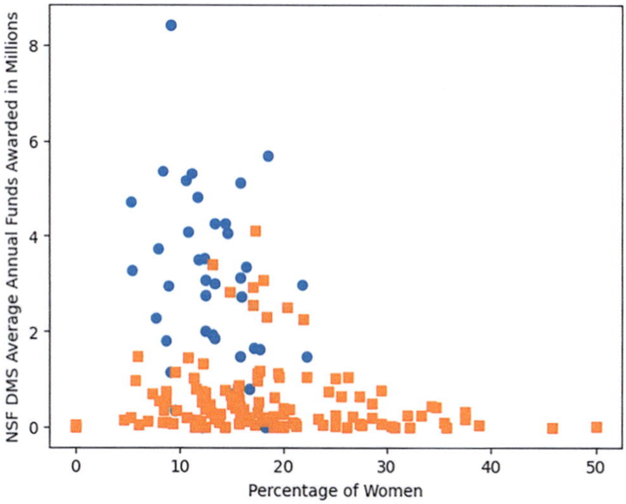

Fig. 5 Average annual grant funding for mathematics departments by the percentage of women faculty in the department. The upper quartile (by prestige) departments are represented by blue circles, while the remainder are orange squares

Another metric that can be used to quantify inequality in NSF DMS funding received by mathematics departments in the United States is the Gini coefficient, a well-known measure of inequality often used to characterize wealth inequality on a scale from 0 to 1. A uniform distribution of wealth would result in a Gini coefficient value of 0, and a case where one individual holds all the wealth would result in a Gini coefficient of 1. The Gini coefficient of NSF DMS funding for the mathematics departments in our dataset is 0.63. For more information on the Gini coefficient, refer to [20].

We also analyzed the full DMS-funded portfolio (excluding fellowships given to people instead of institutions) to avoid a possible sampling effect due to our focus on PhD-granting institutions. In this more comprehensive dataset, the top 20% hold 86.1% of all DMS funding. The Gini coefficient of this distribution is 0.80. Thus, the larger set of all DMS funding recipients demonstrates greater inequality than the subset of PhD-granting institutions.

To address RQ3, we plotted the annual grant funding received by mathematics departments against the percentage of women in those departments in Fig. 5. We note an interesting effect in Fig. 5, where none of the 29 departments (which are all non-elite with respect to Wapman-prestige) with at least 25% women received more than \$1.1M in average annual funding from the DMS. We conducted a Kendall tau test to determine if there is an association between annual funding received from DMS and the percentage of women in mathematics departments and found a significant negative relationship ($\tau = -0.22$, $p < 0.001$).

In addition, we conducted a Kendall tau test for a relationship between DMS funding per faculty and percentage of women to account for variance in department sizes; this was also significant ($\tau = -0.21$, $p < 0.001$). In other words, a higher percentage of women in a mathematics department is associated with less annual DMS funding.

5 Discussion

In this section, we shall discuss the results presented above, which demonstrate that gender-based inequalities exist in faculty composition with respect to gender and the distribution of federal funding to mathematical sciences PhD-granting institutions in the United States.

5.1 Gender-Based Inequality in the Mathematical Sciences

The data shows that almost all PhD-granting institutions have mathematics departments that are composed of faculty that are disproportionately male. In fact, not a single mathematics department represented in this dataset was majority women. We found that the underrepresentation of women is more pronounced among elite mathematics departments (recall that we defined "elite" departments as those in the upper quartile of departments in the prestige ranking generated by Wapman et al.). We believe in the fundamental principle that mathematical talent is distributed equally among all groups of people who do mathematics. In the context of this chapter, we therefore assume an equal distribution of mathematical talent among men and women. Under that assumption, the results presented here highlight the idea that even if mathematical talent is evenly distributed, the opportunities to deploy, use, and leverage that talent in a mathematical sciences department in a PhD-granting institution are not.

Our analysis of NSF DMS funding identifies inequality in the amount of financial support for mathematics PhD-granting departments depending on Wapman-prestige. Pareto models, also popularized as the "20/80" economic model, predict that approximately 80% of assets are held, gained, or earned by only 20% of the population being studied [29]. We found that in the elite institutions, the top 25% in our dataset by Wapman-prestige ranking garnered 65% of the total funds given to the subset of PhD-granting institutions with a Wapman-prestige ranking. When we examine all NSF DMS funding, the top 20% of awardees receive 86% of all funds, with a Gini coefficient of 0.8. This result demonstrates a larger inequality than the classic "20/80" proportion.

5.2 *Inequity = Inequality + Power*

We argue here that the gender-based inequalities presented earlier in this chapter are in part due to underlying mechanisms and societal processes in the institutions and systems present in the mathematical sciences in the United States. We draw on Hasty et al.'s anthropological definition of ***inequity*** as "the unequal distribution of resources due to an unjust power imbalance. It is a type of inequality caused by this unequal distribution, often as a result of injustices against historically excluded groups of people" [13].

Earlier in this chapter, we quantified and documented the "unequal distribution of resources" from the NSF awarded to mathematics departments at PhD-granting institutions in the United States. Now we examine the underlying mechanisms that result in these unequal allocations. The procedures and policies that NSF uses to determine which principal investigators and which institutions receive funding may actually reinforce inequity in the mathematical sciences [1]. NSF uses a process they call merit review, where anonymous reviewers are asked to read, rate, and review funding proposals submitted to the agency. Analysis of panel decisions in science funding in the Netherlands indicated that gender-based bias was moderated by a two-stage review system [1]. In this system, the first stage of reviews is a filter pass in which reviews favored proposals led by men; however, the second-stage final reviews resulted in equal funding. A 2020 analysis of NSF funding by division noted that submission rates by women to the MPS (Mathematical and Physical Sciences) directorate were increasing but were still one of the lowest in the agency, on par with the CISE (Computer and Information Science and Engineering) directorate [30]. Rissler et al. also note that the submission rate discrepancy cannot be explained by the proportion of women at institutions of various Carnegie classification types alone. However, our results suggest that considering the Carnegie "Very High Research Activity" institutions as a single group is too coarse and that important differences in gender representation among these institutions exist and are mediated by Wapman-prestige (see Sect. 4).

The NSF merit review process incorporates mechanisms that favor institutional prestige. NSF panels are instructed to assess the "intellectual merit" and "broader impacts" of all proposals under five elements [28]. Two of these elements in particular may bolster inequity in the mathematical sciences. First, reviewers are asked, "How well qualified is the individual, team, or organization?" This question is likely to skew reviewers toward considering institutional reputations since the information provided about qualifications typically includes individuals' institutional affiliation(s). Second, reviewers are asked, "Are there adequate resources available to the PI (either at the home organization or through collaborations) to carry out the proposed activities?" This second question is especially likely to skew reviewers to more positively rate proposals from well-resourced institutions.

The cliché "The rich get richer" is a colloquial distillation of how systems disproportionately allocate resources towards prestige. Prestigious and well-resourced institutions often provide their faculty with significant advantages in the grant award process. For example, well-resourced universities often have significant funds internally for pilot projects that strengthen submissions to NSF. They have grant departments that assist in the writing and administration of grants, funded by high indirect cost rates. They also have research support teams devoted to data gathering and processing, as well as communications teams devoted to disseminating the results. We also note that expectations in obtaining grant funds vary widely between departments and institutions and are often higher at the elite institutions (those in the top quartile using Wapman-prestige) in our dataset. This likely affects the number of proposals that are submitted by different institutions along with how well they are reviewed.

In short, the effect we are seeing in the unequal allocation of resources in the mathematical sciences community via NSF DMS funding is likely a result of a complicated collection of processes that reinforce and exacerbate the status quo. Hasty's definition of inequity applied to gender-based inequity in the mathematical sciences in the United States requires us to both document gender-based inequalities *and* find evidence that those inequalities are the result of gender-based injustice. In addition to the processes of gender bias in NSF proposal reviews and how institutional prestige biases the merit review process, we also note the long history of erasure, injustice, and exclusion of women from mathematics [31]. We therefore argue that we have not only quantified and documented gender-based inequalities in the mathematical sciences but have also shown the existence of gender-based inequity in the mathematical sciences.

5.3 Limitations

There are a number of limitations that accompany the research presented in this chapter that we want to highlight below. The primary limitation is that there is a paucity of publicly available, comprehensive, self-reported demographic data about the mathematics community. The data we used was part of a dataset shared publicly by Wapman et al. [36] after they had processed it, and the raw data was not available to us. Their methodology of determining Wapman-prestige means that only PhD-granting departments are represented in the prestige data; a large number of faculty in the mathematical sciences who are at community colleges and predominantly undergraduate institutions are not included. Additionally, because of the way disciplines in the mathematical sciences are defined in the Wapman et al. dataset, faculty who are in "Mathematics and Statistics" departments are counted in "Mathematics" and again in "Statistics."

We reiterate here that the dataset [36] consisted of tenured and tenure-track faculty who were in the same department for at least 5 of the years between 2011 and 2020. As a result, we do not have an accurate count of department sizes (which, in any event, vary over time) to fully rule this out as a confounding factor. However, we did examine funding per faculty member in our analysis above to approximately account for this.

Another important limitation is the way gender is ascribed to individuals in the Wapman et al. dataset. Recall that this dataset included gender, but only a small percentage were determined by the individuals themselves, with the rest inferred based on names. While we acknowledge this practice is common, there are multiple issues with name-based gender inference. There is some degree of selection bias in which names can be ascribed to a particular (binary) gender; we note, again, that in the dataset used here, 15% of entries were omitted due to an inability to confidently ascribe them a gender. Furthermore, the reduction of gender to a binary erases the experiences of gender-diverse mathematical scientists from this work.

We also lament the lack of race/ethnicity in the data; there are important questions to be answered about the interaction of race/ethnicity and inequity in the mathematical sciences. However, Chen et al. [7] explores systemic mechanisms exhaustively in their article "Systemic racial disparities in funding rates at the National Science Foundation." This study thoroughly addresses the systemic reinforcing of racial bias within panels and funding distribution decisions of the NSF specifically. Chen et al. also note that a gender analysis alone masks intersectional issues. They point to a study of NIH grants that found that women of color were funded at lower rates than white women [30] and points out the lack of data availability to further investigate this at NSF. A similar argument can be made about the absence of available data about LGBTQ+ identity. The Wapman et al. dataset is limited to only tenured or tenure-track faculty. Without comprehensive demographic data, an intersectional analysis involving multiple identity characteristics is not possible. As discussed below, we hope other researchers will collect or generate additional data that can be used to address important outstanding questions about the mathematical sciences discipline.

6 Future Directions

There are many other directions in which the research presented here could be extended in the future. It is important to study the questions addressed in this chapter with respect to other dimensions of diversity, particularly marginalized social identities such as race/ethnicity, sexual orientation, national origin, and disability status, among others. This future work should be done in a way that allows analysis using intersections of multiple identity characteristics.

The addition of geographic location to the analysis of the gender diversity of PhD-granting institutions as well as of the distribution of federal funding is a possible direction of future research.

Another future direction is to expand this work to a wider range of institutions and faculty appointments. A study encompassing all types of institutions, and particularly community colleges, minority-serving institutions, and primarily undergraduate institutions, is necessary. Additionally, future work should investigate related questions about all types of faculty employed at these institutions, especially the increasing percentage of non-tenure track faculty.

This work's primary goal has been to quantify and document gender-based inequalities in the mathematical sciences. Future work could involve developing mathematical models that are informed by existing data documenting inequality in the mathematical sciences in order to examine, expose, and explicate the mechanisms that create and maintain this imbalance. For example, the data [3, 12, 14] showing the underrepresentation of women in new hires versus their underrepresentation in tenure-stream positions in PhD-granting institutions raises some interesting questions that could also be addressed in future research.

We conclude by inviting interested researchers to join us in the ongoing MetaMath project to use mathematics and data science to analyze the mathematical sciences discipline itself in order to promote social justice and enhance equity in the mathematical sciences.

Acknowledgments This material is based upon work supported by the National Science Foundation under Grant No. DMS-1929284 while all authors were in residence at the Institute for Computational and Experimental Research in Mathematics in Providence, RI, during the Data Science and Social Justice: Networks, Policy, and Education program. The content is solely the responsibility of the authors and does not necessarily represent the official views of the National Science Foundation. The authors thank Phil Chodrow and Victor Piercey for fruitful conversations that pushed this work forward and Sam Zhang for contributions to a first draft of this work.

Author Contributions All the listed co-authors contributed equally to the creation of this article; the order of attribution is alphabetical, as is customary in mathematics and is not intended to demonstrate any distinction in credit or effort.

Competing Interests Buckmire acknowledges sabbatical support provided by the Office of the Dean of the College at Occidental College in Los Angeles, California.

Diaz Eaton was supported in part by the Bates Enhanced Sabbatical Fund and Faculty Professional Development Fund.

Hibdon was supported, in part, by the National Institutes of Health's National Cancer Institute Grant Numbers U54CA202995, U54CA202997, and U54CA203000. The content is solely the responsibility of the authors and does not necessarily represent the official views of the National Institutes of Health.

Kauba is partially supported by the National Science Foundation South Carolina LSAMP Bridge to Doctorate Fellowship HRD-2005030.

Pabón is partially supported by the National Science Foundation under Awards DMS-2108839 and DMS-1450182.

Roca was partially supported by the National Science Foundation under Award DUE-2337055.

Vindas-Meléndez was partially supported by the National Science Foundation under Award DMS-2102921.

References

1. Bol, T., de Vaan, M., van de Rijt, A.: Gender-equal funding rates conceal unequal evaluations. Res. Policy **51**(1), 104399 (2022). https://doi.org/10.1016/j.respol.2021.104399.
2. Brisbin, A., Whitcher, U.: Women's representation in mathematics subfields: evidence from the arxiv (2015). preprint arXiv:1509.07824
3. Buckmire, R.:"Who Does the Math?": On the diversity and demographics of the mathematics community in the USA. In: Improving Applied Mathematics Education, pp. 1–12. Springer, Berlin (2021)

4. Buckmire, R., Eaton, C.D., Hibdon Joseph E, J., Kauba, J., Lewis, D., Ortega, O., Pabón, J., Roca, R., Vindas-Meléndez, A.R.: Data for "quantifying and documenting inequities in PhD-granting mathematical sciences departments in the United States" (2024). http://osf.io/bxh24
5. Buckmire, R., Diaz Eaton, C., Hibdon, J.E., Kinnaird, K.M., Lewis, D., Libertini, J.M., Ortega, O., Roca, R., Vindas Meléndez, A.R.: On Definitions of "Mathematician". J. Hum. Math. 13(2), 8–38 (2023). https://doi.org/10.5642/jhummath.ZRUZ1463. Available at: https://scholarship.claremont.edu/jhm/vol13/iss2/4
6. Buckmire, R., Hibdon Jr, J.E., Lewis, D., Ortega, O., Pabón, J.L., Roca, R., Vindas-Meléndez, A.R..: The Mathematics of Mathematics: Using Mathematics and Data Science to Analyze the Mathematical Sciences Community and Enhance Social Justice. La Matematica, 110–125 (2025). https://doi.org/10.1007/s44007-024-00146-6
7. Chen, C.Y., Kahanamoku, S.S., Tripati, A., Alegado, R.A., Morris, V.R., Andrade, K., Hosbey, J.: Meta-Research: systemic racial disparities in funding rates at the National Science Foundation. eLife **11**, e83071 (2022). https://doi.org/10.7554/eLife.83071
8. Clauset, A., Arbesman, S., Larremore, D.B.: Systematic inequality and hierarchy in faculty hiring networks. Sci. Adv. **1**(1), e1400005 (2015)
9. De Bacco, C., Larremore, D.B., Moore, C.: A physical model for efficient ranking in networks. Sci. Adv. **4**(7), eaar8260 (2018)
10. FitzGerald, C., Huang, Y., Leisman, K.P., Topaz, C.M.: Temporal dynamics of faculty hiring in mathematics. Hum. Soc. Sci. Commun. **10**(1), 247 (2023). https://doi.org/10.1057/s41599-023-01708-9.
11. Fortunato, S., Bergstrom, C.T., Börner, K., Evans, J.A., Helbing, D., Milojević, S., Petersen, A.M., Radicchi, F., Sinatra, R., Uzzi, B., et al.: Science of science. Science **359**(6379), eaao0185 (2018)
12. Golbeck, A.L., Barr, T.H., Rose, C.A.: Report on the 2017–2018 new doctorate recipients. Not. Am. Math. Soc. **67**(8), 1200–1206 (2020)
13. Hasty, J., Lewis, D.G., Snipes, M.M.: Introduction to Anthropology. OpenStax (2022)
14. Jahan, N., Barr, T.H., Rose, C.A., Macias, V.P.: Academic recruitment, hiring, and attrition during 2018–2019. Not. Am. Math. Soc. **69**(6), 1057–1061 (2022)
15. Kawakatsu, M., Chodrow, P.S., Eikmeier, N., Larremore, D.B.: Emergence of hierarchy in networked endorsement dynamics. Proc. Natl. Acad. Sci. **118**(16), e2015188118 (2021)
16. Lerman, K., Yu, Y., Morstatter, F., Pujara, J.: Gendered citation patterns among the scientific elite. Proc. Natl. Acad. Sci. **119**(40), e2206070119 (2022)
17. Medina, H.A.: Doctorate degrees in mathematics earned by Blacks, Hispanics/Latinos, and Native Americans: a look at the numbers. Not. Am. Math. Soc. **51**(7), 772–775 (2004)
18. Mihaljević-Brandt, H., Santamaría, L., Tullney, M.: The effect of gender in the publication patterns in mathematics. PLoS One **11**(10), e0165367 (2016)
19. Morgan, A.C., LaBerge, N., Larremore, D.B., Galesic, M., Brand, J.E., Clauset, A.: Socioeconomic roots of academic faculty. Nat. Hum. Behav. **6**(12), 1625–1633 (2022)
20. Mukhopadhyay, N., Sengupta, P.P.: Gini Inequality Index: Methods and Applications. CRC Press, Boca Raton (2021)
21. Myers, S.A., Mucha, P.J., Porter, M.A.: Mathematical genealogy and department prestige. Chaos: An Interdiscip. J. Nonlinear Sci. **21**(4), 041104 (2011)
22. National Academies of Sciences, Engineering, and Medicine and others: Promising Practices for Addressing the Underrepresentation of Women in Science, Engineering, and Medicine: Opening Doors. National Academies Press (2020)
23. National Center for Science and Engineering Statistics (NCSES): Diversity and STEM: Women, Minorities, and Persons with Disabilities 2023 (Special Report NSF 23-315). National Science Foundation (2023). https://ncses.nsf.gov/wmpd
24. National Research Council: An Assessment of Research-Doctorate Programs in the United States: Mathematical and Physical Sciences, vol. 1. National Academies Press (1982)
25. National Research Council: Research doctorate programs in the United States: Continuity and Change. National Academies Press (1995)

26. National Science Foundation: National Science Foundation Awards Search Website. https://www.nsf.gov/awardsearch/
27. National Science Foundation: About mathematical sciences (DMS) (2023). https://www.nsf.gov/mps/dms/about.jsp
28. National Science Foundation: Proposal & Award Policies & Procedure Guide (2023)
29. Pareto, V.: Cours d'économie politique, vol. 1. Librairie Droz (1964)
30. Rissler, L.J., Hale, K.L., Joffe, N.R., Caruso, N.M.: Gender differences in grant submissions across science and engineering fields at the NSF. Bioscience **70**(9), 814–820 (2020)
31. Saunders, J., National Academies of Sciences, Engineering, and Medicine, et al.: Promising Practices for Addressing the Underrepresentation of Women in Science, Engineering, and Medicine: Opening Doors. National Academies Press (US) (2020)
32. Schlenker, J.M.: The prestige and status of research fields within mathematics (2020). preprint arXiv:2008.13244
33. Schmaling, K.B., Gallo, S.A.: Gender differences in peer reviewed grant applications, awards, and amounts: a systematic review and meta-analysis. Res. Integrity Peer Rev. **8**(1), 2 (2023)
34. Topaz, C.M., Sen, S.: Gender representation on journal editorial boards in the mathematical sciences. PLoS One **11**(8), e0161357 (2016)
35. Vitulli, M.A.: Gender differences in first jobs for new us PhDs in the mathematical sciences. Not. AMs **65**(3) (2018)
36. Wapman, K.H., Zhang, S., Clauset, A., Larremore, D.B.: Data for "Quantifying hierarchy and dynamics in US faculty hiring and retention" (2022). https://doi.org/10.5281/zenodo.6941651
37. Wapman, K.H., Zhang, S., Clauset, A., Larremore, D.B.: Quantifying hierarchy and dynamics in US faculty hiring and retention. Nature **610**(7930), 120–127 (2022)

Appendix A WiSDM 2023: Projects and Participants

WiSDM is a "Women in Data Science and Mathematics" Research Collaboration Workshop that is intended to take place every two years with the goal of bringing together, for one-week, women at all stages of their careers, from graduate students to senior researchers, to collaborate on problems in data science. WiSDM 2017 and WiSDM 2019 workshops were held at the Institute for Computational and Experimental Research in Mathematics (ICERM), Brown University. After a hiatus due to the COVID-19 pandemic, the workshops were reinitiated in 2023, when WiSDM 2023 was held at the Institute for Pure & Applied Mathematics (IPAM), on the campus of the University of California, Los Angeles, during August 7–11. This third edition was organized by top researchers in diverse fields of mathematics and included 42 participants. A summary of the projects in WiSDM 2023, together with a list of project leads with current affiliations and participants with their affiliations at the time of the workshop, is included next.

C. Garcia-Cardona, H. Lee (eds.), *Advances in Data Science*, Association for Women in Mathematics Series 37, https://doi.org/10.1007/978-3-031-87804-6

A.1 WiSDM 2023 Organizing Committee

Andrea Bertozzi: University of California, Los Angeles
Kathryn Leonard: Occidental College
Deanna Needell: University of California, Los Angeles
Linda Ness: Rutgers University

A.2 Project Descriptions

A.2.1 Optimizing NLP Embedding Techniques for Embedded Systems

Description Text embeddings are a way of transforming language into numerical representations that can be used in deep learning architectures for language translation and generation, text summarization, and sentiment analysis for a plethora of natural language processing (NLP) use cases. In NLP, embeddings are typically generated to represent semantic relationships between words or phrases; however, the size of the embedding is usually limited by the temporal and computational constraints imposed by model training and inferencing requirements. Embedded systems, i.e., programmable devices used to perform specific tasks in computationally limited remote environments, typically impose the most stringent computational resources with the goal of optimizing output for a specific task. More and more, embedded systems applications call for online NLP tasks to build a common operating picture in tactical environments. The goal of this project is to generate a representative number of embedded use cases that require on-board NLP and to outline prescriptive methods for optimal text embedding generation that will fulfill the requirements of the embedded system while meeting or exceeding the processing limitations imposed by the computational constraints of each use case.

Leads: Karolyn Babalola (Booz Allen Hamilton)

Participants: Sanchita Ghosh (Texas Tech University), Arnaja Mitra (The University of Texas at Dallas), Chathurangi Pathiravasan (John Hopkins University), and Jing Qin (University of Kentucky)

A.2.2 Geometric Signatures of (Hierarchical) Data

Description Building trees to represent or to fit distances is a critical component of phylogenetic analysis, metric embeddings, approximation algorithms, and computational biology. It is, however, a challenging problem; indeed, many of the tree fitting problem formulations are hard (in a formal sense). Much of the previous algorithmic work has focused on generic metric spaces (i.e., those with no a priori constraints). These spaces do not capture the nature of datasets, especially those datasets that capture some sense of hierarchy. This project will explore two types of geometric signatures of (hierarchical) data and graphs, delta-hyperbolicity and average delta-hyperbolicity. We will compute these quantities for a variety of important test datasets and devise faster, approximate algorithms along the way.

Leads: Anna Gilbert (Yale University)

Participants: Katarzyna Jankiewicz (University of California, Santa Cruz), Manasa Kesapragada (University of California, Santa Cruz), Marzieh Khodaei (Florida State University), Anna Konstorum (Institute for Defense Analysis), and Nazia Riasat (North Dakota State University)

A.2.3 Dimension Reduction and Machine Learning for Tensors

Description Data is now not only everywhere but in such vast quantities that it makes computing quite challenging and often impossible. Moreover, the structure of data is often complicated and multi-modal. For this reason, the algebraic tensor structure has become important in data science and computational methods. There are several tensor dimension reduction techniques that do not require the tensor to be transformed to a vector or matrix, and these can be used for machine learning and reconstruction tasks. In this project, we will study these techniques and develop new methods that work in the dimension reduced space directly. Applications range from imaging to medicine, and we will apply our approaches to both real and synthetic problems.

Leads: Deanna Needell (University of California, Los Angeles), with Jamie Haddock (Harvey Mudd College)

Participants: Alejandra Castillo (Oregon State University), Iryna Hartsock (University of Florida), Paulina Hoyos Restrepo (The University of Texas at Austin), Lara Kassab (University of California, Los Angeles), Alona Kryshchenko (California State University Channel Islands), Kamila Larripa (California State Polytechnic University, Humboldt), Shambhavi Suryanarayanan (Princeton University), and Karamatou Yacoubou Dijma (Wellesley College)

A.2.4 Graph-Based Active Learning

Description This project will be about semi-supervised active learning using a graph approach. Graph-based machine learning algorithms use pairwise comparisons between pieces of data to construct a similarity graph. This project will focus on the "active learning" problem in which specific data points are selected for labels as part of the training data. The algorithm selects the points, and a "human in the loop" labels the data. We will consider a variety of high-dimensional remote sensing data such as hyperspectral images, LIDAR, and SAR data.

Leads: Andrea Bertozzi (University of California, Los Angeles), with Harlin Lee (University of North Carolina at Chapel Hill)

Participants: Malvina Bozhidarova (University of Nottingham), Shuang Li (University of California, Los Angeles), Anna Ma (University of California, Irvine), Namrata Nadagouda (Geoergia Institute of Technology), and Joana Perdomo (Raytheon)

A.2.5 Feature Learning and Optimization Techniques for Machine Learning Tasks

Description Graph-based techniques, which embed datasets into a weighted similarity graph with vertices and edges, form powerful and popular approaches for their ability to capture the structure of the data and pairwise information. However, the success of graph-based approaches depends greatly on the quality of the features of the data used to construct the graph and the computational complexity for dealing with high-dimensional data. This project aims to integrate quality feature learning into graph-based methods to facilitate data classification task. One example of procedures that can be used to obtain high-quality features is autoencoders, which can be of various structures and levels of complexity; such approaches are unsupervised and thus do not require any labeled data. The research will also develop advanced optimization-based models such as auction dynamics learning methods, maximum flow, and spectral approaches for scalable computational efficiency. This project will be supplemented with applications in data science, such as hyperspectral and medical imaging.

Leads: Yifei Lou (University of North Carolina at Chapel Hill), with Cristina Garcia-Cardona (Los Alamos National Laboratory)

Participants: Haiyan Cheng (Willamette University), Weihong Guo (Case Western Reserve University), Sara Hahner (Fraunhofer Institute for Scientific Computing and Algorithms), Yuan Liu (Wichita State University), Michela Marini (University of Houston), and Sui Tang (University of California, Santa Barbara)

A.2.6 *Geometric Supervised Dimension Reduction with Path Metrics*

Description This project will explore new approaches for constructing kernel matrices, a critical task for manifold learning and neural style transfer (i.e., using deep neural networks to learn and transfer the style of an image/audio to another). In supervised learning, a major challenge is how to efficiently incorporate the information carried by the response variables into the kernel matrix. Existing methods based on the Euclidean dissimilarity between pairwise responses are well explored but unsuitable for nonlinear data. This project will explore the benefits of using geometric measures on both the input features and the response variables. In particular, we will focus on the power-weighted shortest path metric, which is a data-driven metric enjoying a rich geometric framework and desirable properties for clustering and dimension reduction. The resulting kernel matrix will be built based on the combination of local gradient information of the labels with power-weighted shortest path distances to stretch the data in directions useful for prediction. The resulting kernels will be evaluated in the context of data visualization and style learning using generative adversarial networks.

Leads: Anna Little (University of Utah) and Rongrong Wang (Michigan State University)

Participants: Jannatul Chhoa (University of Houston), Longxiu Huang (Michigan State University), Aimee Maurais (Massachusetts Institute of Technology), Kirsten Morris (University of Nebraska–Lincoln), Maria van der Walt (Westmont College), and Geetika Verma (RMIT University, Melbourne)

MIX
Papier aus verantwortungsvollen Quellen
Paper from responsible sources
FSC® C105338

If you have any concerns about our products,
you can contact us on
ProductSafety@springernature.com

In case Publisher is established outside the EU,
the EU authorized representative is:
Springer Nature Customer Service Center GmbH
Europaplatz 3, 69115 Heidelberg, Germany

Printed by Libri Plureos GmbH
in Hamburg, Germany